Differentiable Manifolds

PURE AND APPLIED MATHEMATICS

A Series of Monographs and Textbooks

COORDINATOR OF THE EDITORIAL BOARD

S. Kobayashi

UNIVERSITY OF CALIFORNIA AT BERKELEY

DIFFERENTIABLE MANIFOLDS

YOZO MATSUSHIMA

Department of Mathematics
University of Notre Dame
Notre Dame, Indiana

Translated by E.T. Kobayashi

MARCEL DEKKER, INC. New York 1972

Preface

The theory of differentiable manifolds and Lie groups is a basis of the study of differential geometry and differential topology, and the recent development in various branches of mathematics shows that this theory is one of the cornerstones of the edifice of modern mathematics.

The intention of this book is to provide an introduction to the theory of differential manifolds and Lie groups. The book is designed as an advanced undergraduate course or an introductory graduate course and assumes a knowledge of the elements of algebra (vector spaces, groups), point set topology, and some amount of basic analysis.

This book arose from courses given by the author at Osaka University for senior undergraduate students and is a translation of a text published in Japanese in 1965 by Shokabo, Tokyo.

The basic materials (vector spaces, topological spaces, functions of several variables) which are indispensable for the rigorous understanding of the book are gathered in Chapter I. Besides these materials we assume a knowledge of the elements of function theory of one complex variable for the understanding of few sections concerning the complex manifolds and the complex differential forms. However the reader who is not familiar with complex analysis may skip these sections.

I would like to thank Professor H. Ozeki and Professor K. Okamoto of Osaka University who have read the Japanese manuscript and given me various valuable comments.

iii

I am indebted to late Professor E. Kobayashi of New Mexico State University for translating this book and for his contribution to the various improvements of the original text. Professor E. Kobayashi deceased just after completing the translation and I would like to dedicate this volume to his memory.

Notre Dame, Indiana   Y. MATSUSHIMA
February, 1972

Contents

Chapter III. **Differential Forms and Tensor Fields**   **123**

Chapter IV. **Lie Groups and Homogeneous Spaces**   **176**

Differentiable Manifolds

Introduction

We gather here in this chapter, the preliminary material necessary for this book. We explain topological spaces in §1, vector spaces in §2, and the differential calculus of functions of several variables in §3 and §4. This book is written under the assumption that the reader is quite familiar with the material gathered in this chapter. We have quoted reference books for each section of this chapter. Those who are not quite familiar with the material collected here are recommended to acquaint themselves with it through those books first, in order to proceed efficiently through this book afterwards.

Explanation of the notation. We shall use, without explanation, the notation ($\in$, $\cup$, $\cap$, . . .) commonly used in set theory. $\mathbf{R}$ denotes the set of all real numbers and $\mathbf{C}$ the set of all complex numbers. For the numbering of the coordinates, we shall use superscripts on letters, such as x^1, x^2, . . . , x^n. That is, x^i does not mean the ith power of x, but means rather the ith coordinate. For powers of x, we shall always put parentheses around x and write $(x)^i$.

§1. Topological Spaces*

A   Let X be a set, $\mathfrak{O}$ a family† of subsets of X satisfying

(1.1) X itself and the empty set $\varnothing$ both belong to $\mathfrak{O}$.
(1.2) If O_1, $O_2 \in \mathfrak{O}$ then $O_1 \cap O_2 \in \mathfrak{O}$.
(1.3) If $\mathfrak{S}$ is an arbitrary subset of $\mathfrak{O}$, then the union of the subsets of X

---

*Cf. J. Dugundji, *Topology*, Allyn and Bacon, 1965.

† A set, whose elements are sets, is often called a family of sets.

belonging to $\mathfrak{S}$ also belongs to $\mathfrak{O}$, i.e.,

$$\bigcup_{O \in \mathfrak{S}} O \in \mathfrak{O}.$$

We then say that a *topology* is defined in X and we call the subsets of X belonging to $\mathfrak{O}$ *open sets* of this topology. A set X where a topology is defined is called a *topological space*.

In a topological space X, for a point x in X, we call any open set containing x a *neighborhood** of x. Let $\mathfrak{U}(x)$ denote the set of all neighborhoods of x. For the families $\mathfrak{U}(x)$ $(x \in X)$ of subsets of X, the following holds:

(2.1) $\mathfrak{U}(x)$ is not empty. If $U \in \mathfrak{U}(x)$, then $x \in U$.
(2.2) For U_1, $U_2 \in \mathfrak{U}(x)$, there is a $U_3 \in \mathfrak{U}(x)$ such that $U_3 \subset U_1 \cap U_2$.
(2.3) If $U \in \mathfrak{U}(x)$, $y \in U$, then there is a $V \in \mathfrak{U}(y)$ such that $V \subset U$.

In fact, (2.1) is clear from the definition of $\mathfrak{U}(x)$ and from (1.1). If U_1, $U_2 \in \mathfrak{U}(x)$, set $U_1 \cap U_2 = U_3$. Then U_3 is an open set by (1.2) and contains x. Hence $U_3 \in \mathfrak{U}(x)$ and (2.2) holds. If $U \in \mathfrak{U}(x)$ and $y \in U$, then as U is an open set containing y, we have $U \in \mathfrak{U}(y)$. Hence, letting $V = U$, we have (2.3).

Conversely, suppose for each point x of a set X, a family $\mathfrak{U}(x)$ of subsets of X is defined, and suppose for these families the conditions (2.1), (2.2), and (2.3) are satisfied. A subset O of X will be called an open set of X if either O is the empty set or O is nonempty and for an arbitrary x in O there is a $U \in \mathfrak{U}(x)$ such that $U \subset O$. If we let $\mathfrak{O}$ be the family of open sets of X, then from (2.3) we have $\mathfrak{U}(x) \subset \mathfrak{O}$. Moreover it is easy to show that $\mathfrak{O}$ satisfies the conditions (1.1), (1.2), and (1.3). Hence if for each point x of X a family $\mathfrak{U}(x)$ of subsets of X is defined, and if they satisfy conditions (2.1), (2.2), and (2.3), then X becomes a topological space and the sets belonging to $\mathfrak{U}(x)$ are the neighborhoods of x. We then call $\mathfrak{U}(x)$ the *fundamental neighborhood system* of x in the topological space X.

Problem 1. Verify that the $\mathfrak{O}$ defined above satisfies (1.1), (1.2), and (1.3).

In a topological space X, if for two arbitrary points x and y, $x \neq y$, there exist a neighborhood U of x and a neighborhood V of y such that $U \cap V = \varnothing$, then we call X a *Hausdorff space*.

Let X, Y be topological spaces with fundamental neighborhood systems $\mathfrak{U}(x)$ $(x \in X)$, $\mathfrak{U}(y)$ $(y \in Y)$. For an arbitrary (x, y) in the direct product

---

*Also called an open neighborhood. In this book a neighborhood is always an open set.

set $X \times Y$,* let $\mathfrak{U}(x, y) = \{U \times V | U \in \mathfrak{U}(x), V \in \mathfrak{U}(y)\}$. Then the family $\mathfrak{U}(x, y)$ of subsets satisfies (2.1), (2.2), and (2.3). Hence a topology is defined on $X \times Y$. The topological space thus defined is called the *product space* of the topological spaces X and Y.

> *Problem 2.* Show that $\mathfrak{U}(x, y)$ satisfies the three conditions. Show also that the product space $X \times Y$ of two Hausdorff spaces X and Y is again a Hausdorff space.

Let Y be a subset of the topological space X and let $\mathfrak{O}_Y = \{O \cap Y | O \in \mathfrak{O}\}$. The family $\mathfrak{O}_Y$ of subsets of Y can easily be shown to satisfy conditions (1.1), (1.2), and (1.3), and thus a topology is defined in Y. The topological space Y thus defined is called a *topological subspace* of X.

If a family $\{U_\alpha\}_{\alpha \in A}$ of open sets of the topological space X has the property $X = \cup_{\alpha \in A} U_\alpha$, then $\{U_\alpha\}_{\alpha \in A}$ is called an *open covering* of X.

For an arbitrary open covering $\{U_\alpha\}_{\alpha \in A}$ of X, if one can choose a finite number of indices $\alpha_1, \ldots, \alpha_m \in A$ such that $X = U_{\alpha_1} \cup \ldots \cup U_{\alpha_m}$, then the topological space X is said to be *compact*.

If Y is a subset of the topological space X, and if, considered as a subspace of X, Y is compact, then we call Y a *compact subset* of X.

B   Let X and Y be sets. A correspondence f which associates to each element x of X an element $f(x)$ of Y is called a *mapping* or a *map* from X into Y. If, for any two distinct elements x and y of X, the corresponding elements $f(x)$ and $f(y)$ are distinct, i.e., if $f(x) \neq f(y)$ for $x \neq y$, then f is called a *one-to-one map* or an *injection* from X into Y. We also say that f is *injective*. If, for an arbitray element y of Y, there is always an element x of X such that $f(x) = y$, then f is called an *onto map* or a *surjection*. We also say that f is *surjective*. When f is an injection and a surjection, we call f a *bijection*. We also say that f is *bijective*.†

Now let f be a mapping from a topological space X to a topological space Y. For an arbitrary open set O of Y, if the *inverse image* of O by f, $f^{-1}(O) = \{x \in X | f(x) \in O\}$, is always an open set of X, then we call f a *continuous* map from X to Y.

If f is a bijection from X to Y, and if f and its inverse f^{-1} [i.e., the map from Y to X defined by $f^{-1}(y) = x$ if $y = f(x)$] are both continuous maps, then we call f a *homeomorphic map* or simply a *homeomorphism* from X

---

*The set formed by the pairs (x, y), where x is an element of X and y is an element of Y, is called the direct product set of X and Y, and is denoted by $X \times Y$.

†We have listed these terminologies because they are frequently used these days. Since we do not wish to tax a beginner's memory with terminology, we shall not make use of them hereafter in this book. However, they are convenient terms, and we hope that those who have become accustomed to them will use them freely.

onto Y. If there is a homeomorphism from X onto Y, then we say that X and Y are *homeomorphic*.

Problem 3. The continuity of $f: X \to Y$ is equivalent to the following condition: for an arbitrary x in X and an arbitrary neighborhood V of $f(x)$, there is a neighborhood U of x such that $f(U) \subset V$. Prove this.

Problem 4. If f is a continuous map from X to Y, and g is a continuous map from Y to Z, then the composition $g \circ f$ from X to Z (defined by $(g \circ f)(x) = g(f(x))$ for $x \in X$) is continuous. If f is a continuous map from X into Y, and A is a subset of X, then the restriction $f|A$ (defined by $(f|A)(x) = f(x)$ for $x \in A$) is continuous map from A into Y. Prove these.

   C   Let X be a topological space. If there are two open sets O_i ($i = 1, 2$) of X satisfying $O_i \neq \varnothing$ ($i = 1, 2$), $O_1 \cap O_2 = \varnothing$ and $X = O_1 \cup O_2$, then X is said to be disconnected. If X is not disconnected, then X is said to be *connected*. That is, X is connected if the following condition is satisfied: if O_1, O_2 are two open subsets of X satisfying $O_1 \cap O_2 = \varnothing$ and $X = O_1 \cup O_2$, then either O_1 or O_2 is the empty set. Let Y be a subset of the topological space X. If Y, considered as a topological subspace of X, is connected, then we call Y a *connected subset* of X.

   If Y_α ($\alpha \in A$) are connected subsets of X, and if $\cap_{\alpha \in A} Y_\alpha$ is not empty, then the union $Y = \cup_{\alpha \in A} Y_\alpha$ is also connected. In fact, suppose Y is not connected. Then there are open sets U, V of X satisfying $U \cap Y \neq \varnothing$, $V \cap Y \neq \varnothing$, $U \cap V \cap Y = \varnothing$ and $Y = (U \cap Y) \cup (V \cap Y)$. Let $U \cap Y_\alpha = U_\alpha$, $V \cap Y_\alpha = V_\alpha$. Then U_α, V_α are open sets of Y_α and $Y_\alpha = U_\alpha \cup V_\alpha$, $U_\alpha \cap V_\alpha = \varnothing$ hold. However, since Y_α is connected, $U_\alpha = \varnothing$ or $V_\alpha = \varnothing$. Now let $A' = \{\alpha \in A | U_\alpha = \varnothing\}$ and $A'' = \{\alpha \in A | V_\alpha = \varnothing\}$. Then $A = A' \cup A''$, and $Y_\alpha = V_\alpha$ if $\alpha \in A'$, and $Y_\alpha = U_\alpha$ if $\alpha \in A''$. Hence $\cap_{\alpha \in A} Y_\alpha = (\cap_{\alpha \in A'} Y_\alpha) \cap (\cap_{\alpha \in A''} Y_\alpha) \subset (Y \cap V) \cap (Y \cap U)$. However, $(Y \cap U) \cap (Y \cap V) = \varnothing$, so we have $\cap_{\alpha \in A} Y_\alpha = \varnothing$, contrary to our hypothesis that $\cap_{\alpha \in A} Y_\alpha$ is nonempty. Hence Y must be connected.

   In particular, the union of all connected subsets of X containing a point x of X is a connected subset of X. This is called the *connected component* of X containing x. This is the maximum connected subset containing x. Let C be the connected component of X containing x. If y is in C, it is easy to see that C is also the connected component of X containing y. If C is a connected subset of X, $x \in C$, and C is the connected component of X containing x, then we call C a *connected component* of X. If C and C' are connected components of X, and if $C \neq C'$, then $C \cap C' = \varnothing$.

   A topological space X is said to be *locally connected* if each point has a fundamental neighborhood system consisting of connected neighborhoods. If X is locally connected, then each connected component of X is an open set of X.

Problem 5. Prove that the connected components of a topological space X are closed sets, and that if X is locally connected, then the connected components of X are open. Here a subset F of a topological space X is said to be *closed* if its complement $F^C = X - F$ is open.

A continuous map φ from a closed interval $[a, b]$ of the real line to a topological space X is called a *continuous curve* in X. For two points x and y of X, a continuous curve φ such that $\varphi(a) = x$ and $\varphi(b) = y$ is called a continuous curve connecting x and y. $x = \varphi(a)$ is called the starting point of φ, and $y = \varphi(b)$ is called the terminal point of φ.

If for any two points x and y in X, there is always a continuous curve in X connecting x and y, then X is said to be *arcwise connected*. If X is arcwise connected, then X is connected.

If X is a topological manifold (II, §1), and if X is connected, then X is arcwise connected.

Problem 6. Look up the definition of a topological manifold and prove the above assertion. Prove also that a topological manifold is locally connected.

D Let X be a set. Suppose for any pair of points x, y of X, there is a real number $d(x, y)$ satisfying the following three conditions:

(1) If $x \neq y$ then $d(x, y) > 0$; $d(x, x) = 0$,
(2) $d(x, y) = d(y, x)$,
(3) $d(x, y) \leqq d(x, z) + d(z, y)$.

Then we call d a *metric* on X. A set X with a metric d is called a *metric space*. For a point x in a metric space X, and for $\varepsilon > 0$, let

$$U(x;\varepsilon) = \{y \in X | d(x, y) < \varepsilon\}.$$

If we let $\mathfrak{U}(x) = \{U(x;\varepsilon) | 0 < \varepsilon < 1\}$, then $\mathfrak{U}(x)$ satisfies (2.1), (2.2), and (2.3). Thus a topology is defined in the metric space X. Hereafter, we shall always assign this topology to a metric space X and consider X as a topological space.

Let $\{x_n | n = 1, 2, \ldots\}$ be a sequence of points in X. If, for a given $\varepsilon > 0$, there is a positive integer N such that $d(x_n, x) < \varepsilon$ holds for $n > N$, then we say that the sequence $\{x_n\}$ converges to a point x in X. We say that a sequence $\{x_n\}$ of points in X is a *Cauchy sequence* if, for a given $\varepsilon > 0$, there is a positive integer N such that $d(x_n, x_m) < \varepsilon$ holds for arbitrary m, n satisfying $m, n > N$. A sequence converging to a point in X is a Cauchy sequence. If an arbitrary Cauchy sequence in X always converges to a point in X, then we say that the metric space X is *complete*.

Problem 7. Make use of the metrics to define the continuity of a map f from a metric space X to a metric space Y.

Problem 8. Let X and Y be metric spaces with metrics d and d', respectively. We say that a map from X to Y is *uniformly continuous* if for a given $\varepsilon > 0$ there is a $\delta > 0$ such that $d'(f(x), f(y)) < \varepsilon$ holds whenever $d(x, y) < \delta$ for any two points x, y of X. If X is compact, show that a continuous map from X to Y is uniformly continuous.

§2. Vector Spaces*

Let the symbol K stand for **R** or **C**. (The reader who is familiar with the notion of a "field" may consider K to be an arbitrary field.)

A   If a set V satisfies the following conditions in (I) and (II), then V is called a *vector space* over K.
(I) For any two elements a, b of V an element $a + b$ of V, called the sum of a and b, is defined, and the following rules hold:

(1.1) $(a + b) + c = a + (b + c)$
(1.2) $a + b = b + a$
(1.3) An element 0 of V, called the zero vector, exists, and for an arbitrary $a \in V$ satisfies

$$a + 0 = a.$$

(1.4) For an arbitrary $a \in V$, there is an element $-a$ of V, called the inverse element of a, which satisfies

$$a + (-a) = 0.$$

(II) For an element a of V, and an element λ of K, an element λa of V, called λ times a, is defined. λa is called the *scalar multiple* of a by λ. For the addition in (I) and the scalar multiplication, the following rules hold:

(2.1) $(\lambda\mu)a = \lambda(\mu a)$ $(\lambda, \mu \in K)$
(2.2) $1a = a$
(2.3) $\lambda(a + b) = \lambda a + \lambda b$
(2.4) $(\lambda + \mu)a = \lambda a + \mu a$

The elements of a vector space over K are called *vectors*, and the elements of K are called *scalars*. We also call K the *cofficient field* of the vector space V. In particular, if $K = $ **R** we call V a *real vector space*, and if $K = $ **C** we call V a *complex vector space*.

---

*Cf. K. Nomizu, *Fundamentals of Linear Algebra*, McGraw-Hill, New York, 1966.

Problem 1. Show $(-1)a = -a$, $0a = \lambda 0 = 0$. Here 0 of $0a$ is the zero element of K.

Let $a_1, \ldots, a_r$ be elements of V with the following property: if, for r scalars* $\lambda^1, \ldots, \lambda^r$, we have $\lambda^1 a_1 + \ldots + \lambda^r a_r = 0$, then $\lambda^1 = \ldots = \lambda^r = 0$. Then $a_1, \ldots, a_r$ are said to be *linearly independent*.

If there are n linearly independent vectors in V, and $n + 1$ or more vectors are never linearly independent, then we say that V has dimension n. When V has dimension n, a set of n linearly independent vectors $\{a_1, \ldots, a_n\}$ is called a *basis* of V. If $\{a_1, \ldots, a_n\}$ is a basis of V, then an arbitrary vector v of V can be written uniquely as

$$v = \lambda^1 a_1 + \lambda^2 a_2 + \ldots + \lambda^n a_n.$$

The n-tuple of numbers $\{\lambda^1, \ldots, \lambda^n\}$ are called the *components* of v with respect to the basis $\{a_1, \ldots, a_n\}$.

We note that K itself is a vector space over K of dimension 1.

B Let V and W be vector spaces over K. If a map φ from V to W satisfies

$$\varphi(a + b) = \varphi(a) + \varphi(b) \quad (a, b \in V)$$
$$\varphi(\lambda a) = \lambda \varphi(a) \quad (\lambda \in K, a \in V),$$

then φ is called a *linear map* or a *linear transformation* from V to W. Let the dimensions of V and W be n and m, respectively, and let $\{a_1, \ldots, a_n\}$ and $\{b_1, \ldots, b_m\}$ be bases of V and W, respectively. As $\varphi(a_i)$ is an element of W, it can be expressed uniquely as

$$\varphi(a_i) = \sum_{j=1}^{m} \alpha_i^j b_j \quad (i = 1, \ldots, n) \quad (\alpha_i^j \in K).$$

Hence an $m \times n$ matrix

$$A = \begin{bmatrix} \alpha_1^1 & \alpha_2^1 & \cdots & \alpha_n^1 \\ \alpha_1^2 & \alpha_2^2 & \cdots & \alpha_n^2 \\ \cdot & & & \\ \cdot & & & \\ \cdot & & & \\ \alpha_1^m & \alpha_2^m & & \alpha_n^m \end{bmatrix}$$

corresponds to the linear map φ. In this book, the (i, j) component of a matrix is denoted most of the time by α_j^i instead of α_{ij}. For $a \in V$, let $a = \lambda^1 a_1 + \ldots + \lambda^n a_n$ and $\varphi(a) = \mu^1 b_1 + \ldots + \mu^m b_m$. Then we have

---

*As we have remarked already, we sometimes put the indices on the letters as superscripts in this book. That is, λ^i is not a power of λ, but i is an index.

$\varphi(a) = \lambda^1 \varphi(a_1) + \ldots + \lambda^n \varphi(a_n) = \sum_{j=1}^{m}(\sum_{i=1}^{n} \alpha_i^j \lambda^i) b_j$ and as $\{b_1, \ldots, b_m\}$ is a basis of W we have

$$\mu^j = \sum_{i=1}^{n} \alpha_i^j \lambda^i \quad (j = 1, \ldots, m)$$

or in matrix form

$$\begin{bmatrix} \mu^1 \\ \cdot \\ \cdot \\ \cdot \\ \mu^m \end{bmatrix} = A \begin{bmatrix} \lambda^1 \\ \cdot \\ \cdot \\ \cdot \\ \lambda^n \end{bmatrix}.$$

C   Let V be a vector space over K. If a subset W of V satisfies the following conditions, then W is called a *vector subspace*, or simply a *subspace*, of V:

(3.1) If $a, b \in W$, then $a + b \in W$
(3.2) If $\lambda \in K, a \in W$, then $\lambda a \in W$.

Problem 2. The subspace W of V is a vector space with respect to the addition and scalar multiplication in V. Show this.

Let $V_1, \ldots, V_k$ be k vector subspaces of V. If, for an arbitrary vector a of V, there are vectors a_i in V_i $(i = 1, \ldots, k)$ such that

(3.3) $a = a_1 + \ldots + a_k$

then V is said to be the *sum* of $V_1, \ldots, V_k$, and we write

$$V = V_1 + \ldots + V_k.$$

Furthermore, if an arbitrary element a of V can be expressed *uniquely* in the form (3.3), i.e., if

$$a = a_1 + \ldots + a_k = a_1' + \cdots + a_k' \quad (a_i, a_i' \in V_i, i = 1, \ldots, k)$$

implies $a_i = a_i'$ $(i = 1, \ldots, k)$, then V is said to be the *direct sum* of $V_1, \ldots, V_k$ and we write

$$V = V_1 + \ldots + V_k \quad \text{(direct sum)}.$$

Problem 3. Let V be the sum of $V_1, \ldots, V_k$. Show that V is the direct sum of $V_1, \ldots, V_k$ if and only if $a_1 + \ldots + a_k = 0, a_i \in V_i (i = 1, \ldots, k)$, implies $a_i = 0 (i = 1, \ldots, k)$.

D   Let V be a vector space over K, and W a subspace of V. For a, b in V, we say that a and b are equivalent (with respect to W), if the condition

$a - b \in W$ is satisfied, and we write $a \sim b$. The relation that a and b are equivalent satisfies the so-called equivalence laws:

(4.1) $a \sim a$
(4.2) If $a \sim b$, then $b \sim a$
(4.3) If $a \sim b$, $b \sim c$, then $a \sim c$.

Problem 4.   Show that (4.1)–(4.3) hold.

In general, if a relation between the elements of a set M is defined, satisfying the equivalence laws above, then we call this relation an *equivalence relation* in M. When an equivalence relation $\sim$ is defined in M, a subset C of M is called an *equivalence class* with respect to this equivalence relation if it has the following property: if a, $b \in C$, then a and b are equivalent, and if $a \in C$ and $b \notin C$, then a and b are not equivalent. Let $a \in M$, and set $C_a = \{b \in M | a \sim b\}$. Then C_a is an equivalence class containing a. Conversely, let C be an arbitrary equivalence class, and let $a \in C$. Then $C = C_a$. If C and C' are two equivalence classes, then $C = C'$ or $C \cap C' = \varnothing$ holds. The set of equivalence classes is denoted by $M/\sim$, and is called the *quotient set* of M with respect to the equivalence relation

Returning to the vector space V, where an equivalence relation $a \sim b$ is defined with respect to a subspace W, the quotient set $V/\sim$ is denoted by V/W. For an arbitrary element a of V, we write the equivalence class containing a as $\pi(a)$. Then $a \sim b$ is the same as $\pi(a) = \pi(b)$, and we have

$$\pi(a) = \{a + w | w \in W\}.$$

Now if $a \sim a'$, $b \sim b'$, and $\lambda \in K$, we can easily see that $a + a' \sim b + b'$ and $\lambda a \sim \lambda a'$ hold. Hence we can define the sum of two elements of V/W and the product of an element of V/W by an element of K by

$$\pi(a) + \pi(b) = \pi(a + b).$$

$$\lambda \pi(a) = \pi(\lambda a).$$

In fact, the definitions above do not depend on the choice of representatives a, b for the equivalence classes $\pi(a)$, $\pi(b)$. If we define addition and scalar multiplication as above, then V/W becomes a vector space over K. The vector space V/W is called the *quotient vector space* of the vector space V by its subspace W.

Problem 5.   Show that the addition and scalar multiplication in V/W both satisfy the conditions for vector spaces. Show that π is a linear map from V onto V/W.

E Let V be a vector space over K. If a map f from V into K satisfies

$$f(a + b) = f(a) + f(b) \qquad (a, b \in V)$$
$$f(\lambda a) = \lambda f(a) \quad (\lambda \in K, a \in V)$$

then f is called a *linear form* or a *linear function* on V. A linear function is nothing but a linear map from V into the 1-dimensional vector space K. Let the set of all linear functions on V be denoted by V^*. For $f, g \in V^*$, $\lambda \in K$, we define maps $f + g$, λf from V into K by

$$(f + g)(a) = f(a) + g(a)$$
$$(\lambda f)(a) = \lambda(f(a)) \qquad (\lambda \in K, a \in V).$$

It is easy to see that $f + g$ and λf are linear functions on V. Hence addition between elements of V^* and scalar multiplication with elements of K and elements of V^* are defined, and with this addition and scalar multiplication, V^* becomes a vector space over K. The vector space V^* is called the *dual space* of V.

Problem 6. What kind of linear function on V is the 0-vector in V^*?

Now let V have dimension n and let $\{a_1, \ldots, a_n\}$ be a basis of V. For an arbitrary element $a = \lambda^1 a_1 + \ldots + \lambda^n a_n$ of V, we set the ith component λ^i of a equal to $f^i(a)$. It is easy to see that the map $a \to f^i(a)$ defines a linear function on V for each $i = 1, \ldots, n$, and satisfies

$$f^i(a_j) = \delta_j^{i*} \qquad (i, j = 1, \ldots, n). \tag{5}$$

$f^1, \ldots, f^n$ is a set of linearly independent vectors of V^*. In fact, if $\alpha_1 f^1 + \ldots + \alpha_n f^n = 0 \,(\alpha_i \in K)$, then for an arbitrary $a \in V$ we have $0 = 0(a) = (\alpha_1 f^1 + \ldots + \alpha_n f^n)(a) = \sum_{i=1}^n \alpha_i f^i(a)$. Hence, in particular, we have $\sum_{i=1}^n \alpha_i f^i(a_j) = 0 \,(j = 1, \ldots, n)$. But by (5), the left member is equal to α_j. Thus we get $\alpha_j = 0 \,(j = 1, \ldots, n)$, and $f^1, \ldots, f^n$ are linearly independent.

Now let f be an arbitrary element of V^*, and set $f(a_i) = \alpha_i$. We shall show $f = \sum_{i=1}^n \alpha_i f^i$. Set $\sum_{i=1}^n \alpha_i f^i = g$. By (5) we have $g(a_i) = \alpha_i = f(a_i)$. Now for $a \in V$, let $a = \sum_{i=1}^n \lambda^i a_i$. Then $g(a) = \sum_{i=1}^n \lambda^i g(a_i) = \sum_{i=1}^n \lambda^i f(a_i) = f(a)$. Hence $f = g$. We have just proved that $\{f^1, \ldots, f^n\}$ is a basis of V^*. Hence the dual space V^* of the n-dimensional vector space V is also n dimensional.

When condition (5) is satisfied for a basis $\{a_1, \ldots, a_n\}$ of V and a basis

*δ_j^i is the symbol used to denote 1 if $i = j$ and 0 if $i \neq j$, and is called the *Kronecker delta*.

$\{f^1, \ldots, f^n\}$ of V^*, $\{f^1, \ldots, f^n\}$ is called a *dual basis* of $\{a_1, \ldots, a_n\}$. For a given basis $\{a_1, \ldots, a_n\}$ of V there is one and only one dual basis $\{f^1, \ldots, f^n\}$. Moreover for an arbitrary $a \in V$ we have

$$a = \sum_{i=1}^n f^i(a)a_i.$$

Now as V^* is also a vector space, we can consider the dual space $(V^*)^*$ of V^*. For $a \in V$ set

$$a'(f) = f(a) \qquad (f \in V^*).$$

It is easy to see that a' is a linear function on V^*. Moreover $a \to a'$ is a linear map from V to $(V^*)^*$. Let $\{a_1, \ldots, a_n\}$ be a basis of V and $\{f^1, \ldots, f^n\}$ the basis dual to $\{a_1, \ldots, a_n\}$. As $a_i'(f^j) = \delta_i^j$, $\{a_1', \ldots, a_n'\}$ is the basis dual to $\{f^1, \ldots, f^n\}$. Hence, in particular, $\{a_1', \ldots, a_n'\}$ is a basis of $(V^*)^*$, and an arbitrary element a' of $(V^*)^*$ is of the form $a' = \sum_{i=1}^n \lambda^i a_i'$. So, if we identify $a = \sum_{i=1}^n \lambda^i a_i$ with a', then we can consider the dual space $(V^*)^*$ of V^* to be V itself.

Now let φ be a linear map from a vector space V over K to a vector space W over K. For $f \in W^*$, we let

$$({}^t\varphi f)(a) = f(\varphi(a)) \quad (a \in V). \tag{6}$$

Then ${}^t\varphi f$ is a linear function on V, i.e., ${}^t\varphi f \in V^*$. Hence by $f \to {}^t\varphi f$, we have a map ${}^t\varphi$ from W^* to V^*. It is easy to check that ${}^t\varphi$ is a linear map. The linear map ${}^t\varphi$ from W^* to V^* defined by (6) is called the *dual map* of the linear map φ from V to W.

Problem 7. Let $\{a_1, \ldots, a_n\}$ and $\{b_1, \ldots, b_m\}$ be bases of V and W, respectively, and $\{f^1, \ldots, f^n\}$ and $\{g^1, \ldots, g^m\}$ bases of V^* and W^* dual to those of V and W, respectively. If A is the matrix of the linear map φ with respect to the given bases, show that the matrix of ${}^t\varphi$ with respect to the dual bases $\{g^1, \ldots, g^m\}$, $\{f^1, \ldots, f^n\}$ is the transpose matrix ${}^tA$ of the matrix A.

F Let V and W be vector spaces over K. If a map α from the direct product set $V \times W$ into K satisfies the conditions

$$\alpha(a + b, a') = \alpha(a, a') + \alpha(b, a')$$
$$\alpha(a, a' + b') = \alpha(a, a') + \alpha(a, b')$$
$$\alpha(\lambda a, a') = \alpha(a, \lambda a') = \lambda\alpha(a, a')$$
$$(a, b \in V; a', b' \in W; \lambda \in K),$$

then we call α a *bilinear form* on $V \times W$. In particular if $W = V$ we call α a *bilinear form* on V.

Let α be a bilinear form on $V \times W$. If $\alpha(x, a') = 0$ for all $x \in V$ implies $a' = 0$, and if $\alpha(a, y) = 0$ for all $y \in W$ implies $a = 0$, then α is called a *nondegenerate bilinear form* on $V \times W$.

Example. For $a \in V$ and $f \in V^*$, let $<a, f> = f(a)$. Then the map $(a, f) \to <a, f>$ is a nondegenerate bilinear form on $V \times V^*$.

Now let α be a bilinear form on V, i.e., on $V \times V$. If, for any $a, b \in V$,

$$\alpha(a, b) = \alpha(b, a)$$

holds, then α is called a *symmetric bilinear form* on V, and if

$$\alpha(a, b) = -\alpha(b, a)$$

holds, then α is called an *alternating* or *skew-symmetric bilinear form* on V.

Problem 8. Show that an arbitrary bilinear form on V can be written uniquely as the sum of a symmetric bilinear form and an alternating bilinear form on V. There the sum of two bilinear forms α and β is defined by $(\alpha + \beta)(a, b) = \alpha(a, b) + \beta(a, b)$.

Problem 9. Let V be finite dimensional and suppose that there is a nondegenerate alternating bilinear form on V. Show that V is even dimensional.

Let the dimension of V be n and let $\{a_1, \ldots, a_n\}$ be a basis of V. Let α be a bilinear form on V and set

$$\alpha(a_i, a_j) = \alpha_{ij} \quad (i, j = 1, \ldots, n).$$

Then we have

$$\alpha(a, b) = \sum_{i, j = 1}^{n} \alpha_{ij} \lambda^i \mu^j \quad \text{for} \quad a = \sum_{i = 1}^{n} \lambda^i a_i, \quad b = \sum_{i = 1}^{n} \mu^i a_i.$$

Problem 10. Show that:
(1) α is symmetric if and only if $\alpha_{ji} = \alpha_{ij}$ $(i, j = 1, \ldots, n)$;
(2) α is alternating if and only if $\alpha_{ji} = -\alpha_{ij}$ $(i, j = 1, \ldots, n)$.

Remark. As we have said before, we shall in general write the components of a matrix as α_j^i, putting indices as superscripts and subscripts. However, for a matrix corresponding to a bilinear form, we shall write α_{ij}.

Problem 11. Show that a bilinear form α on V is nondegenerate if and only if the determinant $\det(\alpha_{ij})$ of the matrix (α_{ij}) corresponding to α is not 0.

Consider now the case where $K = \mathbf{R}$, i.e., where V is a real vector space. A nondegenerate symmetric bilinear form on V is now called an *inner product* on V. That is, a bilinear form α on a real vector space V is an inner product if the two conditions (1) $\alpha(x, y) = \alpha(y, x)$ and (2) $\alpha(x, a) = 0$ for all $a \in V$ implies $x = 0$, hold. If a symmetric bilinear form α on V satisfies

$$\alpha(x, x) > 0 \quad \text{for} \quad x \in V, \quad x \neq 0,$$

then α is called a *positive inner product* on V.

Problem 12.   If α satisfies $\alpha(x, x) > 0$ for $x \in V, x \neq 0$, then show that α is nondegenerate.

Problem 13.   Let α be a bilinear form on V. Show that α is a positive inner product on V if and only if the matrix (α_{ij}) corresponding to α is a positive symmetric matrix.

Now consider the case where $K = \mathbf{C}$, i.e., where V is a complex vector space. If a map from $V \times V$ to $\mathbf{C}$ satisfies the conditions:

$$h(x + y, z) = h(x, z) + h(y, z)$$
$$h(x, y) = \overline{h(y, x)}^* \quad (x, y, z \in V; \lambda \in \mathbf{C})$$
$$h(\lambda x, y) = \lambda h(x, y)$$

then h is called a *hermitian form* on V.

Problem 14.   Show $h(x, \lambda y) = \overline{\lambda} h(x, y)$.

If the hermitian form h satisfies the condition

$$h(x, a) = 0 \text{ for all } a \in V \text{ implies } x = 0,$$

then h is called a *hermitian inner product* on the complex vector space V. If h is a hermitian form, then $h(x, x)$ is a real number. If h satisfies

$$h(x, x) > 0, \qquad x \in V, \quad x \neq 0,$$

then h is called a *positive hermitian inner product*.

§3. The n-Dimensional Real Space $\mathbf{R}^n$ and C^r Functions†

A   Let $\mathbf{R}^n$ denote the set of all n-tuples of real numbers $(p^1, \ldots, p^n)$. For a point $p = (p^1, \ldots, p^n)$ of $\mathbf{R}^n$, let

$$|p| = \underset{i = 1, \ldots, n}{\text{Max}} |p^i|.$$

---

*In general we write $\overline{\alpha}$ for the complex conjugate of a complex number α.

†For §3 and §4, cf. H. Whitney, *Geometric Integration Theory*, Princeton Univ. Press, Princeton, New Jersey, 1957, Ch. II.

For two points p, q in $\mathbf{R}^n$, let $p - q = (p^1 - q^1, \ldots, p^n - q^n)$, and let

$$d(p, q) = |p - q|.$$

d gives a metric on $\mathbf{R}^n$. Hence a topology is defined in $\mathbf{R}^n$, and $\mathbf{R}^n$ becomes a topological space. This topological space is called the *n-dimensional real space*, and is also denoted by $\mathbf{R}^n$. $\mathbf{R}^n$ is a complete metric space.

Problem 1.   For two points p and q in $\mathbf{R}^n$, let $\rho(p, q) = (\sum_{i=1}^n (p^i - q^i)^2)^{1/2}$. Then ρ is also a metric on $\mathbf{R}^n$. Show that the two metrics d and ρ define the same topology in $\mathbf{R}^n$.

For $p \in \mathbf{R}^n$ and $r > 0$, let

$$Q(p;r) = \{q \in R^n| \ |p - q| < r\}.$$

$Q(p; r)$ is called a *cube* of half-width r centered at p. A subset O of $\mathbf{R}^n$ is an open set of $\mathbf{R}^n$ if and only if, for an arbitrary point p of O, there is an $r > 0$ such that $Q(p;r) \subset O$.

For a point $p = (p^1, \ldots, p^n)$ in $\mathbf{R}^n$, let

$$x^i(p) = p^i \qquad (i = 1, \ldots, n).$$

$x^1, \ldots, x^n$ are continuous functions from $\mathbf{R}^n$ to $\mathbf{R} = \mathbf{R}^1$. We call this n-tuple of functions $(x^1, \ldots, x^n)$ the *standard coordinate system* of $\mathbf{R}^n$.

The notation $(x^1, \ldots, x^n)$ denotes the standard coordinate system in $\mathbf{R}^n$, but it is conventional and often convenient to use the same notation $x = (x^1, \ldots, x^n)$ to denote a "general point" of $\mathbf{R}^n$. In this book we shall use $(x^1, \ldots, x^n)$ in either sense, but there should be no danger of confusion.

A map f from a subset A of $\mathbf{R}^n$ to $\mathbf{R}$ is called a function $f(x^1, \ldots, x^n)$ on n variables $x^1, \ldots, x^n$ defined on A (or with domain A).

B $\mathbf{R}^n$ can also be considered as an n-dimensional vector space over $\mathbf{R}$. That is, for points $p = (p^1, \ldots p^n)$, $q = (q^1, \ldots, q^n)$ of $\mathbf{R}^n$ and for $\lambda \in \mathbf{R}$, we define the sum of elements in $\mathbf{R}^n$ and the scalar multiplication by

$$p + q = (p^1 + q^1, \ldots, p^n + q^n), \quad \lambda p = (\lambda p^1, \ldots, \lambda p^n),$$

and thus $\mathbf{R}^n$ becomes an n-dimensional vector space over $\mathbf{R}$.

C   Let f be a function defined on an open set U of $\mathbf{R}^n$ and let $p \in U$. Let q be a point in a neighborhood of $0 = (0, \ldots, 0)$ If $f(p + q) - f(p)$ can be written as

$$f(p + q) - f(p) = \sum_{i=1}^n A_i \cdot q^i + c(q)$$

[here A_i is a constant independent of q, $c(q)$ is a function of q], and if $c(q)/|q|$ approaches 0 as q approaches the origin 0, then we say that f is *totally differentiable* at p. If f is totally differentiable at p then f is differentiable with respect to each variable x^i at p and

$$\frac{\partial f}{\partial x^i}(p^1, \ldots, p^n) = A_i \qquad (i = 1, \ldots, n)$$

holds. If f is totally differentiable at all points of U, then we say that f is totally differentiable on U.

D   Let f be a function defined on an open set U of $\mathbf{R}^n$. If f is differentiable at all points of U with respect to each variable x^i, and if each partial derivative

$$\frac{\partial f}{\partial x^i} \qquad (i = 1, \ldots, n)$$

is continuous on U, then f is called a *continuously differentiable function* or a C^1 *function* on U. If for a positive integer r, the rth order partial derivatives

$$\frac{\partial^{\alpha_1 + \ldots + \alpha_n} f}{(\partial x^1)^{\alpha_1} \ldots (\partial x^n)^{\alpha_n}} \qquad (\alpha_i \geq 0, \alpha_1 + \ldots + \alpha_n \leq r)$$

all exist and are continuous on U, then f is called an *r-times continuously differentiable function* or a C^r *function* on U, or we say that f is (*of class*) C^r on U. If f is of class C^r on U for every positive integer r then we say that f is a C^∞ *function* on U or that f is (*of class*)C^∞ on U. We also say that f is a C^0 *function* on U or that f is (*of class*) C^0 if f is continuous on U.

Problem 2.   If f is of class C^1 on U, show that f is totally differentiable on U.

If f is C^∞ on U, and if for each point p of U the series

$$\sum_{m_1, \ldots, m_n = 0}^{\infty} \frac{1}{m_1! \ldots m_n!} a_{m_1 \ldots m_n}(x^1 - p^1)^{m_1} \ldots (x^n - p^n)^{m_n},$$

where

$$a_{m_1 \ldots m_n} = \frac{\partial^{m_1 + \ldots + m_n} f}{(\partial x^1)^{m_1} \ldots (\partial x^n)^{m_n}}(p),$$

converges absolutely and uniformly to $f(x^1, \ldots, x^n)$ on a sufficiently small neighborhood of p, then f is called an *analytic function* or a C^{ω} *function* on U.

E   Finally we quote an existence theorem on a system of ordinary differential equations.

Let $f^{i}(t, x^{1}, \ldots, x^{n}; a^{1}, \ldots, a^{m})$ $(i = 1, \ldots, n)$ be functions on $n + m + 1$ variables $(t, x^{1}, \ldots, x^{n}; a^{1}, \ldots, a^{m})$, and suppose they are of class C^{r} ($1 \leq r \leq \infty$ or $r = \omega$) in a neighborhood of 0 in $\mathbf{R}^{n+m+1}$. If $(y^{1}, \ldots, y^{n})$ is an arbitrary point sufficiently close to the origin, then the system of ordinary differential equations

$$\frac{dx^{i}}{dt} = f^{i}(t, x^{1}, \ldots, x^{n}; a^{1}, \ldots, a^{m}),$$

containing parameters $(a^{1}, \ldots, a^{m})$, has a unique solution of class C^{r} with respect to t such that $x^{i} = y^{i}$ at $t = 0$. As this solution depends on the initial condition (y^{i}) and the parameter (a^{k}), we write

$$x^{i} = \varphi^{i}(t, y^{1}, \ldots, y^{n}; a^{1}, \ldots, a^{m}) \quad (i = 1, \ldots, n).$$

Then φ^{i} $(i = 1, \ldots, n)$ is of class C^{r} as a function of $n + m + 1$ variables in a neighborhood of the origin in $\mathbf{R}^{n+m+1}$.

§4. The Inverse Function Theorem

Let φ be a map from an open set U of $\mathbf{R}^{n}$ into $\mathbf{R}^{m}$. Let $\varphi(p) = (\varphi^{1}(p), \ldots, \varphi^{m}(p))$ $(p \in U)$, then each φ^{i} is a function defined on U. If each φ^{i} is of class C^{r} ($0 \leq r \leq \omega$) then we say that φ is a C^{r} *map* or that φ is *(of class)C^{r}* from U into $\mathbf{R}^{m}$. If φ is of class C^{1}, then at each point p of U, for an arbitrary $v \in \mathbf{R}^{n}$,

$$\lim_{t \to 0} \frac{1}{t} [\varphi(p + tv) - \varphi(p)]$$

exists and is equal to $(d\varphi^{1}(p + tv)/dt)_{t=0}, \ldots, (d\varphi^{m}(p + tv)/dt)_{t=0}$. We denote the limit by $(d\varphi)_{p}(v)$ and call it the derivative of φ at p in the direction of v. The correspondence

$$(d\varphi)_{p} : v \to (d\varphi)_{p}(v)$$

is a linear map from the vector space $\mathbf{R}^{n}$ to the vector space $\mathbf{R}^{m}$. In fact, we have

$$\left(\frac{d\varphi^i}{dt}(p + tv)\right)_{t=0} = \sum_{j=1}^{n} \frac{\partial\varphi^i}{\partial x^j}(p)v^j.$$

If we write the elements of $\mathbf{R}^n$ and $\mathbf{R}^m$ as column vectors, then we have

$$(d\varphi)_p \begin{bmatrix} v^1 \\ \cdot \\ \cdot \\ \cdot \\ v^n \end{bmatrix} = \begin{bmatrix} u^1 \\ \cdot \\ \cdot \\ \cdot \\ u^m \end{bmatrix},$$

with

$$u^i = \sum_{j=1}^{n} \frac{\partial\varphi^i}{\partial x^j}(p)v^j.$$

We denote the $m \times n$ matrix

$$\begin{bmatrix} \dfrac{\partial\varphi^1}{\partial x^1}(p) & \cdots & \dfrac{\partial\varphi^1}{\partial x^n}(p) \\ \dfrac{\partial\varphi^2}{\partial x^1}(p) & \cdots & \dfrac{\partial\varphi^2}{\partial x^n}(p) \\ \cdot & & \\ \cdot & & \\ \cdot & & \\ \dfrac{\partial\varphi^m}{\partial x^1}(p) & \cdots & \dfrac{\partial\varphi^m}{\partial x^n}(p) \end{bmatrix}$$

by $(J\varphi)_p$ and call it the *Jacobian matrix* of φ. If $m = n$ then the determinant of the square matrix $(J\varphi)_p$ is written as

$$\det (J\varphi)_p = \frac{D(\varphi^1, \ldots, \varphi^n)}{D(x^1, \ldots x^n)_p}$$

and is called the *Jacobian* or the *Jacobian determinant* of the map φ at p.

Problem 1. Let φ be a C^1 map from an open set U of $\mathbf{R}^n$ into an open set V of $\mathbf{R}^m$, and ψ a C^1 map from V into $\mathbf{R}^p$. Show that $\psi \circ \varphi$ is a C^1 map from $\mathbf{R}^n$ into $\mathbf{R}^p$, and that $(J(\psi \circ \varphi))_p = (J\psi)_{\varphi(p)}(J\varphi)_p$.

LEMMA 1.   *Let φ be a C^1 map from an open set U of $\mathbf{R}^n$ into $\mathbf{R}^m$, and let K be a compact subset of U. For a given $\varepsilon > 0$, there is a $\delta > 0$ such that if $p, q \in K$ and $|p - q| < \delta$, then*

$$|(d\varphi)_p(v) - (d\varphi)_q(v)| < \varepsilon|v|$$

holds for an arbitrary $v \in \mathbf{R}^n$.

Proof.   Let the ith component of $(d\varphi)_p(v)$ be denoted by $(d\varphi)_p^i(v)$. Then

$$|(d\varphi)_p^i(v) - (d\varphi_q^i(v)| = \left| \sum_{j=1}^n \frac{\partial \varphi^i}{\partial x^j}(p)v^j - \frac{\partial \varphi^i}{\partial x^j}(q)\, v^j \right|$$

$$\leqq |v| \sum_{j=1}^n \left| \frac{\partial \varphi^i}{\partial x^j}(p) - \frac{\partial \varphi^i}{\partial x^j}(q) \right|$$

holds. Now, since $\partial \varphi^i / \partial x^j$ $(i, j = 1, \ldots, n)$ is continuous on K, and since K is compact, $\partial \varphi^i / \partial x^j$ is uniformly continuous on K. That is, for a given $\varepsilon > 0$, there is a $\delta > 0$ such that for $p, q \in K$ satisfying $|p - q| < \delta$, we have

$$\left| \frac{\partial \varphi^i}{\partial x^j}(p) - \frac{\partial \varphi^i}{\partial x^j}(q) \right| < \frac{\varepsilon}{n} \qquad (i, j = 1, \ldots, n).$$

Hence, if $p, q \in K$, and if $|p - q| < \delta$, then $|(d\varphi)_p^i(v) - (d\varphi)_q^i(v)| < \varepsilon|v|$ $(i = 1, \ldots, n)$ holds, so we get the inequality in Lemma 1.

LEMMA 2.   *Let φ be as in Lemma 1, and let K be a compact subset of U. We assume further that K is a convex set.* *Set*

$$\underset{p \in K}{\text{Max}} \left| \frac{\partial \varphi^i}{\partial x^j}(p) \right| = M_{ij}, \qquad \underset{i, j}{\text{Max}}\; M_{ij} = M.$$

Then, for $p, q \in K$, we have

$$|\varphi(p) - \varphi(q)| \leqq nM|p - q|.$$

Proof.   Let $p, q \in K$, and let $p_t = tq + (1 - t)p$ $(0 \leqq t \leqq 1)$. Since K is convex, $p_t \in K$. Since

$$\varphi^i(q) - \varphi^i(p) = \int_0^1 \frac{d\varphi^i(p_t)}{dt}\, dt,$$

*For two points p, q in $\mathbf{R}^n$ set $\{p_t | p_t = tq + (1 - t)p, 0 \leqq t \leqq 1\} = \overline{pq}$. $\overline{pq}$ is called the line segment connecting p and q. If a subset A of $\mathbf{R}^n$ has the property that $\overline{pq} \subset A$ for any two points p, q in A, then A is called a convex set.

and since

$$\frac{d\varphi^i(p_t)}{dt} = \sum_{j=1}^{n} \frac{\partial \varphi^i}{\partial x^j}(p_t)(q^j - p^j),$$

we have, for $i = 1, \ldots, n$,

$$|\varphi^i(p) - \varphi^i(q)| \leq \left| \int_0^1 \frac{d\varphi^i(p_t)}{dt} dt \right|$$

$$\leq \sum_{j=1}^{n} |p^j - q^j| \int_0^1 \left| \frac{\partial \varphi^i(p_t)}{\partial x^j} \right| dt$$

$$\leq |p - q| \, n \sum_{j=1}^{n} \int_0^1 M \, dt = |p - q| \, nM.$$

Hence we get $|\varphi(p) - \varphi(q)| \leq nM|p - q|$.

Now let φ be a $1:1$ mapping from an open set U of $\mathbf{R}^n$ onto an open set V of $\mathbf{R}^n$. If φ and its inverse are both of class C^r, then we call φ a *diffeomorphic map of class C^r* or a C^r *diffeomorphism* of U onto V. A C^0 diffeomorphism is nothing but a homeomorphism.

THEOREM 1 (*The inverse function theorem*). *Let φ be a C^r map $(1 \leq r \leq \infty)$ from an open set U of $\mathbf{R}^n$ into $\mathbf{R}^n$. If the Jacobian of the map φ at a point p_0 of U is not zero, then φ is a C^r diffeomorphism from a neighborhood of p_0 in $\mathbf{R}^n$ onto a neighborhood of $\varphi(p_0)$ in $\mathbf{R}^n$, i.e., there are neighborhoods $U_0(U_0 \subset U)$ of p_0 and V_0 of $\varphi(p_0)$ such that when φ is restricted to U_0 then φ is a C^r diffeomorphism of U_0 onto V_0.*

Proof. Without loss of generality, we can assume that $p_0 = \varphi(p_0) = 0$. By hypothesis, the Jacobian matrix $(J\varphi)_0$ of φ at 0 is nondegenerate. We shall first prove the theorem under the assumption that $(J\varphi)_0$ is the $n \times n$ identity matrix 1_n. Since $(J\varphi)_0 = 1_n$ we have

$$\frac{\partial \varphi^i(0)}{\partial x^j} = \delta_j^i \qquad (i, j = 1, \ldots, n).$$

For $p \in U$, let

$$g(p) = \varphi(p) - p, \quad g(p) = (g^1(p), \ldots, g^n(p)). \tag{1}$$

Then g is also a C^r map from U into $\mathbf{R}^n$, and since $g^i(x) = \varphi^i(x) - x^i$ we have

$$\frac{\partial g^i}{\partial x^j}(0) = 0 \qquad (i, j = 1, \ldots, n).$$

That is, the Jacobian matrix $(Jg)_0$ of g is the zero matrix. Hence taking $\rho > 0$ small enough, we can have a closed cube $\bar{Q}(0;\rho) = \{x | |x^i| \leq \rho$ $i = 1, \ldots, n\}$ of half-width ρ, centered at 0, and which is contained in U, and such that $|(\partial g^i/\partial x^j)(p)| < 1/2n$ $(i, j = 1, \ldots, n)$ holds for $p \in \bar{Q}(0;\rho)$. Since $\bar{Q}(0;\rho)$ is a compact convex set, applying Lemma 2 we get

$$|g(p) - g(q)| \leq \tfrac{1}{2}|p - q| \qquad \text{for} \quad p, q \in \bar{Q}(0;\rho). \tag{2}$$

From (1) we have $p = \varphi(p) - g(p)$. Hence

$$|p - q| = |\varphi(p) - \varphi(q) - g(p) + g(q)|$$

$$\leq |\varphi(p) - \varphi(q)| + |g(p) - g(q)|.$$

From this and (2) we have

$$|\varphi(p) - \varphi(q)| \geq \tfrac{1}{2}|p - q| \qquad \text{for} \quad p, q \in \bar{Q}(0;\rho). \tag{3}$$

Hence for two points p, q in $\bar{Q}(0;\rho)$, if we assume $\varphi(p) = \varphi(q)$, then by (3), $|p - q| = 0$, i.e., we get $p = q$. Hence φ is a $1:1$ map on $\bar{Q}(0;\rho)$.

Next let us prove that for each point s in $Q(0;\rho/2)$, there is a point p of $Q(0;\rho)$ such that $\varphi(p) = s$. For this, let us define a sequence of points $\{p_k\}$ $(k = 0, 1, 2, \ldots)$ in $Q(0;\rho)$ inductively as follows: set $p_0 = 0$, $p_1 = s - g(p_0), \ldots, p_k = s - g(p_{k-1}), \ldots$ For this definition to make sense, it is necessary that $g(p_k)$ be defined. So we have to show that $p_k \in U$. For this, it suffices to show, by induction on k, that $p_k \in Q(0;\rho)$. First we note that $p_0 \in Q(0;\rho)$. We assume that $p_0, p_1, \ldots, p_{k-1} \in Q(0;\rho)$, and show that $p_k \in Q(0;\rho)$. For this, let us prove that

$$|p_l - p_{l-1}| \leq 2^{-(l-1)}|s| \tag{4}$$

holds for $1 \leq l \leq k$. For $l = 1$, (4) is clear, so let $l > 1$. Then by the definition of p_l, $p_l - p_{l-1} = g(p_{l-1}) - g(p_{l-2})$, so that from (2) we see that $|p_l - p_{l-1}| \leq |p_{l-1} - p_{l-2}|/2$, and one can prove (4) by induction on l. Now

$$|p_k| \leq |p_k - p_{k-1}| + |p_{k-1} - p_{k-2}| + \ldots + |p_1 - p_0| + |p_0|,$$

so by (4) we get $|p_k| \leq \sum_{l=0}^{k-1}(1/2)^l|s| < 2|s|$. But since $s \in Q(0;\rho/2)$, this gives us $|p_k| < r$. This shows $p_k \in Q(0;\rho)$.

By what was said, the sequence $\{p_k\}$ in $Q(0;\rho)$ is defined, but by (4),

$\{p_k\}$ is a Cauchy sequence, so it converges to a point p in $\mathbf{R}^n$. But, as shown above, we have $|p_k| < 2|s|$, and this gives us $|p| \leq 2|s|$, so $p \in Q(0;\rho)$. Now if we let $k \to \infty$ in

$$p_k = s - g(p_{k-1}),$$

then we get $p = s - g(p)$, so $s = p + g(p) = \varphi(p)$. Hence p is the desired point of $Q(0;\rho)$

So far we have proved that φ is $1:1$ on $Q(0;\rho)$, and that $\varphi(Q(0;\rho)) \supset Q(0;\rho/2)$. Now let $U_\rho = \varphi^{-1}Q(0;\rho/2) \cap Q(0;\rho)$, and let φ_0 be the restriction of φ to U_ρ. Then φ_0 is a $1:1$ map of class C^r from U_ρ onto $Q(0;\rho/2)$. Moreover, by (3), for $p, q \in Q(0;\rho/2)$,

$$|\varphi_0^{-1}(p) - \varphi_0^{-1}(p)| \leq 2|p - q| \tag{5}$$

holds. Hence letting $\varphi_0^{-1} = \psi_0$, we see that ψ_0 is a continuous map from $Q(0;\rho/2)$ to U_ρ.

Next we want to show that ψ_0 is of class C^r. The determinant of $(J\varphi)_p$, i.e., the Jacobian of φ at p, is a continuous function of p, and is not 0 at the origin. Hence by taking ρ sufficiently small, one can assume that the Jacobian of φ is not 0 at each point of $Q(0;\rho)$, i.e., that $(J\varphi)_p$ is nonsingular. For $p, p_1 \in U_\rho$, let $\varphi_0(p) = s$ and $\varphi_0(p_1) = s_1$. Then $p = \psi_0(s)$ and $p_1 = \psi_0(s_1)$. Since φ^i is totally differentiable, we have

$$\varphi^i(p) - \varphi^i(p_1) = \sum_{j=1}^{n} \frac{\partial \varphi^i}{\partial x^j}(p_1)(p^j - p_1^j) + C^i(p, p_1) \qquad (i = 1, \ldots, n),$$

where

$$C^i(p, p_1)/|p - p_1| \to 0 \quad \text{as} \quad p \to p_1 \qquad (i = 1, \ldots, n). \tag{6}$$

Writing $C(p, p_1)$ for the column vector with components $C^i(p, p_1)$ $(i = 1, \ldots, n)$, we can write the above expression as

$$s - s_1 = (J\varphi)_{p_1}(\psi_0(s) - \psi_0(s_1)) + C(\psi_0(s), \psi_0(s_1)). \tag{7}$$

Now let $A(s)$ be the inverse matrix of $(J\varphi)_p$ $(p \in U_\rho)$. The entries a_j^i of $A(s)$ are continuous functions of p, but since ψ_0 is continuous, we can think of these entries as continuous functions of $s \in Q(0;\rho/2)$. Letting $C_1(s, s_1) = -A(s_1)C(\psi_0(s), \psi_0(s_1))$, and multiplying $A(s_1)$ on both sides of (7), we obtain

$$\psi_0(s) - \psi_0(s_1) = A(s_1)(s - s_1) + C_1(s, s_1). \tag{8}$$

Observe that

$$\frac{C_1(s, s_1)}{|s - s_1|} = - A(s_1) \frac{C(p, p_1)}{|p - p_1|} \frac{|p - p_1|}{|s - s_1|}. \tag{9}$$

On the other hand, by (3), we have $|s - s_1| \geq |p - p_1|/2$, so that $|p - p_1|/|s - s_1| \leq 2$. Substituting this in the right member of (9), we get

$$\frac{|C_1^i(s, s_1)|}{|s - s_1|} \leq 2 \sum_{j=1}^{n} |a_j^i(s_1)| \frac{|C^j(p, p_1)|}{|p - p_1|}. \tag{10}$$

Now since ψ_0 is continuous, if $s \to s_1$ then $p \to p_1$. Hence, by (6) and (10), we have

$$C_1^i(s, s_1)/|s - s_1| \to 0 \quad \text{as} \quad s \to s_1.$$

Thus setting $\psi_0(s) = (\psi_0^1(s), \ldots, \psi_0^n(s))$, we have, by (8), that each ψ_0^i is totally differentiable, and that

$$A(s_1) = (J\psi_0)_{s_1}.$$

That is

$$\frac{\partial \psi_0^i}{\partial x^j}(s_1) = a_j^i(s_1) \qquad (i, j = 1, \ldots, n). \tag{11}$$

Now a_j^i is a continuous function of s, and s_1 is an arbitrary point of $Q(0; \rho/2)$, so that by (11), the derivatives of ψ_0 are all continuous, and hence ψ_0 is of class C^1. However, since $A(s) = (J\varphi)_{\psi_0(s)}^{-1}$, $a_j^i(s)$ is a rational function of the entries of $(J\varphi)_{\psi_0(s)}$, i.e., of $(\partial \varphi^k/\partial x^l)(\psi_0(s))$ $(k, l = 1, \ldots, n)$. Now $\partial \varphi^k/\partial x^l$ is C^{r-1} and ψ_0 is C^1. So if $r > 1$, then a_j^i is C^1. Then, by (11) again, we can say that ψ_0^i is C^2. Repeating this argument, we can thus conclude that ψ_0^i is C^r. We have finished the proof of the inverse function theorem for the case where $(J\varphi)_0 = 1_n$. If $(J\varphi)_0$ is not the identity matrix 1_n let the inverse matrix of $(J\varphi)_0$ be $B = (b_j^i)$, and let β denote the linear map from $\mathbf{R}^n$ to $\mathbf{R}^n$ given by $(x^i) \to (\sum_j b_j^i x^j)$. If we let

$$\varphi' = \beta \circ \varphi,$$

then $\varphi'(0) = 0$ and $(J\varphi')_0 = 1_n$, so the inverse theorem holds for φ'. But β is a C^r diffeomorphism from $\mathbf{R}^n$ to $\mathbf{R}^n$, so we can easily see that the theorem holds for φ.

Theorem 1 is thus proved.

There are various results that can be derived from the inverse function theorem. Let us prove the following first.

THEOREM 2.   *Let U be a neighborhood of 0 in $\mathbf{R}^{n+k}$, and let φ be a C^r map $(1 \leq r \leq \infty)$ from U into $\mathbf{R}^n$, and suppose $\varphi(0) = 0$. Furthermore, assume that the Jacobian matrix of φ has rank n at 0, say*

$$
\begin{bmatrix}
\dfrac{\partial \varphi^1}{\partial x^1}(0) & \cdots & \dfrac{\partial \varphi^1}{\partial x^n}(0) \\
\cdot & & \\
\cdot & & \\
\cdot & & \\
\dfrac{\partial \varphi^n}{\partial x^1}(0) & \cdots & \dfrac{\partial \varphi^n}{\partial x^n}(0)
\end{bmatrix}
\neq 0.
$$

Then there is a C^r diffeomorphism ψ from a neighborhood V_0 of 0 in $\mathbf{R}^{n+k}$ onto a neighborhood W_0 of 0 in $\mathbf{R}^{n+k}$ such that at every point $p = (p^1, \ldots, p^n, p^{n+1}, \ldots, p^{n+k})$ of V_0,

$$
\varphi^i(\psi(p^1, \ldots, p^{n+k})) = p^i \quad (i = 1, \ldots, n)
$$

holds.

Proof.   Define a mapping from U into $\mathbf{R}^{n+k}$ by

$$
\Phi(p) = (\varphi^1(p), \ldots, \varphi^n(p), p^{n+1}, \ldots, p^{n+k}).
$$

Then $(J\Phi)_0$ is of the form

$$
\begin{bmatrix}
\dfrac{\partial \varphi^1}{\partial x^1}(0) & \cdots & \dfrac{\partial \varphi^1}{\partial x^n}(0) & & \\
\cdot & & & & \\
\cdot & & & * & \\
\cdot & & & & \\
\dfrac{\partial \varphi^n}{\partial x^1}(0) & \cdots & \dfrac{\partial \varphi^n}{\partial x^n}(0) & & \\
& 0 & & & 1_k
\end{bmatrix}
\qquad (1_k \text{ denotes the } k \times k \text{ unit matrix}),
$$

so that the Jacobian of Φ is not 0 at the origin. Hence, by Theorem 1, there is a C^r diffeomorphism ψ from a neighborhood V_0 of 0 in $\mathbf{R}^{n+k}$ onto a neighborhood W_0 of 0 in $\mathbf{R}^{n+k}$, and if $p \in V_0$, then $\Phi(\psi(p)) = p$ holds.

By the definition of Φ, $\Phi(\psi(p)) = (\varphi^1(\psi(p)), \ldots, \varphi^n(\psi(p)), \psi^{n+1}(p),$
$\ldots, \psi^{n+k}(p))$. Hence, if $p \in V_0$, then we have

$$\varphi^i(\psi(p^1, \ldots, p^{n+k})) = p^i \qquad (i = 1, \ldots, n).$$

Theorem 2 is thus proved.

Observe that, in the proof above, comparing the $(n + j)$th coordinates of the equation $\Phi(\psi(p)) = p$, we have

$$\psi^{n+j}(p) = p^{n+j} \qquad (j = 1, \ldots, k).$$

Hence we have

$$\varphi^i(\psi^1(p^1, \ldots, p^{n+k}), \ldots, \psi^n(p^1, \ldots, p^{n+k}), p^{n+1}, \ldots, p^{n+k}) = p^i$$

$(i = 1, \ldots, n)$. So letting p be an arbitrary point with coordinates $(0, \ldots, 0, p^{n+1}, \ldots, p^{n+k})$, and setting

$$g^i(p^{n+1}, \ldots, p^{n+k}) = \psi^i(0, \ldots, 0, p^{n+1}, \ldots, p^{n+k}),$$

we have

$$\varphi^i(g^1(p^{n+1}, \ldots, p^{n+k}), \ldots, g^n(p^{n+1}, \ldots, p^{n+k}), p^{n+1}, \ldots, p^{n+k})$$
$$= 0 \ (i = 1, \ldots, n).$$

We can express this as follows.

THEOREM 3.   *Let $f^1, \ldots, f^n$ be n functions of class C^r defined in a neighborhood of 0 in $\mathbf{R}^{n+k}$, and suppose $f^i(0) = 0$ $(i = 1, \ldots, n)$. If the determinant*

$$\begin{bmatrix} \dfrac{\partial f^1}{\partial x^1}(0) & \cdots & \dfrac{\partial f^n}{\partial x^1}(0) \\ \cdot \\ \cdot \\ \cdot \\ \dfrac{\partial f^1}{\partial x^n}(0) & \cdots & \dfrac{\partial f^n}{\partial x^n}(0) \end{bmatrix}$$

is not zero, then there are n functions $g^1, \ldots, g^n$ on k variables $x^{n+1}, \ldots, x^{n+k}$ such that:

(1) *each $g^i(x^{n+1}, \ldots, x^{n+k})$ is of class C^r on a sufficiently small neighborhood U of $(0, \ldots, 0)$, and*

(2) *$f^i(g^1(x^{n+1}, \ldots, x^{n+k}), \ldots, g^n(x^{n+1}, \ldots, x^{n+k}), x^{n+1}, \ldots, x^{n+k}) = 0$ holds for $x \in U$.*

II

Differentiable Manifolds

A differentiable manifold, roughly speaking, is a topological space, where each point has a neighborhood that is parameterized in such a way that the transformation between two sets of parameters is given by a set of differentiable functions. A function on such a topological space can be considered locally as a function on these parameters. Hence we can discuss the differentiablility of functions and mappings. Using the idea of differentials, we can "linearize" a very small neighborhood of a point on a manifold and consider a tangent space. In this chapter we shall discuss the basic facts of the differential calculus on a differentiable manifold.

§1. The Definition of a Manifold

Let M be a Hausdorff space. If each point of M has a neighborhood homeomorphic to an open set in $\mathbf{R}^n$, then we call M an n-dimensional *topological manifold*. If M is an n-dimensional topological manifold, U an open set of M, homeomorphic to an open set E of $\mathbf{R}^n$ by the homeomorphism $\psi: U \to E$, then we call the pair (U, ψ) a coordinate neighborhood of M. If p is a point of U then $\psi(p)$ is a point of $\mathbf{R}^n$, so $\psi(p)$ is an n-tuple of real numbers. Let the ith coordinate of $\psi(p)$ be $x^i(p)$. Then we have $\psi(p) = (x^1(p), \ldots, x^n(p))$. Since ψ is continuous, each x^i is a real-valued continuous function defined on U. Furthermore, since ψ is $1:1$, if we have $x^i(p) = x^i(q)$ $(i = 1, \ldots, n)$ for two points p, q of U, then $p = q$. That is, the point p of U is determined by the n-tuple of real numbers $(x^1(p), \ldots, x^n(p))$. $(x^1(p), \ldots, x^n(p))$ is called the set of *local coordinates* of the point p of U with respect to the coordinate neighborhood (U, ψ), and the n-tuple $(x^1, \ldots, x^n)$ of functions on U is called the local *coordinate system* on (U, ψ). That is, we have:

The local coordinates of p in U with respect to (U, ψ) equals the coordinates of $\psi(p)$ in $\mathbf{R}^n$.

We use the adjective "local" to indicate that the coordinates are defined only on the part U of M. A local coordinate system is defined on the whole of M only if M is homeomorphic to an open set of $\mathbf{R}^n$.

Since M is an n-dimensional topological manifold, M is covered by a family $\{U_\alpha\}$ of open sets U_α homeomorphic to open sets in $\mathbf{R}^n$. Let A be the set of indices α of the open sets U_α and we write $\{U_\alpha\}_{\alpha \in A}$ for $\{U_\alpha\}$. Let E_α be an open set of $\mathbf{R}^n$ homeomorphic to U_α, and let ψ_α be a homeomorphism from U_α onto E_α (Fig. 2.1). We call the collection $\{U_\alpha, \psi_\alpha\}_{\alpha \in A}$

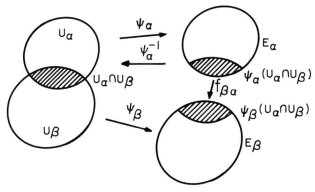

Fig. 2.1

of coordinate neighborhoods, a *coordinate neighborhood system* or an *atlas*, and denote it by $S = \{(U_\alpha, \psi_\alpha)\}_{\alpha \in A}$. If we take a point p of M, then there might be more than one set of local coordinates for p, because, for each U_α containing p, local coordinates are determined as coordinates of $\psi_\alpha(p)$ in $\mathbf{R}^n$. For each $\alpha \in A$, let us denote the local coordinate system on (U_α, ψ_α) by $(x_\alpha^1, \ldots, x_\alpha^n)$. If the intersection $U_\alpha \cap U_\beta$ of U_α and U_β is nonempty, then at each point of $U_\alpha \cap U_\beta$ two coordinate systems $(x_\alpha^1, \ldots, x_\alpha^n)$ and $(x_\beta^1, \ldots, x_\beta^n)$ are defined. Let us investigate the relation between these coordinate systems.

$\psi_\alpha(U_\alpha \cap U_\beta)$ and $\psi_\beta(U_\alpha \cap U_\beta)$ are open sets contained in the open sets E_α and E_β of $\mathbf{R}^n$, respectively. The inverse map ψ_α^{-1} of ψ_α gives a homeomorphism from $\psi_\alpha(U_\alpha \cap U_\beta)$ onto $U_\alpha \cap U_\beta$, and ψ_β gives a homeomorphism from $U_\alpha \cap U_\beta$ onto $\psi_\beta(U_\alpha \cap U_\beta)$. The map from $\psi_\alpha(U_\alpha \cap U_\beta)$ to $\psi_\beta(U_\alpha \cap U_\beta)$ obtained by composing these two homeomorphisms is denoted by $f_{\beta\alpha}$. $f_{\beta\alpha}$ is a homeomorphism, and for any point u in $\psi_\alpha (U_\alpha \cap U_\beta)$ we have

$$f_{\beta\alpha}(u) = \psi_\beta(\psi_\alpha^{-1}(u)). \tag{1}$$

Let $(u^1, \ldots, u^n)$ be the coordinates of u, and $(f_{\beta\alpha}^1(u^1, \ldots, u^n), \ldots, f_{\beta\alpha}^n(u^1, \ldots, u^n))$ the coordinates of $f_{\beta\alpha}(u)$. Then $f_{\beta\alpha}^i(u^1, \ldots, u^n)$ is a continuous function on n variables.

Now if $p \in U_\alpha \cap U_\beta$, then $(x_\alpha^1(p), \ldots, x_\alpha^n(p)) = \psi_\alpha(p)$ and $(x_\beta^1(p), \ldots, x_\beta^n(p)) = \psi_\beta(p)$, and since $\psi_\alpha(p) \in \psi_\alpha(U_\alpha \cap U_\beta)$, setting $u = \psi_\alpha(p)$ in (1), the right member becomes $(x_\beta^1(p), \ldots, x_\beta^n(p))$, while the left member is equal to $(f_{\beta\alpha}^1(x_\alpha^1(p), \ldots, x_\alpha^n(p)), \ldots, f_{\beta\alpha}^n(x_\alpha^1(p), \ldots, x_\alpha^n(p)))$. Hence we get the relations

$$x_\beta^i(p) = f_{\beta\alpha}^i(x_\alpha^1(p), \ldots, x_\alpha^n(p)) \qquad (i = 1, \ldots, n; p \in U_\alpha \cap U_\beta). \tag{2}$$

We shall write this as

$$x_\beta^i = f_{\beta\alpha}^i(x_\alpha^1, \ldots, x_\alpha^n) \qquad (i = 1, \ldots, n).$$

This is the transformation formula between two local coordinate systems $(x_\alpha^1, \ldots, x_\alpha^n)$ and $(x_\beta^1, \ldots, x_\beta^n)$ defined on $U_\alpha \cap U_\beta$.

Let f be a (real-valued) continuous function defined on a neighborhood of p in M. The point p is contained in some coordinate neighborhood U_α, and the points of U_α are determined by their local coordinates, so f can be considered as a continuous function on n variables. If this function on n variables is differentiable at $(x_\alpha^1(p), \ldots, x_\alpha^n(p))$, then we shall say, tentatively, that f is differentiable at p. Suppose f is defined on a neighborhood V of p. Set $V \cap U_\alpha = V_\alpha$, then $p \in V_\alpha$. On the open set $\psi_\alpha(V_\alpha)$ of $\mathbf{R}^n$, define a continuous function $F_\alpha(u)$ $(u = (u^1, \ldots, u^n) \in \psi_\alpha(V_\alpha))$ by $F_\alpha(u^1, \ldots, u^n) = f(\psi_\alpha^{-1}(u))$. The differentiability of f at p, as defined above, means that $F_\alpha(u^1, \ldots, u^n)$ is differentiable at $(x_\alpha^1(p), \ldots, x_\alpha^n(p))$ $(\in \psi_\alpha(V_\alpha))$. Suppose p is also contained in U_β. Set $V \cap U_\beta = V_\beta$, and define a continuous function $F_\beta(v)$ $(v = (v^1, \ldots, v^n) \in \psi_\beta(V_\beta))$ on $\psi_\beta(F_\beta)$ as above by $F_\beta(v^1, \ldots, v^n) = f(\psi_\beta^{-1}(v))$. If $u \in \psi_\beta(V_\alpha \cap V_\beta)$, then $f_{\beta\alpha}(u) = \psi_\beta(\psi_\alpha^{-1}(u))$ is a point in $\psi_\beta(V_\alpha \cap V_\beta)$, and by the definitions of F_α and F_β we have

$$F_\alpha(u^1, \ldots, u^n) = F_\beta(f_{\beta\alpha}^1(u^1, \ldots, u^n), \ldots, f_{\beta\alpha}^n(u^1, \ldots, u^n)). \tag{3}$$

Similarly, if $v \in \psi_\beta(V_\alpha \cap V_\beta)$, then $f_{\alpha\beta}(v) = \psi_\alpha(\psi_\beta^{-1}(v))$ is a point of $\psi_\alpha(V_\alpha \cap V_\beta)$, and we have

$$F_\beta(v^1, \ldots, v^n) = F_\alpha(f_{\alpha\beta}^1(v^1, \ldots, v^n), \ldots, f_{\alpha\beta}^n(v^1, \ldots, v^n)). \tag{3'}$$

From (1) we have $f_{\beta\alpha}(x_\alpha^1(p), \ldots, x_\alpha^n(p)) = (x_\beta^1(p), \ldots, x_\beta^n(p))$. Similarly $f_{\alpha\beta}(x_\beta^1(p), \ldots, x_\beta^n(p)) = (x_\alpha^1(p), \ldots, x_\alpha^n(p))$. We have defined above the differentiability of f at p by choosing a U_α containing p and using the coordinate neighborhood (U_α, ψ_α). The local coordinate systems should all have equal rights, so there is no reason that (U_α, ψ_α) should be privileged. That is, $F_\beta(v^1, \ldots, v^n)$ should also be differentiable at $(x_\beta^1(p), \ldots, x_\beta^n(p))$. However, F_α and F_β are related by (3) and (3'), and as $f_{\alpha\beta}^i, f_{\beta\alpha}^i$ are only continuous functions of $(u^1, \ldots, u^n)$, there is no guarantee that F_β is also differentiable at $(x_\beta^1(p), \ldots, x_\beta^n(p))$, even if F_α is differentiable at $(x_\alpha^1(p), \ldots, x_\alpha^n(p))$, and, conversely, even if F_β is differentiable at $(x_\beta^1(p), \ldots, x_\beta^n(p))$, it is not necessarily true that F_α is differentiable at $(x_\alpha^1(p), \ldots, x_\alpha^n(p))$. This difficulty is removed by imposing the condition that the functions on n variables $f_{\alpha\beta}^i, f_{\beta\alpha}^i$, giving the transformation of local coordinate systems, should be differentiable, for all α and β. Then we can define the differentiability of f independent of the choice of coordinate neighborhoods. In this way we come to the concept of a differentiable manifold.

DEFINITION 1.   A coordinate neighborhood system $S = \{(U_\alpha, \psi_\alpha)\}_{\alpha \in A}$ of an n-dimensional topological manifold M is called a *coordinate neighborhood system of class C^r*, or a *C^r coordinate neighborhood system*, if S has the following property: for any α, β in A such that $U_\alpha \cap U_\beta$ is nonempty, the functions $f_{\alpha\beta}^i, f_{\beta\alpha}^i$ $(i = 1, \ldots, n)$ on n variables, determining the transformation between the local coordinate systems (U_α, ψ_α) and (U_β, ψ_β), are r times $(r \geq 1)$ continuously differentiable on $\psi_\beta(U_\alpha \cap U_\beta)$ and $\psi_\alpha(U_\alpha \cap U_\beta)$, respectively. If $f_{\alpha\beta}^i, f_{\beta\alpha}^i$ are all real analytic, S is called a *coordinate neighborhood system of class C^ω*, or a *C^ω coordinate neighborhood system*. If S is a coordinate neighborhood system of class C^r $(1 \leq r \leq +\infty$ or $r = \omega)$ on M, then we say that S defines a *differentiable structure of class C^r* on the topological manifold M.*

DEFINITION 2.   If an n-dimensional topological manifold M has a coordinate neighborhood system S of class C^r, then we call M an n-dimensional differentiable manifold of class C^r, or a *C^r manifold*. In particular, if $r = \omega$, then we call M an *analytic manifold*.

---

*In the same spirit as a fundamental neighborhood system defines a topology in a topological space.

Remark.   A coordinate neighborhood system of class C^r is a coordinate neighborhood system of class C^s for any s such that $s \leqq r$. Hence a C^r manifold is a C^s manifold for any s such that $s \leqq r$. In particular, an analytic manifold and a C^∞ manifold are C^r manifolds for any r such that $1 \leqq r \leqq \infty$.

§2. Examples of Differentiable Manifolds

1 $\mathbf{R}^n$ is an analytic manifold. In fact, for an open covering take $\mathbf{R}^n$ itself, and let r be the identity mapping from $\mathbf{R}^n$ to itself. Then it is clear that $\{(\mathbf{R}^n, r)\}$ is a C^ω coordinate neighborhood system.

The coordinates $x^1, \ldots, x^n$ of $\mathbf{R}^n$ form a (local) coordinate system for the whole $\mathbf{R}^n$. When $\mathbf{R}^n$ is considered as a differentiable manifold, it is called an *affine space*. Let U be an open set of $\mathbf{R}^n$, and let $f^1, \ldots, f^n$ be real valued C^r functions defined on U. Suppose the map $x \to f(x) = (f^1(x), \ldots, f^n(x))$ $(x \in U)$ from U into $\mathbf{R}^n$ is a 1:1 map, and that the Jacobian $D(f^1, \ldots, f^n)/D(x^1, \ldots, x^n)$ is not 0 at each point of U. Then by the inverse function theorem (I, §1), $f(U)$ is an open set $\mathbf{R}^n$, and $f^{-1}: f(U) \to U$ is of class C^r. This means that $\{(R^n, r), (U, f)\}$ is a C^r coordinate neighborhood system for the affine space. Hence $(f^1, \ldots, f^n)$ is a local coordinate system of the affine space, and is usually called a *curvilinear coordinate system*. In particular, if each $f^i(x)$ is a linear function on x, i.e., if

$$f^i(x) = \sum_{j=1}^{n} a^i_j x^j + b^i \qquad (i = 1, \ldots, n),$$

then $D(f^1, \ldots, f^n)/D(x^1, \ldots, x^n)$ is the determinant $\det(a^i_j)$ of the matrix

$$\begin{bmatrix} a^1_1 & a^1_2 & \cdots & a^1_n \\ a^2_1 & a^2_2 & \cdots & a^2_n \\ \cdot & & & \\ \cdot & & & \\ \cdot & & & \\ a^n_1 & a^n_2 & \cdots & a^n_n \end{bmatrix}$$

and it is not 0 by assumption. In general, let $y^i = \sum_{j=1}^{n} a^i_j x^j + b^i$ $(i = 1, \ldots, n)$. If $\det(a^i_j) \neq 0$, then $(y^1, \ldots, y^n)$ is a (local) coordinate system defined on the whole affine space, and is called an *affine coordinate system*, or a *linear coordinate system*. We also call $(x^1, \ldots, x^n)$ the *standard coodinate system* of the affine space $\mathbf{R}^n$.

Problem 1.   Define a closed set X of $\mathbf{R}^2$ by $X = \{(x, 0) | x \geqq 0\}$. Show that the polar coordinates of $\mathbf{R}^2$ form a local coordinate system of $\mathbf{R}^2$ defined on the open set $\mathbf{R}^2 - X$.

2   Let S^1 be the circle in the xy-plane $\mathbf{R}^2$ centered at the origin and of radius 1(Fig. 2.2). Give S^1 the topology of a subspace of $\mathbf{R}^2$. Let

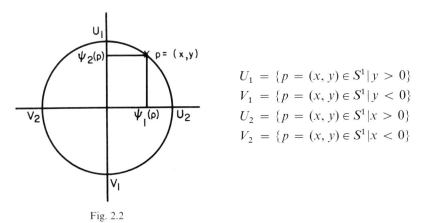

$$U_1 = \{p = (x, y) \in S^1 | y > 0\}$$
$$V_1 = \{p = (x, y) \in S^1 | y < 0\}$$
$$U_2 = \{p = (x, y) \in S^1 | x > 0\}$$
$$V_2 = \{p = (x, y) \in S^1 | x < 0\}$$

Fig. 2.2

Then U_i and V_i are open sets of S^1, and $U_1 \cup U_2 \cup V_1 \cup V_2 = S^1$. Let ψ_i and φ_i be the maps from U_i and V_i, respectively, to the open interval $I = \{t | -1 < t < 1\}$, given by $\psi_1(x, y) = x$, $\varphi_1(x, y) = x$, $\psi_2(x, y) = y$, $\varphi_2(x, y) = y$. The maps ψ_i and φ_i are homeomorphisms. Thus S^1 is a 1-dimensional topological manifold, and $S = \{(U_i, \psi_i), (V_i, \varphi_i)\}_{i = 1, 2}$ is a coordinate neighborhood system. Now $U_1 \cap U_2 = \{p = (x, y) \in S^1 | x, y > 0\}$, so $\psi_1(U_1 \cap U_2)$, $\psi_2(U_1 \cap U_2)$ are both equal to $J = \{t | 0 < t < 1\}$. For $t \in J$, we have $f_{21}(t) = \psi_2(\psi_1^{-1}(t)) = (1 - t^2)^{1/2}$ and $f_{12}(t) = \psi_1(\psi_2^{-1}(t)) = (1 - t^2)^{1/2}$. $f_{21}(t)$ and $f_{12}(t)$ are analytic for $0 < t < 1$. Similarly the other coordinate transformations are also analytic. Thus S^1 is an analytic manifold.

Problem 2.   Using polar coordinates, show that S^1 is a differentiable manifold.

3   The topological space obtained by giving the subspace topology to the set of points $(x^1, \ldots, x^{n+1})$ in $\mathbf{R}^{n+1}$ satisfying $(x^1)^2 + (x^2)^2 + \ldots + (x^{n+1})^2 = r^2$ is denoted by S^n and called the *n-dimensional sphere*. The sphere S^n is an n-dimensional analytic manifold. To verify this we let, as in the case of the circle S^1,

$$U_i = \{p = (x^1, \ldots, x^{n+1}) \in S^n | x^i > 0\},$$
$$V_i = \{p = (x^1, \ldots, x^{n+1}) \in S^n | x^i < 0\} \qquad (i = 1, \ldots, n + 1)$$

and define maps φ_i and ψ_i from U_i and V_i to

$$B^n = \{(y^1, \ldots, y^n) | (y^1)^2 + \ldots + (y^n)^2 < r^2\},$$

the interior of the sphere S^{n-1} in $\mathbf{R}^n$, by

$$p = (x^1, x^2, \ldots, x^{n+1}) \to (x^1, \ldots, x^{i-1}, x^{i+1}, \ldots, x^{n+1}).$$

Then φ_i, ψ_i are both homeomorphisms from U_i, V_i, respectively, onto the open set B^n in $\mathbf{R}^n$, so S^n is an n-dimensional topological manifold. As in the case of the circle, the coordinate transformation functions are analytic. The details are left to be checked by the reader.

4 *Projective spaces.* Remove the origin $0 = (0, \ldots, 0)$ from $\mathbf{R}^{n+1}$, and define an equivalence relation on the set $\mathbf{R}^{n+1} - \{0\}$ as follows. Two points $x = (x^i)$ and $y = (y^i)$ are defined to be equivalent if there is a nonzero real number λ such that $x = \lambda y$, i.e., $x^i = \lambda y^i (i = 1, \ldots, n+1)$. The set of equivalence classes given by this equivalence relation is denoted by P^n. Let $\pi(x)$ denote the equivalence class containing the point x of $\mathbf{R}^{n+1} - \{0\}$. Then π is a map from $\mathbf{R}^{n+1} - \{0\}$ onto P^n. To an element $\pi(x)$ of P^n associate the line tx $(t \in R)$ through the origin of $\mathbf{R}^{n+1}$. This correspondence is $1:1$, so we can regard P^n as the set of all lines passing through the origin in $\mathbf{R}^{n+1}$.

A subset U of P^n is defined to be an open set if the inverse image $\pi^{-1}(U)$ in $\mathbf{R}^{n+1} - \{0\}$ is an open set. In this way a topology is defined on P^n, and P^n becomes a Hausdorff space. Furthermore the projection π from $\mathbf{R}^{n+1} - \{0\}$ onto P^n is continuous.

Problem 3. Prove what was said about the topology of P^n.

Take a point p of P^n and let $p = \pi(x), x = (x^1, \ldots, x^{n+1})$. Because of the way the equivalence was defined, the property that the ith coordinate x^i of x is not 0 does not depend on the representative x for p. Let U_i $(i = 1, \ldots, n+1)$ be the set of all points p in P^n, which are represented by points x in $\mathbf{R}^{n+1} - \{0\}$ whose ith coordinate x^i is not 0.

Since $\pi^{-1}(U_a) = \{x | x^a \neq 0\}$ is an open set in $\mathbf{R}^{n+1} - \{0\}$, U_a is an open set in P^n. If $p \in U_\alpha$ and $p = \pi(x)$, then the ratio x^j/x^a of the jth coordinate x^j of x $(j = 1, \ldots, n+1)$ to the ath coordinate x^a of x does not depend on the choice of the representative x for p. Hence we can define real-valued functions $x_a^1, \ldots, x_a^n$ on U_a by

$$x_a^j(p) = x^j/x^a \quad (j = 1, \ldots, a-1),$$
$$x_a^j(p) = x^{j+1}/x^a \quad (j = a, \ldots, n),$$

where $p = \pi(x)$. Now define a map ψ_a from U_a to $\mathbf{R}^n$ by $\psi_a(p) = (x_a^1(p),$ $\ldots, x_a^n(p))$. Then ψ_a is a homeomorphism from U_a onto the whole of $\mathbf{R}^n$. Hence P^n is an n-dimensional topological manifold, and $\{(U_\alpha, \psi_a)\}_{a=1,\ldots,n+1}$ is a coordinate neighborhood system. If $p \in U_a \cap U_b$, $a < b$, and $p = \pi(x)$, then, since $x^a \neq 0$, $x^b \neq 0$, we have $x_b^a(p) = x^a/x^b \neq 0$ and $x_a^{b-1}(p) = x^b/x^a \neq 0$. Hence $x_b^a(p) = (x_a^{b-1}(p))^{-1}$. Furthermore, as

$$\frac{x^i}{x^b} = \frac{x^i}{x^a} \cdot \left(\frac{x^b}{x^a}\right)^{-1} \quad \text{and} \quad \frac{x^j}{x^a} = \frac{x^j}{x^b} \cdot \left(\frac{x^a}{x^b}\right)^{-1},$$

we have

$$x_b^i(p) = \frac{x_a^i(p)}{x_a^{b-1}(p)} \qquad \text{(if } i < a \text{ or } i \geq b)$$

$$= \frac{1}{x_a^{b-1}(p)} \qquad \text{(if } i = a)$$

$$= \frac{x_a^{i-1}(p)}{x_a^{b-1}(p)} \qquad \text{(if } a < i < b)$$

$$x_a^j(p) = \frac{x_b^j(p)}{x_b^a(p)} \qquad \text{(if } j < a \text{ or } j \geq b)$$

$$= \frac{x_b^{j+1}(p)}{x_b^a(p)} \qquad \text{(if } a \leq j < b - 1)$$

$$= \frac{1}{x_b^a(p)} \qquad \text{(if } j = b - 1).$$

Hence the coordinate transformation functions are rational functions, and the denominators are not 0 on $U_a \cap U_b$, so P^n is an analytic manifold. P^n is called the n-dimensional (real) *projective space*.

Problem 4.   Map the point $(x^1, \ldots, x^{n+1})$ $(\sum (x^i)^2 = 1)$ os S^n to the point $\pi(x)$ of P^n. Show that this map is continuous, and hence that P^n is compact.

5   *Open submanifolds.* Let M be a C^r manifold, D an open set of M, and $\{(U_\alpha, \psi_\alpha)\}_{\alpha \in A}$ a coordinate neighborhood system of class C^r. Set $U_\alpha' = U_\alpha \cap D$, and let ψ_α' be the restriction of ψ_α to U_α'. Then $\{(U_\alpha', \psi_\alpha')\}_{\alpha \in A}$ is a coordinate neighborhood system of class C^r on D, and D is a C^r manifold. D is called an *open submanifold* of M.

6   *Product manifolds.* Let M and N be C^r manifolds of dimension m and n, respectively. Let $\{(U_\alpha, \psi_\alpha)\}_{\alpha \in A}$ and $\{(V_i, \varphi_i)\}_{i \in I}$ be coordinate neighborhood systems of class C^r of M and N, respectively. First, if we give the direct product set $M \times N$ the topology of the product space of M and N, then $M \times N$ becomes a Hausdorff space. $\{U_\alpha \times V_i\}_{(\alpha, i) \in A \times I}$ is an open cover of $M \times N$. Let $\psi_\alpha \times \varphi_i$ be the map that sends the point (p, q) of $U_\alpha \times V_i$ to the point $(\psi_\alpha(p), \varphi_i(p))$ of $\mathbf{R}^{m+n}$. Then $\psi_\alpha \times \varphi_i$ is a homeomorphism from $U_\alpha \times V_i$ onto an open set $\psi_\alpha(U_\alpha) \times \varphi_i(V_i)$ of $\mathbf{R}^{m+n}$. Hence $M \times N$ is an $(m + n)$-dimensional topological manifold. We can check easily that the coordinate transformation functions are of class C^r, so $M \times N$ is a C^r manifold. We call $M \times N$ the *product manifold* of M and N. If $(x^1, \ldots, x^m)$ is the local coordinate system of M on U_α, and $(y^1, \ldots, y^n)$ is the local coordinate system of N on V_i, then by setting $\tilde{x}^i(p, q) = x^i(p), \tilde{y}^k(p, q) = y^k(q) \, (i = 1, \ldots, m; k = 1, \ldots, n)$ for (p, q) in $U_\alpha \times V_i$, $(\tilde{x}^1, \ldots, \tilde{x}^m, \tilde{y}^1, \ldots, \tilde{y}^n)$ becomes a local coordinate system on $U_\alpha \times V_i$. Since it becomes quite cumbersome to put the tildes on every x^i and y^k, we simply write x^i, y^k instead of $\tilde{x}^i, \tilde{y}^k$ most of the time. We can define the product manifold $M_1 \times \cdots \times M_l$ of l many C^r manifolds in the same way.

7 The *n-dimensional torus.* The product $S^1 \times \cdots \times S^1$ of n circles S^1 is an n-dimensional analytic manifold. We denote this by T^n, and call it the *n-dimensional torus* (or *n-torus*).

Problem 5. Show that the surface of revolution obtained by rotating a circle of radius 1 centered at (2, 0) in the xy-plane around the y-axis (the surface of a doughnut) is a 2-dimensional torus.

§3. Differentiable Functions and Local Coordinate Systems

Let M be an n-dimensional C^r manifold. Here we let $1 \leq r \leq \infty$ or $r = \omega$. Let U be an open set of M, and let f be a (real-valued) continuous function defined on U. Let $\{(U_\alpha, \psi_\alpha)\}_{\alpha \in A}$ be a C^r coordinate neighborhood system of M. For a point p in U, we can take a neighborhood V of p sufficiently small so that $V \subset U_\alpha \cap U$ for some index $\alpha \in A$. If the function $(f\psi_\alpha^{-1})(u) \, (u \in \psi_\alpha(V))$, defined on the open set $\psi_\alpha(V)$ of $\mathbf{R}^n$, is of class $C^s \, (s \leq r)$ in a neighborhood of $\psi_\alpha(p)$, then we say that f is of class C^s at the point p in U. As explained in §1, because $\{(U_\alpha, \psi_\alpha)\}_{\alpha \in A}$ is of class C^r, this definition does not depend on the choice of $\alpha \in A$ such that $p \in U_\alpha$. If f is of class C^s at all points of U, then f is called a C^s *function* on U. In particular, the local coordinate functions $x^i_\alpha \, (i = 1, \ldots, n)$ on each U_α are C^r functions on U.

Now suppose f is of class C^s in a neighborhood of a point p in M. If

$p \in U_\alpha$, we can write

$$f(\psi_\alpha^{-1}(u)) = F_\alpha(u^1, \ldots, u^n) \tag{1}$$

in a neighborhood of $\psi_\alpha(p)$. Here F_α is a C^s function on n variables $u^1, \ldots, u^n$. If we let $q = \psi_\alpha^{-1}(u)$, then since $(u^1, \ldots, u^n) = (x_\alpha^1(q), \ldots, x_\alpha^n(q))$, we can write (1) as

$$f(q) = F_\alpha(x_\alpha^1(q), \ldots, x_\alpha^n(q)).$$

We abbreviate this as

$$f = F_\alpha(x_\alpha^1, \ldots, x_\alpha^n). \tag{1'}$$

Also we write $\partial f(q)/\partial x_\alpha^i$ for $(\partial F_\alpha/\partial u^i)_{u = \psi_\alpha(q)}$.

Now suppose that p is also in U_β and let

$$f = F_\beta(x_\beta^1, \ldots, x_\beta^n) \tag{2}$$

in a neighborhood of p. If we set $(u^1, \ldots, u^n) = (x_\alpha^1(q), \ldots, x_\alpha^n(q))$ for $q \in U_\alpha \cap U_\beta$, then using (3) of §1, we have

$$F_\alpha(u^1, \ldots, u^n) = F_\beta(f_{\beta\alpha}^1(u), \ldots, f_{\beta\alpha}^n(u)).$$

Hence we have

$$\frac{\partial f}{\partial x_\alpha^i}(q) = \frac{\partial F_\alpha}{\partial u^i} = \sum_{j=1}^{n} \frac{\partial F_\beta}{\partial v^j}(f_{\beta\alpha}(u)) \frac{\partial f_{\beta\alpha}^j}{\partial u^i}.$$

But by (2) of §1, we have $x_\beta^j(q) = f_{\beta\alpha}^j(x_\alpha^1(q), \ldots, x_\alpha^n(q))$, hence

$$\frac{\partial f_{\beta\alpha}^j}{\partial u^i} = \frac{\partial x_\beta^j}{\partial x_\alpha^i}(q) \quad \text{and} \quad \frac{\partial F_\beta}{\partial v^j}(f_{\beta\alpha}(u)) = \frac{\partial f}{\partial x_\beta^j}(q).$$

Hence

$$\frac{\partial f}{\partial x_\alpha^i} = \sum_{j=1}^{n} \frac{\partial f}{\partial x_\beta^j} \frac{\partial x_\beta^j}{\partial x_\alpha^i} \tag{3}$$

holds on $U_\alpha \cap U_\beta$. Here $\partial f/\partial x_\alpha^i$ denotes the function that takes the value $\partial f(q)/\partial x_\alpha^i$ at the point q. Similarly for $\partial f/\partial x_\beta^j$ and $\partial x_\beta^j/\partial x_\alpha^i$. Similarly we have

$$\frac{\partial f}{\partial x_\beta^k} = \sum_{l=1}^{n} \frac{\partial f}{\partial x_\alpha^l} \frac{\partial x_\alpha^l}{\partial x_\beta^k} \tag{3'}$$

holds on $U_\alpha \cap U_\beta$. In particular, if we let $f = x_\alpha^k$ in (3) and $f = x_\beta^j$ in (3'), then since $\partial x_\alpha^k/\partial x_\alpha^i = \delta_i^k$, $\partial x_\beta^j/\partial x_\beta^k = \delta_k^j$, we have

$$\sum_{j=1}^{n} \frac{\partial x_\alpha^k}{\partial x_\beta^j} \frac{\partial x_\beta^j}{\partial x_\alpha^i} = \delta_i^k, \qquad \sum_{l=1}^{n} \frac{\partial x_\beta^j}{\partial x_\alpha^l} \frac{\partial x_\alpha^l}{\partial x_\beta^k} = \delta_k^j \qquad (4)$$

on $U_\alpha \cap U_\beta$. Define the $n \times n$ matrix $A_{\alpha\beta}$ by

$$A_{\alpha\beta} = \begin{bmatrix} \dfrac{\partial x_\alpha^1}{\partial x_\beta^1} & \dfrac{\partial x_\alpha^1}{\partial x_\beta^2} & \cdots & \dfrac{\partial x_\alpha^1}{\partial x_\beta^n} \\ \cdot & \cdot & & \cdot \\ \cdot & \cdot & & \cdot \\ \cdot & \cdot & & \cdot \\ \dfrac{\partial x_\alpha^n}{\partial x_\beta^1} & \dfrac{\partial x_\alpha^n}{\partial x_\beta^2} & \cdots & \dfrac{\partial x_\alpha^n}{\partial x_\beta^n} \end{bmatrix}$$

Then (4) means

$$A_{\alpha\beta} A_{\beta\alpha} = A_{\beta\alpha} A_{\alpha\beta} = 1_n \quad \text{(where } 1_n \text{ is the } n \times n \text{ unit matrix).}$$

That is, the matrix $A_{\beta\alpha}$ is the inverse matrix of $A_{\alpha\beta}$. In particular, det $A_{\alpha\beta} \neq 0$ at all points of $U_\alpha \cap U_\beta$.

Now let $f^1, \ldots, f^n$ be C^s functions defined on a neighborhood U of p. For $q \in U \cap U_\alpha$, let

$$\frac{D(f^1, \ldots, f^n)}{D(x_\alpha^1, \ldots, x_\alpha^n)_q} = \det\left(\frac{\partial f^i}{\partial x_\alpha^j}(q)\right)_{i,j=1,\ldots,n},$$

and call it the functional determinant of $(f^1, \ldots, f^n)$ at q with respect to the local coordinate system $(x_\alpha^1, \ldots, x_\alpha^n)$. If $q \in U_\alpha \cap U_\beta \cap U$, then by (3) we have

$$\frac{D(f^1, \ldots, f^n)}{D(x_\beta^1, \ldots, x_\beta^n)_q} = (\det A_{\alpha\beta}(q)) \frac{D(f^1, \ldots, f^n)}{D(x_\alpha^1, \ldots, x_\alpha^n)_q}.$$

Hence the property that $D(f^1, \ldots, f^n)/D(x_\alpha^1, \ldots, x_\alpha^n)_p \neq 0$ does not depend on the choice of U_α containing p.

DEFINITION 1. Let M be an n-dimensional C^r manifold and let $f^1, \ldots, f^n$ be n functions of class C^s ($s \leq r$) defined on a neighborhood of a point p on M, satisfying $D(f^1, \ldots, f^n)/D(x_\alpha^1, \ldots, x_\alpha^n)_p \neq 0$ where $p \in U_\alpha$. Then we call $(f^1, \ldots, f^n)$ a *local coordinate system* around p.

For example, since $D(x_\alpha^1, \ldots, x_\alpha^n)/D(x_\alpha^1, \ldots, x_\alpha^n)_p = 1$ holds for all points $p \in U_\alpha$, $(x_\alpha^1, \ldots, x_\alpha^n)$ is a C^r local coordinate system around each point p in U. Since $D(f^1, \ldots, f^n)/D(x_\alpha^1, \ldots, x_\alpha^n)_q$ is a continuous function of q, if it is not 0 at p, then it is not 0 at each point q of a sufficiently small neighborhood of p. Hence, if $(f^1, \ldots, f^n)$ is a local coordinate system around p, then it is a local coordinate system around each point q in a sufficiently small neighborhood of p.

Now, as before, in a neighborhood of the point $\psi_\alpha(p) = (x_\alpha^1(p), \ldots, x_\alpha^n(p))$ in $\mathbf{R}^n$, we let

$$f^i(\psi_\alpha^{-1}(u)) = F_\alpha^i(u^1, \ldots, u^n) \quad (i = 1, \ldots, n).$$

Then by the definition of $\partial f^i / \partial x_\alpha^j$, we have

$$\frac{D(f^1, \ldots, f^n)}{D(x_\alpha^1, \ldots, x_\alpha^n)_p} = \frac{D(F_\alpha^1, \ldots, F_\alpha^n)}{D(u^1, \ldots, u^n)_{\psi_\alpha(p)}} \neq 0.$$

Hence by the inverse function theorem, the mapping $u \to F(u) = (F_\alpha^1(u), \ldots, F_\alpha^n(u))$ is a homeomorphism from a neighborhood of $\psi_\alpha(p)$ in $\mathbf{R}^n$ onto a neighborhood of $F(\psi_\alpha(p))$ in $\mathbf{R}^n$, and the inverse map F^{-1} is also of class C^s. If we let U be a sufficiently small neighborhood of p contained in U_α, then the map $q \to f(q) = (f^1(q), \ldots, f^n(q))$ is a homeomorphism from U onto a neighborhood of $f(p) = F(\psi_\alpha(p))$ in $\mathbf{R}^n$. Hence (U, f) is a coordinate neighborhood of the topological manifold M, and $(f^1, \ldots, f^n)$ is a local coordinate system on U.

On the other hand, $f = F \circ \psi_\alpha$ and F^{-1} is a C^s map from $f(U)$ to $\psi_\alpha(U)$, so if we set $F^{-1}(v) = (G_\alpha^1(v), \ldots, G_\alpha^n(v))$ for $v \in f(U)$, then $G_\alpha^i(v^1, \ldots, v^n)(i = 1, \ldots, n)$ are C^s functions (Fig. 2.3). As $F^{-1}(f(q)) = \psi_\alpha(q)$ for $q \in U$, we have

$$x_\alpha^i(q) = G_\alpha^i(f^1(q), \ldots, f^n(q)).$$

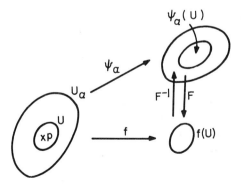

Fig. 2.3

Also from the definition of F_α^i we have

$$f^i(q) = F_\alpha^i(x_\alpha^1(q), \ldots, x_\alpha^n(q)).$$

These two equations show that the transformation between the two local coordinate systems $(x_\alpha^1, \ldots, x_\alpha^n)$ and $(f^1, \ldots, f^n)$ is of class C^s. Hence the coordinate neighborhood system $\{(U_\alpha, \psi_\alpha), (U, f)\}$ obtained by adjoining (U, f) to the C^r coordinate neighborhood system $\{(U_\alpha, \psi_\alpha)\}_{\alpha \in A}$ is of class C^s.

DEFINITION 2.   Let M be n-dimensional C^r manifold ($1 \leq r \leq \infty$ or $r = \omega$) and let $S = \{(U_\alpha, \psi_\alpha)\}_{\alpha \in A}$ be a C^r coordinate neighborhood system of M. Let (U, ψ) be a coordinate neighborhood of M (as a topological manifold). If the coordinate neighborhood system obtained by adjoining (U, ψ) to S is of class C^s ($s \leq r$), then we call (U, ψ) a C^s coordinate neighborhood of the C^r manifold M. Set $\psi(q) = (x^1(q), \ldots, x^n(q))$. The n functions $(x^1, \ldots, x^n)$ on U of class C^s are said to form a C^s local coordinate system on U of the C^r manifold M.

From what was proved above, we see that if $(f^1, \ldots, f^n)$ is a C^s local coordinate system around p, then $(f^1, \ldots, f^n)$ is a local coordinate system on a sufficiently small neighborhood U of p of M.

If $(x^1, \ldots, x^n)$ is a C^s local coordinate system on U, then, as before, a C^s function f defined around a point p of U can be written as

$$f(q) = F(x^1(q), \ldots, x^n(q)).$$

Here F is a C^s function of n variables $(u^1, \ldots, u^n)$. We shall abbreviate the above equation and write

$$f = F(x^1, \ldots, x^n).$$

We also write $(\partial f/\partial x^i)(q)$ for $(\partial F/\partial u^i)(x^1(q), \ldots, x^n(q))$, and we write $\partial f/\partial x^i$ for the function that takes value $(\partial f/\partial x^i)(q)$ at the point q.

Remark. The subordinance and the equivalence of differentiable structures. Let M be a topological manifold and let us consider two coordinate neighborhood systems $S = \{(U_\alpha, \psi_\alpha)\}_{\alpha \in A}$ and $T = \{(V_i, \varphi_i)\}_{i \in I}$. Suppose S is of class C^r and T is of class C^s ($s \leq r$). S and T define differentiable structures of classes C^r and C^s, respectively, on M. Let us write these C^r and C^s manifolds as (M, S) and (M, T), respectively. If for each $i \in I$, (V_i, φ_i) is a C^s coordinate neighborhood system of the C^r manifold (M, S), then we say that the differentiable structure of class C^s given by T is subordinate to the differentiable structure of class C^r given by S, and we write this as $T < S$. If $r = s$ and $T < S$ and $S < T$, then we say that S and T are equivalent and that S and T define the same C^r differentiable structure on M. If the topological manifold M has a countable basis,* then we can prove that any

*For topological manifolds with countable bases, see §15.

C^r differentiable structure (for $1 \leqq r < \infty$) on M is subordinate to some C^∞ differentiable structure on M. There is an example by Milnor of a topological manifold which has inequivalent differentiable structures.

§4. Differentiable Mappings

Let M and M' be manifolds* of dimensions n and m, respectively, and let φ be a continuous map from M into M'. Let $p \in M$ and let f be a C^∞ function defined on a neighborhood V of $\varphi(p)$ on M'. Since φ is continuous, we can take a neighborhood U of p such that $\varphi(U) \subset V$. Define a function $\varphi^* f$ on the neighborhood U of p by

$$(\varphi^* f)(q) = f(\varphi(q))$$

Since f and φ are both continuous, $\varphi^* f$ is also a continuous function. If, for an arbitrary C^∞ function f defined on a neighborhood of $\varphi(p)$, $\varphi^* f$ is of class C^∞ in a neighborhood of p, then we say that the mapping φ is C^∞ at the point p, or that φ is differentiable at the point p. If φ is differentiable at all points of M, then φ is called a *differentiable map* or a C^∞ *map* from M to M'.

Let (U, ψ) and (V, η) be coordinate neighborhoods of M and M' such that $\varphi(U) \subset V$, and let $(x^1, \ldots, x^n)$ and $(y^1, \ldots, y^m)$ be local coordinate systems on U and V, respectively. We can express the continuous functions $\varphi^* y^i$ on U as

$$\varphi^* y^i = \varphi^i(x^1, \ldots, x^n) \quad (i = 1, \ldots, m). \tag{1}$$

Here the φ^i's are continuous functions on n variables.

Then the differentiablility of φ at the point p in U is equivalent to the functions $\varphi^1, \ldots, \varphi^m$ being of class C^∞ in a neighborhood of $(x^1(p), \ldots, x^n(p))$.

Problem 1   Show that the map from S^n to P^n in Problem 4 of §2 is differentiable. Show that the identity map from S^n into $\mathbf{R}^{n+1}$ is differentiable.

Problem 2.   Define a C^r map from M into M'.

If a differentiable map φ from a manifold M to a manifold M' has the property that it is $1:1$ and onto and its inverse φ^{-1} is also differentiable, then φ is called a *diffeomorphic map or a diffeomorphism* from M onto M'.

---

*From now on, unless otherwise stated, our manifolds are of class C^∞, and we simply call them manifolds.

Example 1. Let $(x^1, \ldots, x^n)$ be the standard coordinate system in the affine space $\mathbf{R}^n$ (§2). If a map φ from $\mathbf{R}^n$ to $\mathbf{R}^n$ satisfies

$$\varphi^* x^i = \sum_{j=1}^{n} a_j^i x^j + b^i \qquad (i = 1, \ldots, n),$$

then we call φ an *affine map* of $\mathbf{R}^n$. In particular, if $b^i = 0$ $(i = 1, \ldots, n)$, then φ is called a *linear map*. If $a_j^i = \delta_j^i$ $(i, j = 1, \ldots, n)$, then φ is called a *parallel translation*. The affine map φ is a diffeomorphism if and only if $\det(a_j^i) \neq 0$, and when this is the case, φ is called an *affine transformation*.

Example 2. Let φ be a linear map of $\mathbf{R}^{n+1}$ given by $\varphi^* x^i = \sum_{j=1}^{n+1} a_j^i x^j$ $(i = 1, \ldots, n+1)$. If $\det(a_j^i) \neq 0$, then φ maps $\mathbf{R}^{n+1} - \{0\}$ onto $\mathbf{R}^{n+1} - \{0\}$. Furthermore, if $x = \lambda y(\lambda \in \mathbf{R})$, then $\varphi(x) = \lambda \varphi(y)$. Let π be the map from $\mathbf{R}^{n+1} - \{0\}$ to the projective space P^n. Then, by setting $\bar{\varphi}(\pi(x)) = \pi(\varphi(x))$, a map $\bar{\varphi}$ from P^n to P^n is defined. $\bar{\varphi}$ is called a *projective transformation* of P^n. A projective transformation is a diffeomorphism of P^n.

Problem 3. Prove that a projective transformation is a diffeomorphism of P^n.

The set of all affine transformations of $\mathbf{R}^n$ forms a group called the *affine transformation group* of $\mathbf{R}^n$. The set of all projective transformations forms a group called the *projective transformation group* of P^n.

Problem 4. Let Z be the set of all matrices of the form $\lambda 1_{n+1}$, where 1_{n+1} is the $(n+1) \times (n+1)$ unit matrix, and λ is a nonzero real number. Then Z is a normal subgroup of $GL(n+1, \mathbf{R})$. Prove that the projective transformation group of P^n is isomorphic to $GL(n+1, \mathbf{R})/Z$.

Differentiable curves. Let (a, b) be an open interval of $\mathbf{R}^1$. $\mathbf{R}^1$ is a 1-dimensional manifold and (a, b) is an open submanifold of $\mathbf{R}^1$. A differentiable map φ from (a, b) into a manifold M is called a *differentiable curve* of M defined on (a, b).

Remark. A differentiable curve is a map φ and is not the image $\varphi((a, b))$ of the interval (a, b) by φ. However, we do sometimes speak about, for example, a point $\varphi(t)$ on the curve φ.

If φ is a map from a closed interval $[a, b]$ into a manifold M, and if there is an open interval (a', b') containing $[a, b]$ and a differentiable map φ' from (a', b') into M such that $\varphi'(t) = \varphi(t)$, for $t \in [a, b]$, then we call φ a differentiable curve defined on $[a, b]$. For intervals $[a, b)$ and $(a, b]$, we define differentiable curves similarly. If φ is a continuous map from $[a, b]$ into M such that, by taking a finite number of points $a_0, a_1, \ldots, a_m$ on $[a, b]$, where $a = a_0 < a_1 < a_2 < \cdots < a_m = b$, the restrictions of φ to each of the subintervals $[a_0, a_1], [a_1, a_2], \ldots, [a_{m-1}, a_m]$ are differentiable curves, then we call φ a *piecewise differentiable curve* (Fig. 2.4).

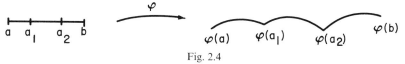

Fig. 2.4

Let φ be a differentiable curve of M defined on (a, b). Let $t_0 \in (a, b)$, and let $(x^1, \ldots, x^n)$ be a local coordinate system of M in a neighborhood of $\varphi(t_0)$, and set

$$(\varphi * x^i)(t) = \varphi^i(t) \quad (i = 1, \ldots, n).$$

Each $\varphi^i(t)$ is a C^∞ function in a neighborhood of t_0. The n-dimensional vector given by

$$\left(\left(\frac{d\varphi^1}{dt} \right)_{t=t_0}, \ldots, \left(\frac{d\varphi^n}{dt} \right)_{t=t_0} \right)$$

is called the tangent vector to the curve φ at $\varphi(t_0)$ with respect to the local coordinate system $(x^1, \ldots, x^n)$.

Problem 5. If $(x^i) = (x^1, \ldots, x^n)$ and $(\bar{x}^i) = (\bar{x}^1, \ldots, \bar{x}^n)$ are two local coordinate systems in a neighborhood of $\varphi(t_0)$, and $(\xi^1, \ldots, \xi^n)$ and $(\bar{\xi}^1, \ldots, \bar{\xi}^n)$ are tangent vectors to φ at $\varphi(t_0)$ with respect to (x^i) and $(\bar{x}^i)$, respectively, then show that

$$\bar{\xi}^i = \sum_{j=1}^n \frac{\partial \bar{x}^i}{\partial x^j} (\varphi(t_0)) \xi^j,$$

$$\xi^i = \sum_{j=1}^n \frac{\partial x^i}{\partial \bar{x}^j} (\varphi(t_0)) \bar{\xi}^j.$$

Problem 6. Define a C^r curve for a manifold M.

§5. Tangent Vectors and Tangent Spaces, the Riemannian Metric

Let M be a manifold and p a point on M. We consider the set of all real-valued C^∞ functions, each defined on some neighborhood of p, and denote it by $\mathfrak{F}(p)$. If $f, g \in \mathfrak{F}(p)$, then $f + g$ and $f \cdot g$ are defined on the intersection of the neighborhood where f is defined and the neighborhood where g is defined; λf is defined on the neighborhood where f is defined. If for each $f \in \mathfrak{F}(p)$, there corresponds a real number $v(f)$ satisfying

(1) $v(\lambda f + \mu g) = \lambda v(f) + \mu v(g)$

                                  $(\lambda, \mu \in \mathbf{R}; \ f, g \in \mathfrak{F}(p))$

(2) $v(fg) = v(f)g(p) + f(p)v(g)$

then the map $v : \mathfrak{F}(p) \to \mathbf{R}$ is called a *tangent vector* of M at p.

We should note that the value $v(f)$, for $f \in \mathfrak{F}(p)$, depends only on the local behavior of f. More precisely, if $f, g \in \mathfrak{F}(p)$, and if f and g coincide on some neighborhood V of p, then $v(f) = v(g)$. In fact, from (1), it suffices to show that $v(h) = 0$ for any $h \in \mathfrak{F}(p)$ which is 0 on V. To show this, we pick $\varphi \in \mathfrak{F}(p)$ such that $\varphi = 1$ on $M - V$ and $\varphi(p) = 0$. The existence of such a φ will follow from Lemma 1 of §10. We now observe that $h = \varphi \cdot h$, and apply (2). We get $v(h) = v(\varphi \cdot h) = v(\varphi)h(p) + \varphi(p)v(h) = 0$.

For tangent vectors v, v' of M at p, and for $\lambda \in \mathbf{R}$, we define the sum $v + v'$ and the scalar multiple λv by

$$(v + v')(f) = v(f) = v(f) + v'(f), \quad (\lambda v)(f) = \lambda(v(f)) \quad (f \in \mathfrak{F}(p)).$$

Then $v + v'$ and λv are also tangent vectors of M at p. Hence, defining the sum and the scalar multiple of tangent vectors at p in this manner, the set of all tangent vectors at p becomes a vector space over $\mathbf{R}$. We write this vector space as $T_p(M)$, and call it the *tangent vector space* or *tangent space* of M at p.

Let $(x^1, \ldots, x^n)$ be a local coordinate system on U, and at a point p of U let

$$\left(\frac{\partial}{\partial x^i}\right)_p f = \frac{\partial f}{\partial x^i}(p) \qquad (i = 1, \ldots, n).$$

Then $(\partial/\partial x^i)_p$ is a tangent vector at p. In the proof of the following theorem we shall show that $\{(\partial/\partial x^1)_p, \ldots, (\partial/\partial x^n)_p\}$ is a basis of $T_p(M)$.

THEOREM *If M is a manifold of dimension n, then the tangent space $T_p(M)$ is also of dimension n.*

Proof. Let $(x^1, \ldots, x^n)$ be a local coordinate system on a neighborhood U of p. To show that $T_p(M)$ is of dimension n, it suffices to show that $\{(\partial/\partial x^1)_p, \ldots, (\partial/\partial x^n)_p\}$ is a basis of $T_p(M)$. The vectors $(\partial/\partial x^1)_p$, $\ldots, (\partial/\partial x^n)_p$ are linearly independent. In fact, since $(\partial/\partial x^i)_p x^j = \delta_i^j$, we have $\sum_{i=1}^n \lambda^i(\partial/\partial x^i)_p x^j = \lambda^j$. Hence, if $\sum_{i=1}^n \lambda^i(\partial/\partial x^i)_p = 0$, then we have

$$\lambda^j = \sum_{i=1}^n \lambda^i \left(\frac{\partial}{\partial x^i}\right)_p x^j = 0(x^j) = 0,$$

and this shows that $(\partial/\partial x^1)_p, \ldots, (\partial/\partial x^n)_p$ are linearly independent. Next, if $v \in T_p(M)$, we shall prove that v can be written as

$$v = \sum_{i=1}^n v(x^i)\left(\frac{\partial}{\partial x^i}\right)_p. \tag{1}$$

First, if λ is a constant, then $v(\lambda) = 0$. In fact, since $\lambda = \lambda \cdot 1$, by condition (1) for tangent vectors, we have $v(\lambda) = \lambda v(1)$. On the other hand, by (2) we have $v(\lambda) = v(\lambda) \cdot 1 + \lambda \cdot v(1)$. Hence we have $v(\lambda) = 0$. Now, if

f is a C^∞ function defined on a neighborhood of p, f is written as

$$f(q) = F(x^1(q), \ldots, x^n(q))$$

on a small neighborhood of p. Here $F(u^1, \ldots, u^n)$ is a C^∞ function defined on a neighborhood of $(x^1(p), \ldots, x^n(p))$, and can be expanded as

$$F(u^1, \ldots, u^n) = F(x^1(p), \ldots, x^n(p)) + \sum_{i=1}^{n} F_i(u^1, \ldots, u^n)(u^i - x^i(p))$$

on some neighborhood of $(x^1(p), \ldots, x^n(p))$, where the $F_i(u^1, \ldots, u^n)$)'s are C^∞ functions on this neighborhood. In fact,

$$
\begin{aligned}
F(u^1, &\ldots, u^n) \\
&= F(x^1(p), \ldots, x^n(p)) \\
&\quad + \int_0^1 \frac{dF}{dt}(t(u^1 - x^1(p)) + x^1(p), \ldots, t(u^n - x^n(p)) + x^n(p))\, dt \\
&= F(x^1(p), \ldots, x^n(p)) + \sum_{i=1}^{n} \int_0^1 \frac{\partial F}{\partial s^i} \frac{ds^i}{dt}\, dt,
\end{aligned}
$$

where we let $s^i = t(u^i - x^i(p)) + x^i(p)$. Hence it suffices to let

$$F_i(u^1, \ldots, u^n) = \int_0^1 \frac{\partial F}{\partial s^i}\, dt.$$

We note that $F_i(x(p)) = \partial F(x(p))/\partial u^i = \partial f(p)/\partial x^i$. Restricting f to some neighborhood of p, we write

$$f = f(p) + \sum_{i=1}^{n} F_i(x^1, \ldots, x^n)(x^i - x^i(p)).$$

Now apply v to this equation. Using (1), (2) and $v(\lambda) = 0$, we get

$$v(f) = \sum_{i=1}^{n} F_i(x(p))v(x^i) = \sum_{i=1}^{n} \frac{\partial f}{\partial x^i}(p)v(x^i).$$

This shows that $v(f) = (\sum_{i=1}^{n} v(x^i)(\partial/\partial x^i)_p)(f)$ holds for all $f \in \mathfrak{F}(p)$, and (1) is proved. Thus $\{(\partial/\partial x^1)_p, \ldots, (\partial/\partial x^n)_p\}$ is a basis of $T_p(M)$, and hence $T_p(M)$ is an n-dimensional vector space.

For a tangent vector $v \in T_p(M)$, let $\xi^i = v(x^i)$, and call the n-dimensional vector $(\xi^1, \ldots, \xi^n)$ the components of v with respect to the local coordinate system $(x^1, \ldots, x^n)$. We then have

$$v = \sum_{i=1}^{n} \xi^i \left(\frac{\partial}{\partial x^i}\right)_p.$$

Let $(x^1, \ldots, x^n)$ and $(\bar{x}^1, \ldots, \bar{x}^n)$ be two local coordinate systems at p. Then $\{(\partial/\partial x^1)_p, \ldots, (\partial/\partial x^n)_p\}$ and $\{(\partial/\partial \bar{x}^1)_p, \ldots, (\partial/\partial \bar{x}^n)_p\}$ are both bases of $T_p(M)$ and we have the transformation rules

$$\left(\frac{\partial}{\partial \bar{x}^i}\right)_p = \sum_{j=1}^{n} \frac{\partial x^j}{\partial \bar{x}^i}(p)\left(\frac{\partial}{\partial x^j}\right)_p$$

$$\left(\frac{\partial}{\partial x^i}\right)_p = \sum_{j=1}^{n} \frac{\partial \bar{x}^j}{\partial x^i}(p)\left(\frac{\partial}{\partial \bar{x}^j}\right)_p.$$

(2)

The components (ξ^i), $(\bar{\xi}^i)$ of $v \in T_p(M)$ with respect to (x^i), $(\bar{x}^i)$, respectively, satisfy the transformation rules

$$\bar{\xi}^i = \sum_{j=1}^{n} \frac{\partial \bar{x}^i}{\partial x^j} \xi^j, \qquad \xi^i = \sum_{j=1}^{n} \frac{\partial x^i}{\partial \bar{x}^j} \bar{\xi}^j.$$

(3)

Problem 1. Prove the rules (2) and (3).

Now let φ be a differentiable curve on M, defined on an open interval (a, b), and let $t_0 \in (a, b)$. Take a local coordinate system $(x^1, \ldots, x^n)$ of M, and let $((d\varphi^1/dt)_{t=t_0}, \ldots, (d\varphi^n/dt)_{t=t_0})$ be the tangent vector to φ at $\varphi(t_0)$ with respect to $(x^1, \ldots, x^n)$(§4). Let

$$v = \sum_{i=1}^{n} \left(\frac{d\varphi^i}{dt}\right)_{t=t_0} \left(\frac{\partial}{\partial x^i}\right)_{\varphi(t_0)}.$$

Then v is a tangent vector to M at $\varphi(t_0)$ and does not depend on the choice of the local coordinate system $(x^1, \ldots, x^n)$. In fact, if $(\bar{x}^1, \ldots, \bar{x}^n)$ is another local coordinate system at $\varphi(t_0)$, and if we let $\bar{\varphi}^i = \varphi^* \bar{x}^i$, then

$$\sum_{i=1}^{n} \left(\frac{d\varphi^i}{dt}\right)_{t=t_0} \left(\frac{\partial}{\partial x^i}\right)_{\varphi(t_0)} = \sum_{i=1}^{n} \left(\frac{d\bar{\varphi}^i}{dt}\right)_{t=t_0} \left(\frac{\partial}{\partial \bar{x}^i}\right)_{\varphi(t_0)}.$$

(4)

We call v the tangent vector of the differentiable curve φ at $\varphi(t_0)$. If the tangent vector at each point of the differentiable curve φ is not zero, and if $\varphi(t) \neq \varphi(t')$ for $t \neq t'$, then φ is called a *regular differentiable curve*.

Problem 2.   Prove (4) above.

Example.   Let us consider the special case where M is the affine space $\mathbf{R}^n$. $\mathbf{R}^n$ can also be considered as an n-dimensional vector space. To avoid confusion, we shall write $V(\mathbf{R}^n)$ to denote $\mathbf{R}^n$ considered as a vector space. Let $(x^1, \ldots, x^n)$ be the standard coordinate system of $\mathbf{R}^n$. To a tangent vector v at a point p in $\mathbf{R}^n$, associate the components $(\xi^1, \ldots, \xi^n) \in V(\mathbf{R}^n)$ of v with respect to $(x^1, \ldots, x^n)$. By this correspondence $T_p(M)$ and $V(\mathbf{R}^n)$ become isomorphic. If v and v' are tangent vectors at p and p', respectively, and if the components of v and v' are equal, then we say that the vector v at p is parallel to the vector v' at p'. For two points p and q of $\mathbf{R}^n$, the tangent vector at p with components $(x^1(q) - x^1(p), \ldots, x^n(q) - x^n(p))$ is denoted by $\vec{pq}$, and is called the vector with base point p and end point q. By the isomorphism between $T_p(M)$ and $V(\mathbf{R}^n)$, if we consider $\vec{pq}$ as a vector in $V(\mathbf{R}^n)$, then we have $\vec{pq} = -\vec{qp}$, $\vec{pq} + \vec{qr} = \vec{pr}$.

Problem 3.   Let $T(\mathbf{R}^n)$ denote the set of all tangent vectors of the affine space $\mathbf{R}^n$. Prove that the relation that two vectors are parallel is an equivalence relation, and that the equivalence classes form a vector space isomorphic to $V(\mathbf{R}^n)$.

Problem 4.   We say that $n + 1$ points $p_0, p_1, \ldots, p_n$ of $\mathbf{R}^n$ are affinely independent if the n vectors $\vec{p_0 p_1}, \ldots, \vec{p_0 p_n}$ are linearly independent. If $(p_0, p_1, \ldots, p_n)$ and $(q_0, q_1, \ldots, q_n)$ are both affinely independent, show that there is a unique affine transformation φ satisfying $\varphi(p_i) = q_i$ ($i = 0, 1, \ldots, n$).

Riemannian metric.   Suppose that at each point p of a manifold M a positive inner product $g_p(u, v)$ ($u, v \in T_p(M)$) on $T_p(M)$ is given. For a local coordinate system $(x^1, \ldots, x^n)$ on U, we let $g_{ij}(q) = g_q((\partial/\partial x^i)_q, (\partial/\partial x^j)_q)$ for $q \in U$. Each $g_{ij}(q)$ ($i, j = 1, \ldots, n$) is a function on U, and the matrix $(g_{ij}(q))_{i,j=1,\ldots,n}$ is a positive definite symmetric matrix. If $(\bar{x}^1, \ldots, \bar{x}^n)$ is another coordinate system on U, let $\bar{g}_{ij}(q) = g_q((\partial/\partial \bar{x}^i)_q, (\partial/\partial \bar{x}^j)_q)$, then by (2) we have

$$\bar{g}_{ij}(q) = \sum_{k,l=1}^{n} \frac{\partial x^k}{\partial \bar{x}^i}(q) \frac{\partial x^l}{\partial \bar{x}^j}(q) g_{kl}(q),$$

$$g_{ij}(q) = \sum_{k,l=1}^{n} \frac{\partial \bar{x}^k}{\partial x^i}(q) \frac{\partial \bar{x}^l}{\partial x^j}(q) \bar{g}_{kl}(q).$$

Hence, if the functions g_{ij} on U are all of class C^r, then so are the functions $\bar{g}_{ij}$, and conversely. That is, the property that the g_{ij}'s are all of class C^r does not depend on the choice of the local coordinate system. If all the

g_{ij}'s are of class C^r on a neighborhood of each point of M, then the map $g:p \to g_p$ which assigns a positive inner product g_p to $T_p(M)$ for each point p of M, is called a *Riemannian metric of class C^r*. g_p is called the value of g at p and the g_{ij}'s are called the components of g with respect to $(x^1, \ldots, x^n)$. For simplicity, we shall restrict ourselves to Riemannian metrics of class C^∞, and call a Riemannian metric of class C^∞ simply a Riemannian metric. When a Riemannian metric is given, we define the length $\|v\|$ of a tangent vector v at p by

$$\|v\|^2 = g_p(v, v).$$

If v has components $(\xi^1, \ldots, \xi^n)$ with respect to $(x^1, \ldots, x^n)$, then

$$\|v\|^2 = \sum_{i, j = 1}^{n} g_{ij}(p)\xi^i\xi^j.$$

The Riemannian metric is often denoted by the symbol

$$ds^2 = \sum_{i, j} g_{ij} \, dx^i \, dx^j.$$

(See Remark in §6.) A manifold with a Riemannian metric is called a *Riemannian manifold*.

Let φ be a differentiable curve on a Riemannian manifold M defined on an open interval (a, b), and let v_t be the tangent vector of φ at $\varphi(t)$. Then $t \to \|v_t\|$ is a continuous function of t. If $a < c < d < b$, then we call

$$L(\varphi; c, d) = \int_c^d \|v_t\| \, dt$$

the length of φ between $\varphi(c)$ and $\varphi(d)$. For a differentiable curve φ defined on a closed interval $[a, b]$, there is a differentiable curve φ' defined on an open interval (a', b') containing $[a, b]$ such that $\varphi(t) = \varphi'(t)$ for $t \in [a, b]$. We let $L(\varphi) = L(\varphi'; a, b)$, and call $L(\varphi)$ the length of φ. If φ is a piecewise differentiable curve, then it is a sum of a finite number of differentiable curves defined on closed intervals, and its length is defined as the sum of the lengths of these differentiable curves.

Problem 5. Let φ and ψ be differentiable curves on a Riemannian manifold M, defined on open intervals (a, b) and (c, d), respectively. If there is a diffeomorphism θ from (a, b) onto (c, d) such that $\psi(\theta(t)) = \varphi(t)$ and $d\theta(t)/dt > 0$ hold for all $t \in (a, b)$, then we say that the two curves are equivalent. Prove that this is an equivalence relation. Prove also that if φ and ψ are equivalent, then the length of φ between $\varphi(a')$ and $\varphi(b')$ is equal to the length of ψ between $\psi(\theta(a'))$ and $\psi(\theta(b'))$.

Remark. For two points p, q on a Riemannian manifold M, let $C_{p,q}$ denote the set of all piecewise differentiable curves connecting p and q. For $\varphi \in C_{p,q}$, let $L(\varphi)$ be the length of φ between p and q. If we let $d(p, q)$ be the infimum of $L(\varphi)$ as φ runs through $C_{p,q}$, then d is a metric on M. That is, d satisfies (1) $d(p, q) \geq 0$, and $d(p, q) = 0$ if and only if $p = q$, (2) $d(p, q) = d(q, p)$, (3) $d(p, q) \leq d(p, r) + d(r, q)$. Moreover, the metric d on M defines the same topology as that on M. We omit the proofs of these statements* because they require more preparation. However, the conditions for d to be a metric are easily verified except $d(p, q) = 0$ if and only if $p = q$, so the reader should try it.

Example 1. Let $(x^1, \ldots, x^n)$ be the standard coordinate system on $\mathbf{R}^n$. Let $u, v \in T_p(\mathbf{R}^n)$, and let the components of u and v be (ξ^i) and (η^i). Let $g_p(u, v) = \sum_{i=1}^n \xi^i \eta^i$. Then $p \to g_p$ gives a Riemannian metric on $\mathbf{R}^n$. We denote this metric by $(dx^1)^2 + (dx^2)^2 + \ldots + (dx^n)^2$. The Riemannian manifold obtained from $\mathbf{R}^n$ with this metric is called the n-dimensional Euclidean space, and will be denoted by $\mathbf{E}^n$ from now on.

Problem 6. For any two points x, y of $\mathbf{E}^n$, prove that $d(x, y) = (\sum_{i=1}^n (x^i - y^i)^2)^{1/2}$.

Example 2. Let H denote the set of all points in the (x, y)-plane with positive y coordinates, and call it the upper half-plane. H is an open submanifold of $\mathbf{R}^2$. Let $p = (x_0, y_0) \in H$, and let the components of u, $u' \in T_p(H)$ with respect to (x, y) be (ξ, η), (ξ', η'), respectively. We can define a Riemannian metric g on H by letting $g_p(u, u') = (\xi\xi' + \eta\eta')/y_0^2$. We denote this Riemannian metric by $((dx)^2 + (dy)^2)/y^2$. If we consider H to be the set of all complex numbers z with positive imaginary part $I(z)$, then using the notation $dz = dx + i\,dy$, $d\bar{z} = dx - i\,dy$, we can write formally $(dx)^2 + (dy)^2 = dz\,d\bar{z}$, and we can write the Riemannian metric as $dz\,d\bar{z}/I(z)^2$. The Riemannian manifold H is called the *hyperbolic plane*.

§6. The Differential of a Function and Critical Points

Let f be a differentiable function defined on an open set U of a manifold M. For $p \in U$, and for an arbitrary $v \in T_p(M)$, we let

$$(df)_p(v) = v(f). \tag{1}$$

Then $(df)_p$ is a mapping from $T_p(M)$ to $\mathbf{R}$ and satisfying $(df)_p(v + v') = (df)_p(v) + (df)_p(v')$ and $(df)_p(\lambda v) = \lambda(df)_p(v)$ for $v, v' \in T_p(M)$ and $\lambda \in \mathbf{R}$. That is, $(df)_p$ is a linear function on the vector space $T_p(M)$, and hence

*For the proof see: . Milnor, "Morse Theory," *Annals of Math. Studies*, No. 51, (1963), Part II; S. Kobayashi‐   . Nomizu, *Foundations of Differential Geometry*, Vol. 1, Wiley-Interscience, New York, 1963, Ch. IV.

an element of the dual space $T_p^*(M)$ of $T_p(M)$. The linear function $(df)_p$ is called the *differential* of f at p.

If $(x^1, \ldots, x^n)$ is a local coordinate system on a neighborhood of p, then

$$(df)_p\left(\frac{\partial}{\partial x^i}\right)_p = \frac{\partial f}{\partial x^i}(p), \tag{2}$$

and, in particular,

$$(dx^j)_p\left(\frac{\partial}{\partial x^i}\right)_p = \delta_i^j. \tag{3}$$

Hence $\{(dx^1)_p, \ldots, (dx^n)_p\}$ is a basis of $T_p^*(M)$ dual to the basis $\{(\partial/\partial x^1)_p, \ldots, (\partial/\partial x^n)_p\}$ of $T_p(M)$. We can express $(df)_p$ as a linear combination of the $(dx^i)_p$'s, and write $(df)_p = \sum_j \lambda_j (dx^j)_p$. Then

$$(df)_p\left(\frac{\partial}{\partial x^i}\right)_p = \sum_j \lambda_j (dx^j)_p\left(\frac{\partial}{\partial x^i}\right)_p = \sum_j \lambda_j \delta_i^j = \lambda_i,$$

so λ_i is equal to $\partial f(p)/\partial x^i$, and we have

$$(df)_p = \sum_{i=1}^n \frac{\partial f}{\partial x^i}(p)(dx^i)_p. \tag{4}$$

Remark. We can give a meaning to the notation $\sum_{i,j} g_{ij}\, dx^i\, dx^j$ for the Riemannian metric as follows. Since the basis $\{(\partial/\partial x^i)_p\}$ of $T_p(M)$ and the basis $\{(dx^i)_p\}$ of $T_p^*(M)$ are dual to each other, for a tangent vector $v \in T_p(M)$, $\{(dx^i)_p(v)\}$ is the set of components of v with respect to $(x^1, \ldots, x^n)$, and hence we can write $\|v\|^2 = \sum_{i,j} g_{ij}(p)(dx^i)_p(v)(dx^j)(v)$. We can consider this to be abbreviated to $\sum_{i,j} g_{ij}\, dx^i\, dx^j$.

Problem 1. Show that $(df^1)_p, \ldots, (df^n)_p$ are linearly independent if and only if $(f^1, \ldots, f^n)$ is a local coordinate system around p.

DEFINITION 1. Let f be a C^∞ function on a manifold M and let p be a point on M. If $f(p) = 0$, then the point p is called a *zero point* of f. If $(df)_p \neq 0$, then the point p is called a regular point of f, and $f(p)$ is called a regular value of f. If $(df)_p = 0$, then the point p is called a *critical point* of f, and $f(p)$ is called a *critical value* of f.

Problem 2. Define the local maximal value and the local minimal value of f and show that they are critical values.

If p is a regular point of f, then there is a local coordinate system $(y^1, \ldots, y^n)$ around p such that $y^i(p) = 0$ $(i = 1, \ldots, n)$ and $f = f(p) + y^1$

hold. In fact, if $(x^1, \ldots, x^n)$ is a local coordinate system around p, since $(df)_p \neq 0$, by (4) we have $\partial f(p)/\partial x^i \neq 0$ for some i. By renumbering, if necessary, we can take $i = 1$ and assume $\partial f(p)/\partial x^1 \neq 0$. Set $y^1 = f - f(p)$, $y^i = x^i - x^i(p)$ $(i = 2, \ldots, n)$. Then $y^j(p) = 0$ $(j = 1, \ldots, n)$ and

$$\frac{D(y^1, \ldots, y^n)}{D(x^1, \ldots, x^n)_p} = \det \begin{bmatrix} \dfrac{\partial f}{\partial x^1}(p) & \dfrac{\partial f}{\partial x^2}(p) & \cdots & \dfrac{\partial f}{\partial x^n}(p) \\ 0 & 1 & \cdots & 0 \\ \cdot & \cdot & & \cdot \\ \cdot & \cdot & & \cdot \\ \cdot & \cdot & & \cdot \\ 0 & 0 & \cdots & 1 \end{bmatrix} = \frac{\partial f}{\partial x^1}(p) \neq 0,$$

so $(y^1, \ldots, y^n)$ is a local coordinate system around p, and by the definition of y^1 we have $f = f(p) + y^1$.

Next let p be a critical point of f. Let (ξ^i), (η^i) be, respectively, the components of $u, v \in T_p(M)$ with respect to the local coordinate system (x^i). Define a symmetric bilinear form $H_f(u, v)$ on $T_p(M)$ by

$$H_f(u, v) = \sum_{i,j=1}^n \frac{\partial^2 f}{\partial x^i \partial x^j}(p)\xi^i\eta^j.$$

Since p is a critical point of f, the definition of H_f does not depend on the choice of the local coordinate system. In fact, if $(x^1, \ldots, \bar{x}^n)$ is another local coordinate system around p and if $(\bar{\xi}^i)$, $(\bar{\eta}^i)$ are, respectively, the components of u, v with respect to it, then since

$$\frac{\partial f}{\partial \bar{x}^j} = \sum_k \frac{\partial f}{\partial x^k} \frac{\partial x^k}{\partial \bar{x}^j},$$

we have

$$\frac{\partial^2 f}{\partial \bar{x}^i \partial \bar{x}^j} = \sum_{k,l} \frac{\partial^2 f}{\partial x^l \partial x^k} \frac{\partial x^l}{\partial \bar{x}^i} \frac{\partial x^k}{\partial \bar{x}^j} + \sum_k \frac{\partial f}{\partial x^k} \frac{\partial^2 x^k}{\partial \bar{x}^i \partial \bar{x}^j}.$$

But since $\partial f(p)/\partial x^k = 0$ $(k = 1, \ldots, n)$, we get

$$\frac{\partial^2 f}{\partial \bar{x}^i \partial \bar{x}^j}(p) = \sum_{k,l} \frac{\partial^2 f}{\partial x^l \partial x^k}(p) \frac{\partial x^l}{\partial \bar{x}^i}(p) \frac{\partial x^k}{\partial \bar{x}^j}(p).$$

But we have the relations (3) of §5 between (ξ^i), $(\bar{\xi}^i)$ and between (η^i),

$(\bar{\eta}^i)$, so we get

$$\sum_{i,j=1}^{n} \frac{\partial^2 f}{\partial \bar{x}^i \partial \bar{x}^j}(p)\bar{\xi}^i\bar{\eta}^j = \sum_{k,l=1}^{n} \frac{\partial^2 f}{\partial x^l \partial x^k}(p)\xi^l\eta^k.$$

This shows that the definition of $H_f(u, v)$ does not depend on the choice of the local coordinate system.

The symmetric bilinear form H_f is called the *Hessian* of f at the critical point p. If the Hessian is nondegenerate, i.e., if the matrix $\partial^2 f(p)/\partial x^i \partial x^j$ is a nonsingular matrix, then we call p a *nondegenerate critical point* of f.

In general, if $G(u, v)$ is a nondegenerate symmetric bilinear form on an n-dimensional real vector space V $(u, v \in V)$, then by choosing a suitable base $\{e_1, \ldots, e_n\}$ for the vector space V, we can have for any $u = \sum_{i=1}^{n} \xi^i e^i$:

$$G(u, u) = (\xi^1)^2 + \ldots + (\xi^r)^2 - (\xi^{r+1})^2 - \ldots - (\xi^n)^2.$$

We call the pair (r, s) (where $r + s = n$) the signature of $G(u, v)$. The pair (r, s) does not depend on the choice of the base $\{e_1, \ldots, e_n\}$, and only depends on G.* We call $s(= n - r)$ the *index* of G.

DEFINITION 2. If p is a nondegenerate critical point of f, then the index of the Hessian H_f of f at p is called the *index* of the nondegenerate critical point p.

THEOREM(Morse). *Let p be a nondegenerate critical point of f. Then there is a local coordinate system $(y^1, \ldots, y^n)$ on a neighborhood of p satisfying $y^i(p) = 0$ $(i = 1, \ldots, n)$ and*

$$f = f(p) + (y^1)^2 + \ldots + (y^r)^2 - (y^{r+1})^2 - \ldots - (y^n)^2.$$

Here $n - r$ is equal to the index of p.

Proof. Suppose $(y^1, \ldots, y^n)$ satisfies the condition in the theorem. Then we have

$$\frac{\partial^2 f}{\partial y^i \partial y^j} = 0 \quad (i \neq j), \qquad \frac{\partial^2 f}{\partial (y^i)^2} = 2 \quad (1 \leq i \leq r),$$

$$\frac{\partial^2 f}{\partial (y^i)^2} = -2 \quad (r + 1 \leq i \leq r + s).$$

If we let a base $\{e_1, \ldots, e_n\}$ of $T_p(M)$ be given by $e_i = (1/\sqrt{2})(\partial/\partial y^i)_p$,

*Cf. S. MacLane and G. Birkhoff, *Algebra*, Macmillan, New York, 1967, 386–388.

then we have

$$H_f(u, u) = (\xi^1)^2 + \ldots + (\xi^r)^2 - (\xi^{r+1})^2 - \ldots - (\xi^{r+s})^2$$

$$(u = \textstyle\sum_i \xi^i e_i),$$

so s is the index of H_f at p of f, i.e., the index of the nondegenerate critical point p.

Now we shall prove that there is such a local coordinate system.

LEMMA 1.   *Let $a_{ij}(u^1, \ldots, u^n)$ $(i, j = 1, \ldots, n)$ be C^∞ functions defined on a neighborhood U of the origin $(0, \ldots, 0)$ of $\mathbf{R}^n$ satisfying $a_{ij}(u) = a_{ji}(u)$, $\det A(u) \neq 0$ for $u \in U$, where we set $A(u) = (a_{ij}(u))$. Then there exist $n \times n$ nonsingular matrices $T(u) = (t_{ij}(u))$ such that*

(1) *$t_{ij}(u)$ are C^∞ functions defined on some neighborhood V $(V \subset U)$ of $(0, \ldots, 0)$, and*
(2)

$$
{}^t T(u) A(u) T(u) =
\begin{bmatrix}
\varepsilon_1 & & & & \\
 & \varepsilon_2 & & & \\
 & & \cdot & & \\
 & & & \cdot & \\
 & & & & \cdot \\
 & & & & & \varepsilon_n
\end{bmatrix}, \qquad \varepsilon_i = \pm 1
$$

holds at each point $u \in V$.

Proof of Lemma 1.   Let $A_u(\xi, \eta) = \sum_{ij} a_{ij}(u) \xi^i \eta^j$ be the bilinear form corresponding to the symmetric matrix $A(u)$. Assuming that matrices $T(u)$ satisfying (1) and (2) exist, we let the ith column vector of $T(u)$ be denoted by $e_i(u)$. Then as $T(u)$ is nonsingular, $e_1(u), \ldots, e_n(u)$ are linearly independent for any fixed u, and (2) is equivalent to $A_u(e_i(u), e_j(u)) = \varepsilon_j \delta_{ij}$.

Conversely, if there are n-dimensional vector-valued functions $e_1(u), \ldots, e_n(u)$ defined on some neighborhood $V (\subset U)$ of (0) satisfying (a) each component of $e_i(u)$ is a C^∞ function of u, (b) $e_1(u), \ldots, e_n(u)$ are linearly independent for each fixed u, and (c) $A_u(e_i(u), e_j(u)) = \varepsilon_j \delta_{ij}$ $(\varepsilon_i = \pm 1)$, then the $n \times n$ matrix $T(u)$ formed by these n column vectors satisfies conditions (1) and (2). Hence it suffices to show the existence of the vector-valued functions $e_1(u), \ldots, e_n(u)$ satisfying (a), (b), and (c).

Let e_j be the n-dimensional vector whose jth component is 1 and all other components 0 $(j = 1, \ldots, n)$. Let us determine a vector $e_1'(u)$, whose components are n functions $b_i(u)$ ($i = 1, \ldots, n$), and such that the following three conditions are satisfied: (i) Each $b_i(u)$ is of class C^∞ in a neighborhood $V_1 (\subset U)$ of (0), (ii) $A_u(e_1'(u), e_j) = 0$ ($j = 2, \ldots, n$), and (iii) $A_u(e_1'(u), e_1'(u)) \neq 0$ for $u \in V_1$. Now (ii) is equivalent to

(ii')
$$\sum_{i=1}^{n} a_{ji}(u)b_i(u) = 0 \qquad (j = 2, \ldots, n).$$

As the $(n - 1) \times n$ matrix $(a_{ji}(0))_{j = 2, \ldots, n; \ i = 1, \ldots, n}$ has rank $n - 1$, we can assume, without loss of generality, that the determinant of the $(n - 1) \times (n - 1)$ matrix $A'(0) = (a_{ji}(0))_{2 \leq i, j \leq n}$ is not 0.

By taking a neighborhood V_1 of (0) sufficiently small, we can also assume that the determinant of $A'(u) = (a_{ji}(u))_{2 \leq i, j \leq n}$ is not 0 on V_1. Now set $b_1(u) = 1$, then (ii') can be written as

$$\sum_{i=2}^{n} a_{ji}(u)b_i(u) = -a_{j1}(u) \qquad (j = 2, \ldots, n). \tag{5}$$

The determinant of the coefficients is not 0 for $u \in V_1$, so, using Cramer's rule, we get the solution $(b_2(u), \ldots, b_n(u))$ of (5), and we see that $b_2(u), \ldots, b_n(u)$ are C^∞ functions on V_1. Since $b_1(u) = 1$, we have

$$A_u(e_1'(u), e_1'(u)) = \sum_{i, j = 1}^{n} a_{ij}(u)b_i(u)b_j(u)$$

$$= a_{11}(u) + 2 \sum_{j=2}^{n} a_{1j}(u)b_j(u) + \sum_{i, j = 2}^{n} a_{ij}(u)b_i(u)b_j(u)$$

and, because

$$\sum_{j=2}^{n} a_{ij}(u)b_j(u) = -a_{i1}(u) \qquad (i = 2, \ldots, n),$$

this gives us

$$A_u(e_1'(u), e_1'(u)) = a_{11}(u) + \sum_{j=2}^{n} a_{1j}(u)b_j(u). \tag{6}$$

But the first component of the column vector $A(u)e_1'(u)$ is $a_{11} + \sum_{j=2}^{n} a_{1j}(u)b_j(u)$, and by (5) the other components are 0. On the other hand, $A(u)$ is nonsingular and $e_1'(u) \neq 0$, so $A(u)e_1'(u) \neq 0$. Hence by (6) we have $A_u(e_1'(u), e_1'(u)) \neq 0$. We have determined $b_i(u)$ so that (i), (ii), and (iii) are satisfied. Now let $c(u)$ be the square root of the absolute value of $A_u(e_1'(u), e_1'(u))$, and set $e_1(u) = [1/c(u)]e_1'(u)$. Then we have $A_u(e_1(u), e_j) = 0$ ($j = 2, \ldots, n$), $A_u(e_1(u), e_1(u)) = \pm 1$ and $e_1(u)$ is C^∞ on V_1. We can now prove the lemma by induction on the dimension n.

LEMMA 2.   *Let φ be a C^∞ function defined on a convex* neighborhood W of $(0, \ldots, 0)$ and suppose $\varphi(0) = 0$. Then we can write*

$$\varphi(u) = \sum_{i=1}^{n} \varphi_i(u)u^i,$$

where the $\varphi_i(u)$'s are C^∞ functions on W satisfying $\partial\varphi(0)/\partial u^i = \varphi_i(0)$.

Proof.   If $(u^1, \ldots, u^n) \in W$, then since W is convex, we have $(tu^1, \ldots, tu^n) \in W$ for $0 \leq t \leq 1$. Hence, writing $x^i = tu^i$ $(i = 1, \ldots, n)$, we have

$$\varphi(u^1, \ldots, u^n) = \int_0^1 \frac{d\varphi}{dt}(tu^1, \ldots, tu^n)dt$$

$$= \sum_{i=1}^{n} \int_0^1 \frac{\partial\varphi}{\partial x^i}(tu^1, \ldots, tu^n)u^i \, dt,$$

and it suffices to set $\varphi_i(u) = \int_0^1 \partial\varphi(tu^1, \ldots, tu^n)/\partial x^i dt$.

The proof of the theorem (continued). We can assume $f(p) = 0$ from the beginning, because if necessary we can replace f by $f - f(p)$. Take a local coordinate system $(x^1, \ldots, x^n)$ around p such that $x^i(p) = 0$ $(i = 1, \ldots, n)$. Assume that f is defined on $W = \{q \mid |x^i(q)| < a \; (i = 1, \ldots, n)\}$. Then by Lemma 2 we can write

$$f = \sum_{i=1}^{n} \varphi_i x^i,$$

where φ_i are of class C^∞ on W and $\partial f(p)/\partial x^i = \varphi_i(p)$. But since $(df)_p = 0$, we have $\varphi_i(p) = 0$. So, again by Lemma 2, we get $\varphi_i = \sum_{j=1}^{n} \varphi_{ij} x^j$, where the φ_{ij}'s are of class C^∞ on W and $\varphi_{ij}(p) = \partial\varphi_i(p)/\partial x^j$. Set $a_{ij} = \frac{1}{2}(\varphi_{ij} + \varphi_{ji})$. Then $a_{ij} = a_{ji}$ and we have

$$f = \sum_{i,j=1}^{n} a_{ij} x^i x^j. \tag{7}$$

Using (7) to compute $\partial^2 f(p)/\partial x^i \partial x^j$, we get $2a_{ij}(p) = \partial^2 f(p)/\partial x^i \partial x^j$. Since p is a nondegenerate critical point of f, the matrix $(a_{ij}(p))$ is nonsingular. Hence, if we choose a sufficiently small neighborhood U of p, then $A(q) = (a_{ij}(q))$ is a nonsingular matrix for each q in U. Hence, by Lemma 1, there exist a neighborhood V of p and nonsingular matrices $T(q) =$

---

*This means that the line segment connecting any two points in W lies completely in W.

$(t^i_j(q))$ $(q \in V)$ such that the $t^i_j(q)$'s are of class C^∞ on V and

$$\sum_{i,j=1}^{n} a_{ij}(q)t^i_k(q)t^j_l(q) = \delta^k_l \varepsilon_k \qquad (\varepsilon_k = \pm 1) \qquad (8)$$

holds. Let s^i_j be the (i, j)th entry of the inverse $T^{-1}(q)$ of the matrix $T(q)$, and set

$$y^i = \sum_{k=1}^{n} s^i_k x^k \qquad (i = 1, \ldots, n). \qquad (9)$$

Then y^i is of class C^∞ on V, and we have $x^i = \sum_k t^i_k y^k$. Substituting this in (7), we have, by (8),

$$f = \sum_{i=1}^{n} \varepsilon_i (y^i)^2.$$

But, by (9), we have

$$\partial y^i(q)/\partial x^k = s^i_k(q).$$

Hence we have $D(y^1, \ldots, y^n)/D(x^1, \ldots, x^n)_p = T^{-1}(p)$, and $(y^1, \ldots, y^n)$ is the desired local coordinate system around p.

§7. The Differential of a Map

Let φ be a differentiable map from a manifold M into a manifold M'. If f is a differentiable function defined in a neighborhood of $\varphi(p)$ $(p \in M)$, then $\varphi^* f = f \circ \varphi$ is a differentiable function defined in a neighborhood of p. For $v \in T_p(M)$, we set

$$((\varphi_*)_p v)(f) = v(\varphi^* f). \qquad (1)$$

We have

$$((\varphi_*)_p v)(f + g) = v(\varphi^* f + \varphi^* g) = v(\varphi^* f) + v(\varphi^* g)$$
$$= ((\varphi_*)_p v)f + ((\varphi_*)_p v)g,$$
$$((\varphi_*)_p v)(\lambda f) = \lambda((\varphi_*)_p v)(f),$$
$$((\varphi_*)_p v)(fg) = ((\varphi_*)_p v)f \cdot g(\varphi(p)) + f(\varphi(p)) \cdot ((\varphi_*)_p v)g,$$

so $(\varphi_*)_p v$ is a tangent vector to M' at $\varphi(p)$. Hence $(\varphi_*)_p$ is a map from $T_p(M)$ into $T_{\varphi(p)}(M')$, and it is easy to see that this map is a linear map. $(\varphi_*)_p$ is called the *differential* of the differentiable map φ at p.* Let $(x^1, \ldots, x^n)$ and $(y^1, \ldots, y^m)$ be local coordinate systems at p and $\varphi(p)$, respectively, and set

$$\varphi^* y^i = \varphi^i.$$

*When it is clear that u is a tangent vector at p, instead of writing $(\varphi_*)_p u$, we shall sometimes write $\varphi_* u$.

Then we have

$$\left(\varphi_*\right)_p\left(\frac{\partial}{\partial x^i}\right)_p = \sum_{j=1}^m \frac{\partial \varphi^j}{\partial x^i}\,(p)\left(\frac{\partial}{\partial y^j}\right)_{\varphi(p)}. \tag{2}$$

Note that if φ is the identity map of M, then $(\varphi_*)_p$ is the identity map of $T_p(M)$. If φ is a differentiable map from M to M', and ψ is a differentiable map from M' to M'', then $\psi \circ \varphi$ is a differentiable map from M to M'', and

$$((\psi \circ \varphi)_*)_p = (\psi_*)_{\varphi(p)} \circ (\varphi_*)_p.$$

In particular, if φ is a diffeomorphism from M onto M', then $\varphi^{-1} \circ \varphi$ and $\varphi \circ \varphi^{-1}$ are the identity maps of M and M', respectively, so $(\varphi_*)_p$ is a 1–1 linear map from $T_p(M)$ onto $T_{\varphi(p)}(M')$, and its inverse map $(\varphi_*)_p^{-1}$ coincides with $((\varphi^{-1})_*)_{\varphi(p)}$. Hence the dimension of $T_p(M)$ is equal to the dimension of $T_{\varphi(p)}(M')$. Hence if there is a diffeomorphism from M onto M', then M and M' have the same dimension.

Problem 1. Let t be the standard coordinate for $\mathbf{R}^1$, and let (a, b) be an open interval of $\mathbf{R}^1$. Prove that the tangent vector to the differentiable curve φ on M at $\varphi(t_0)$, where φ is defined on (a, b), is the tangent vector $(\varphi_*)_{t_0}(\partial/\partial t)_{t_0}$.

Problem 2. If f is a differentiable function on M, then f is a differentiable map from M to $\mathbf{R}^1$. Prove that $(f_*)_p v = (df)_p(v)(\partial/\partial t)_{f(p)}$ for $v \in T_p(M)$.

Now $(\varphi_*)_p : T_p(M) \to T_{\varphi(p)}(M)$ determines a linear map from the dual space $T_{\varphi(p)}{}^*(M')$ of $T_{\varphi(p)}(M')$ to the dual space $T_p^*(M)$ of $T_p(M)$ (i.e., the dual map of $(\varphi_*)_p$. (Cf. 1, §2.) We denote this map by $(\varphi^*)_p$. For a differentiable function f defined in a neighborhood of $\varphi(p)$, we have

$$(\varphi^*)_p(df)_{\varphi(p)} = (d(\varphi^*f))_p. \tag{3}$$

$(\varphi^*)_p$ is called the *dual differential* of φ at p.

Problem 3 Prove formula (3).

The rank of the linear map $(\varphi_*)_p : T_p(M) \to T_{\varphi(p)}(M)$ is called the *rank* of the differentiable map φ from M to M' at the point p, and is denoted by $\operatorname{rank}_p(\varphi)$. By definition, $\operatorname{rank}_p(\varphi)$ is equal to the dimension of the image space $(\varphi_*)_p(T_p(M))$, and is equal to the rank of the matrix $(\partial \varphi^i(p)/\partial x^j)$ defined by (2). So, if n and m are the dimensions of M and M', respectively, then $\operatorname{rank}_p(p) \leqq \operatorname{Min}(n, m)$. The maximum value of $\operatorname{rank}_p(\varphi)$, as p runs through all the points of M, is denoted by $\operatorname{rank}(\varphi)$, and is called the rank of the map φ. We have $\operatorname{rank}(\varphi) \leqq \operatorname{Min}(n, m)$.

Problem 4. Show that the set of points p of M such that $\operatorname{rank}_p(\varphi) = \operatorname{rank}(\varphi)$ is an open set of M.

THEOREM *If at some point p of M, we have $\operatorname{rank}_p(\varphi) = n (= \dim M)$, then we have $\operatorname{rank}(\varphi) = n$ and $n \leq m (= \dim M')$, and we can find local coordinate systems $(x^1, \ldots, x^n)$ and $(y^1, \ldots, y^m)$ around p and $\varphi(p)$, respectively, such that*

$$y^i \circ \varphi = x^i \quad (i = 1, \ldots, n), \quad y^j \circ \varphi = 0 \quad (j = n + 1, \ldots, m).*$$

Proof. In general we have $\operatorname{rank}(\varphi) \leq \operatorname{Min}(n, m)$, but by hypothesis we have $\operatorname{rank}_p(\varphi) = n$ at p, so $\operatorname{rank}(\varphi) = n$ and $n \leq m$ hold. Let $(x^1, \ldots, x^n)$ and $(y^1, \ldots, y^m)$ be local coordinate systems around p and $\varphi(p)$, respectively, and suppose $x^k(p) = y^j(\varphi(p)) = 0$ ($k = 1, \ldots, n$; $j = 1, \ldots, m$). Let $\varphi^i = y^i \circ \varphi$. As the $m \times n$ matrix

$$\begin{bmatrix} \dfrac{\partial \varphi^1}{\partial x^1}(p) & \cdots & \dfrac{\partial \varphi^1}{\partial x^n}(p) \\ \cdot & & \\ \cdot & & \\ \cdot & & \\ \dfrac{\partial \varphi^m}{\partial x^1}(p) & \cdots & \dfrac{\partial \varphi^m}{\partial x^n}(p) \end{bmatrix}$$

has rank n, by renumbering the y^i's if necessary, we can assume that the $n \times n$ matrix $(\partial \varphi^i(p)/\partial x^j)_{i, j = 1, \ldots, n}$ has nonzero determinant. Let $\bar{x}^i = \varphi^i$ ($i = 1, \ldots, n$). Then, as $D(\bar{x}^1, \ldots, \bar{x}^n)/D(x^1, \ldots, x^n)_p \neq 0$, $(\bar{x}^1, \ldots, \bar{x}^n)$ is a local coordinate system around p, and we have $y^i \circ \varphi = \bar{x}^i$ ($i = 1, \ldots, n$) and $\partial \varphi^i(p)/\partial x^j = \partial \bar{x}^i(p)/\partial x^j = \delta_j^i$. Now write $y^j \circ \varphi$ ($j > n$) as a function of $\bar{x}^1, \ldots, \bar{x}^n$: $(y^j \circ \varphi)(q) = F^j(\bar{x}^1(q), \ldots, \bar{x}^n(q))$ ($j = n + 1, \ldots, m$). Here the F^j's are C^∞ functions on a neighborhood of $(0, \ldots, 0)$. Let

$$\bar{y}^i = y^i \quad (i = 1, \ldots, n), \quad \bar{y}^j = y^j - F^j(y^1, \ldots, y^n)$$
$$(j = n + 1, \ldots, m).$$

Then $D(\bar{y}^1, \ldots, \bar{y}^m)/D(y^1, \ldots, y^m)_{\varphi(p)} = 1$, so $(\bar{y}^1, \ldots, \bar{y}^m)$ is a local coordinate system around $\varphi(p)$. For $1 \leq i \leq n$, we have $y^i \circ \varphi = y^i \circ \varphi = \bar{x}^i$, and for $n + 1 \leq j \leq m$, we have

$$\bar{y}^j \circ \varphi = y^j \circ \varphi - F^j(y^1 \circ \varphi, \ldots, y^n \circ \varphi)$$
$$= y^j \circ \varphi - F(\bar{x}^1, \ldots, \bar{x}^n) = 0,$$

so $(\bar{x}^1, \ldots, \bar{x}^n)$ and $(\bar{y}^1, \ldots, \bar{y}^m)$ are the desired local coordinate systems.

---

*We write $y^i \circ \varphi$ instead of $\varphi^* y^i$.

Problem 5.   If at a point p of M, we have $\text{rank}_p(\varphi) = m$, then $n \geq m$ holds, and we can choose local coordinate systems $(x^1, \ldots, x^n)$ and $(y^1, \ldots, y^m)$ around p and $\varphi(p)$, respectively, so that

$$y^i \circ \varphi = x^i \quad (i = 1, \ldots, m)$$

is satisfied.

Problem 6.   Define $\varphi : \mathbf{R}^2 \to \mathbf{R}^3$ by $y^1 = (x^1)^2$, $y^2 = x^1 \cdot x^2$, $y^3 = x^2$, and find the rank of φ at each point p of $\mathbf{R}^2$.

Problem 7.   Define $\varphi : \mathbf{R}^3 \to \mathbf{R}^3$ by $y^1 = x^1 \cdot x^2$, $y^2 = x^2$, $y^3 = x^3$. Restrict φ to S^2 to obtain the map ψ from S^2 to $\mathbf{R}^3$. Find the rank of ψ at each point of S^2.

§8. Sard's Theorem

Let M and M' be n-dimensional manifolds, and let φ be a differentiable map from M to M'. If $\text{rank}_p(\varphi) = n$, then the point p is called a *regular point* of φ. If $\text{rank}_p(\varphi) < n$, then the point p is called a *critical point* of φ. A point q of M' is called a *critical value* of φ if there is a critical point p of φ such that $\varphi(p) = q$. A point q of M' is called a *regular value* of φ if it is not a critical value. Hence, if q is a regular value of φ, either q does not belong to $\varphi(M)$ or q belongs to $\varphi(M)$ and all points in $\varphi^{-1}(q)$ are regular points of φ.

Sard's theorem tells us that there are not too many critical values in M'. To express mathematically the term "not too many," we shall define the notion of a subset of measure 0 of a manifold.*

As is well known, a subset A of $\mathbf{R}^n$ is said to have (Lebesgue) measure 0 if, for a given $\varepsilon > 0$, there are finite or countably many cubes Q_1, $Q_2, \ldots$, such that $A \subset \cup_i Q_i$ and $\sum_i m(Q_i) < \varepsilon$. Here $m(Q_i)$ denotes the volume of the cube Q_i.

LEMMA 1.   *Let φ be a C^1 map from $\mathbf{R}^n$ into $\mathbf{R}^n$, and let A be a subset of $\mathbf{R}^n$ of measure 0. Then $\varphi(A)$ also has measure 0.*

Proof.   Let $Q(r)$ be the cube of half-width r and centered at the origin, that is, let $Q(r) = \{(u^1)| \ |u^1| < r\}$. Let $\{r_k\}$ $(k = 1, 2, \ldots)$ be a sequence of real numbers such that $r_k \to +\infty$, and set $A_k = A \cap \overline{Q}(r_k)$. Then we have $A = \cup_k A_k$, $\varphi(A) = \cup_k \varphi(A_k)$.So it suffices to show that each $\varphi(A_k)$ has measure 0. Hence we may assume from the beginning that $A \subset \overline{Q}(r)$ for some r, and prove that $\varphi(A)$ has measure 0. Let $\varphi(x) = ((\varphi^1(x), \ldots,$

---

*If φ maps M to a single point q of M', then every point of M is a critical point of φ, but q is the only critical value of φ. So the set of critical points can be "big."

$\varphi^n(x)$), and let

$$\underset{\substack{x \in \overline{Q}(r) \\ i, j = 1, \ldots, n}}{\text{Max}} \left| \frac{\partial \varphi^i}{\partial x^j} \right| = L.$$

Then, for $x, y \in \overline{Q}(r)$, we have

$$|\varphi^i(x) - \varphi^i(y)| \leq Ln \cdot \max|x^j - y^j|.$$

Hence, if $Q = Q(x_0; s)$ is a cube contained in $\overline{Q}(r)$, then $\varphi(Q)$ is contained in $Q' = Q(\varphi(x_0); Lns)$. Thus we have $m(\varphi(Q)) \leq (Ln)^n m(Q)$. Now, since A has measure 0, for an arbitrary $\varepsilon > 0$, there exists a sequence $\{Q_i\}$ of cubes such that $A \subset \cup_i Q_i$ and $\sum_i m(Q_i) < \varepsilon$. Here we can take r sufficiently large and assume that $Q_i \subset \overline{Q}(r)$. Then we have $\varphi(A) \subset \cup_i Q_i'$ and $m(Q_i') = (Ln)^n m(Q_i)$. Hence $\sum_i m(Q_i') < (Ln)^n \varepsilon$ and thus $\varphi(A)$ also has measure 0.

Now let $\{U_\alpha, \varphi_\alpha\}_{\alpha \in A}$ be a coordinate neighborhood system on an n-dimensional manifold M. A subset A of M is said to have measure 0 if $\varphi_\alpha(U_\alpha \cap A)$ has measure 0 in $\mathbf{R}^n$ for each $\alpha \in A$. Using Lemma 1, it is easy to show that this definition does not depend on the choice of the coordinate neighborhood system on M (and the reader should carry out the proof).

It is immediate from this definition that a subset A of M which has measure 0 contains no open subset of M. Hence, its complement A^c is a dense subset of M.

LEMMA 2. *Let M and M' be manifolds of dimensions n and m, respectively. Let φ be a differentiable map from M to M'. Then*

(1) *if $n < m$, then $\varphi(M)$ is a subset of measure 0 of M';*
(2) *if $n = m$, and if A is a subset of M of measure 0, then $\varphi(A)$ also has measure 0.*

Proof (2) is clear from Lemma 1 and the definition of measure 0. If $n < m$, set $M_1 = M \times \mathbf{R}^{m-n}$, and define $\psi: M_1 \to M'$ by $\psi(p, x) = \varphi(p)$ ($p \in M, x \in \mathbf{R}^{m-n}$). Then ψ is differentiable. Fix $x_0 \in \mathbf{R}^{m-n}$, and set $A = M \times \{x_0\}$. A is a subset of measure 0 in M_1, and we have $\psi(A) = \varphi(M)$. Hence, by (2), $\varphi(M)$ has measure 0.

THEOREM(Sard). *Let M and M' both be manifolds of dimension n, and let φ be a differentiable map from M to M'. We suppose that M has a countable base.* Then the set K of the critical values of φ is a subset of measure 0 of M'.*

---

*We mean that M as a topological space has a base consisting of a countable number of open subsets. Cf. §15.

Proof. Let C be the set of critical points of φ in M. Then we have $K = \varphi(C)$. Let $\{(V_\alpha, \psi_\alpha)\}_{\alpha \in A}$ be a coordinate neighborhood system on M'. Since M has a countable base, we can find a countable coordinate neighborhood system $\{(U_i, \psi_i)\}_{i = 1, 2, \ldots}$ of M such that

(1) each $\psi_i(U_i)$ is a cube in $\mathbf{R}^n$ with center at 0;
(2) for each i there is an $\alpha \in A$ such that $\varphi(U_i) \subset V_\alpha$.*

Set $C_i = U_i \cap C$, $\varphi(C_i) = K_i$. Then $K = \cup_{i=1}^\infty K_i$, so it suffices to show that each K_i has measure 0 (because the countable union of sets of measure 0 is again of measure 0).

Let α be such that $\varphi(U_i) \subset V_\alpha$. Then we can consider U_i to be a cube centered at the origin of $\mathbf{R}^n$, and V_α to be an open set in $\mathbf{R}^n$. Furthermore, we can consider φ restricted to U_i as a differentiable map from a cube in $\mathbf{R}^n$ to $\mathbf{R}^n$. Hence, it suffices to prove the following: If φ is a differentiable map from $\mathbf{R}^n$ to $\mathbf{R}^n$, $Q = Q(0; r)$ a cube centered at the origin of $\mathbf{R}^n$, C the set of critical points contained in Q, then $\varphi(C)$ has measure 0 in $\mathbf{R}^n$. Now, as before, set $\varphi(x) = (\varphi^1(x), \ldots, \varphi^n(x))$. Then for $x, y \in Q$, we have, by the mean value theorem,

$$\varphi^i(y) - \varphi^i(x) = \sum_{j=1}^n \frac{\partial \varphi^i}{\partial x^j}(x_0)(y^j - x^j),$$

where x_0 is a point on the line segment connecting x and y. Hence, as before, if we set

$$L = \underset{\substack{x \in \overline{Q} \\ i, j = 1, \ldots, n}}{\mathrm{Max}} \left| \frac{\partial \varphi^i}{\partial x^j}(x) \right|,$$

then we have

$$|\varphi^i(y) - \varphi^i(x)| \leqq Ln \max_j |y^j - x^j|. \tag{1}$$

Fix $x \in Q$, and define an affine map φ_x from $\mathbf{R}^n$ to $\mathbf{R}^n$ by

$$\varphi_x^i(y) = \varphi^i(x) + \sum_{j=1}^n \frac{\partial \varphi^i}{\partial x^j}(x)(y^j - x^j).$$

Then, for $x, y \in Q$, we have

$$\varphi^i(y) - \varphi_x^i(y) = \sum_{i=1}^n \left(\frac{\partial \varphi^i}{\partial x^j}(x_0) - \frac{\partial \varphi^i}{\partial x^j}(x) \right)(y^j - x^j).$$

*This can be done because M is a Lindelöf space (§15).

Now, since $\partial\varphi^i/\partial x^j$ is uniformly continuous on Q, there is a function b of a real variable r such that $b(r) > 0$, $b(r) \to 0$ as $r \to 0$, and such that

$$|\varphi^i(y) - \varphi_x^i(y)| \leqq b(\|x - y\|)\max_j |y^j - x^j| \qquad (2)$$

holds for x, $y \in Q$. Here $\|x - y\|$ is the distance between x and y.

Now let $x \in C$. Then since $\det(\partial\varphi^i(x)/\partial x^j) = 0$, all $\varphi_x(y)$ ($y \in \mathbf{R}^n$) are contained in an $(n - 1)$-dimensional hyperplane Π_x passing through $\varphi(x)$. Take $\varepsilon > 0$ sufficiently small, and let $\bar{Q}_\varepsilon$ be a closed cube of half-width $\varepsilon/2$ and containing x. For $y \in \bar{Q}_\varepsilon$, we have from (1) that $\varphi(y) \in \bar{Q}(\varphi(x);$ $\varepsilon L')(L' = nL)$, and from (2) that the distance from $\varphi(y)$ to Π_x does not exceed $b(n^{1/2}\varepsilon)\, n^{1/2}\varepsilon$. Let D_x be the domain bounded by two hyperplanes parallel to Π_x, each being of distance $b(n^{1/2}\varepsilon)\, n^{1/2}\varepsilon$ to Π_x. Then we have $\varphi(y) \in \bar{Q}(\varphi(x);\varepsilon L') \cap D_x$. As the diameter of $\bar{Q}(\varphi(x);\varepsilon L')$ is $2n^{1/2}\varepsilon L'$ the volume of $\bar{Q}(\varphi(x);\varepsilon L') \cap D_x$ does not exceed $A\varepsilon^n b(n^{1/2}\varepsilon)$, where $A = (2n^{1/2})^n L'^{n-1}$ does not depend on the critical point x. Thus, we have found that if a closed cube $\bar{Q}_\varepsilon$ of half-width $\varepsilon/2$ contains a critical point x, then $\varphi(\bar{Q}_\varepsilon)$ is contained in a polyhedron whose volume does not exceed $A\varepsilon^n b(n^{1/2}\varepsilon)$. We now divide $\bar{Q} = \bar{Q}(0;r)$ into $(2m)^n$ many cubes of half-width $r/2m$. If one of these cubes $\bar{Q}_i$ contains a critical point, then $\varphi(\bar{Q}_i)$ is contained in a polyhedron P_i whose volume does not exceed $A(r/m)^n b$ $\cdot (n^{1/2}r/m)$. Hence, in particular, $\varphi(C)$ is covered by the union of these polyhedra, and the volume of the union does not exceed $A(2r)^n b(n^{1/2}r/m)$. Because $b(n^{1/2}r/m) \to 0$ as $m \to +\infty$, we conclude that $\varphi(C)$ has measure 0.

§9. The Motion in a Riemannian Manifold

Let M and M' be Riemannian manifolds with Riemannian metrics g and g', respectively. If a differentiable map φ from M to M' satisfies, for any point p of M and any vector u of $T_p(M)$, the equality

$$\|\varphi_* u\| = \|u\|, \qquad (1)$$

then we call φ an *isometry* from M to M'.

If a diffeomorphism φ of a Riemannian manifold M onto itself is an isometry from M to M, then we call φ a *motion* of M. The set of all motions of M forms a group called the group of motions of the Riemannian manifold M.

Remark. For a fixed point p of M, that (1) holds for all $u \in T_p(M)$ is equivalent to

$$g_{\varphi(p)}(\varphi_* u, \varphi_* v) = g_p(u, v) \qquad \text{for all} \quad u, v \in T_p(M).$$

Example 1. Let $\mathbf{E}^n$ be an n-dimensional Euclidean space and $(x^1, \ldots, x^n)$ the standard coordinate system. A motion φ of $\mathbf{E}^n$ is an affine transformation, and if we let

$$\varphi^* x^i = \sum_j a^i_j x^j + b^i,$$

then the matrix (a^i_j) is an orthogonal matrix. Conversely, if φ is an affine transformation, and if the matrix (a^i_j) is an orthogonal matrix, then φ is a motion in $\mathbf{E}^n$.

Proof. Let

$$\varphi^* x^i = \varphi^i(x^1, \ldots, x^n) \qquad (i = 1, \ldots, n).$$

We have, then, $\varphi_*(\partial/\partial x^i)_p = \sum_k (\partial \varphi^k/\partial x^i)(p)(\partial/\partial x^k)_{\varphi(p)}$. The Riemannian metric g of $\mathbf{E}^n$ satisfies $g_q((\partial/\partial x^i)_q, (\partial/\partial x^j)_q) = \delta_{ij}$ at each point q.

Since φ is a motion, we have $g_{\varphi(p)}(\varphi_*(\partial/\partial x^i)_p, \varphi_*(\partial/\partial x^j)_p) = \delta_{ij}$, from which we get

$$\sum_s \frac{\partial \varphi^s}{\partial x^i} \frac{\partial \varphi^s}{\partial x^j} = \delta_{ij}. \tag{2}$$

Hence the matrix $(\partial \varphi^k/\partial x^i)$ is an orthogonal matrix. Differentiating (2) with respect to x^k, we get

$$\sum_s \left(\frac{\partial^2 \varphi^s}{\partial x^i \partial x^k} \frac{\partial \varphi^s}{\partial x^j} + \frac{\partial \varphi^s}{\partial x^i} \frac{\partial^2 \varphi^s}{\partial x^j \partial x^k} \right) = 0, \qquad (i, j, k = 1, \ldots, n).$$

Let the left-hand member of this equation be denoted by L_{ikj}, and compute $L_{ikj} + L_{kji} - L_{jik}$. One finds it equal to

$$\sum_s \frac{\partial \varphi^s}{\partial x^i} \frac{\partial^2 \varphi^s}{\partial x^j \partial x^k}.$$

However, $L_{ikj} = L_{kji} = L_{jik} = 0$. Hence we get

$$\sum_s \frac{\partial \varphi^s}{\partial x^i} \frac{\partial^2 \varphi^s}{\partial x^j \partial x^k} = 0.$$

Multiplying $\partial \varphi^t/\partial x^i$ on both sides of this equation, summing on i, and using $\sum_i (\partial \varphi^t/\partial x^i)\partial \varphi^s/\partial x^i = \delta_{ts}$, we get

$$\frac{\partial^2 \varphi^t}{\partial x^j \partial x^k} = 0 \qquad (t, j, k = 1, \ldots, n).$$

Hence, $\varphi^i(x)$ is a linear function in $x^1, \ldots, x^n$, and φ becomes an affine transformation. If we let $\varphi^i(x) = \sum_j a_j^i x^j + b^i$, then, since $\partial \varphi^i / \partial x^j = a_j^i$, (a_j^i) is an orthogonal matrix. Conversely, it is easy to see that if φ is an affine transformation and (a_j^i) an orthogonal matrix, then φ is a motion.

Example 2. Let H be the hyperbolic plane. Considering the points of H as complex numbers with positive imaginary parts, we can view a diffeomorphism φ of H to be a complex-valued function defined on the upper half-plane H, and set $z' = \varphi(z)$. If $\varphi(z)$ is an analytic function of z, then we call φ a holomorphic transformation of H. The set of all holomorphic transformations of H forms a group. We denote this group by $A(H)$. The group of motions of H is denoted by $I(H)$. As is known in the theory of complex variables, the group $A(H)$ coincides with the group of linear fractional transformations $z \to (az + b)/(cz + d)$ (where a, b, $c, d \in \mathbf{R}$, $\begin{vmatrix} a & b \\ c & d \end{vmatrix} = 1$).* On the other hand, $\theta : z \to -\bar{z}$ is a diffeomorphism of H, and one can easily verify that $\theta \in I(M)$ and $\theta^2 = 1$. We have the following relation between $I(H)$ and $A(H)$:

THEOREM. *The group $A(H)$ is a normal subgroup of $I(H)$ of index 2, and $I(H) = A(H) + \theta A(H)$ holds.*

Proof. Let φ be a diffeomorphism of H and let $u(x, y)$ and $v(x, y)$ be, respectively, the real and imaginary parts of $\varphi(z)$. That is, $u(x, y) = \varphi^* x$ and $v(x, y) = \varphi^* y$. The condition, with respect to the Riemannian metric g of H, for $\varphi \in I(H)$, is

$$\left(\frac{\partial u}{\partial x}\right)^2 + \left(\frac{\partial v}{\partial x}\right)^2 = h^2, \qquad \left(\frac{\partial u}{\partial y}\right)^2 + \left(\frac{\partial v}{\partial y}\right)^2 = h^2,$$

$$\frac{\partial u}{\partial x}\frac{\partial u}{\partial y} + \frac{\partial v}{\partial x}\frac{\partial v}{\partial y} = 0. \tag{3}$$

Here, $h = v(x, y)y^{-1}$. This means that in a 2-dimensional vector space, $(h^{-1}\,\partial u/\partial x, h^{-1}\,\partial v/\partial x)$ and $(h^{-1}\,\partial u/\partial y, h^{-1}\,\partial v/\partial y)$ are two mutually orthogonal vectors of lengths 1. However, $(h^{-1}\,\partial v/\partial x, -h^{-1}\,\partial u/\partial x)$ is also a vector orthogonal to $(h^{-1}\,\partial u/\partial x, h^{-1}\,\partial v/\partial x)$ and of length 1. Thus, we have $(h^{-1}\,\partial v/\partial x, -h^{-1}\,\partial u/\partial x) = \pm(h^{-1}\,\partial u/\partial y, h^{-1}\,\partial v/\partial y)$, and hence $\partial u/\partial y = \pm\partial v/\partial x$, $-\partial u/\partial x = \pm\partial v/\partial y$. Here, if we have the negative sign, then (u, v) satisfies the Cauchy-Riemann differential equations, and hence

---

*Cf. H. Cartan, *Elementary Theory of Analytic Functions of One or Several Complex Variables*, Addison-Wesley, Reading, Massachusetts, 1963, p. 182.

$\varphi(z) = u + iv$ is a holomorphic function, so $\varphi \in A(H)$. On the other hand, if we have the positive sign, then $(-u, v)$ satisfies the Cauchy-Riemann differential equations, so $-\overline{\varphi(z)} = -u + iv$ is a holomorphic function of z. Since $-\overline{\varphi(z)} = \theta(\varphi(z))$, we have $\theta \circ \varphi \in A(H)$. But $\theta^2 = 1$, so $\varphi \in \theta A(H)$. That is, if φ is a motion of H, we have $\varphi \in A(H)$ or $\varphi \in \theta A(H)$. Conversely, if $\varphi \in A(H)$, then, as we have said, φ is a linear fractional transformation, and we can write

$$\varphi(z) = \frac{az + b}{cz + d} \qquad (a, b, c, d \in \mathbf{R}, \quad \begin{vmatrix} a & b \\ c & d \end{vmatrix} = 1). \tag{4}$$

Since $d\varphi/dz = \partial u/\partial x + i\,\partial v/\partial x = \partial v/\partial y - i\,\partial v/\partial y$, we have

$$\left(\frac{\partial u}{\partial x}\right)^2 + \left(\frac{\partial v}{\partial x}\right)^2 = \left(\frac{\partial u}{\partial y}\right)^2 + \left(\frac{\partial v}{\partial y}\right)^2 = \left|\frac{d\varphi}{dz}\right|^2. \tag{5}$$

From (4) we get

$$\frac{d\varphi}{dz} = \frac{1}{(cz + d)^2}, \qquad v(x, y) = \frac{y}{|cz + d|^2}.$$

Hence $|d\varphi/dz|^2 = (v(x, y)y^{-1})^2 = h^2$, and u, v satisfy the first two equations of (3). Since u, v also satisfy the Cauchy-Riemann equations $\partial u/\partial x = \partial v/\partial y$, $\partial u/\partial y = -\partial v/\partial x$, they also satisfy the last equation of (3). Hence φ is a motion. Hence $A(H)$ is a subgroup of $I(H)$, and we have $I(H) = A(H) + \theta A(H)$. As θ is clearly not holomorphic, $\theta \notin A(H)$, so $A(H)$ is a subgroup of $I(H)$ of index 2, and hence a normal subgroup.

From what we have said above, we can conclude that if $\varphi \in I(H)$, then

$$\varphi(z) = \frac{az + b}{cz + d} \quad \text{or} \quad \varphi(z) = \frac{-a\bar{z} - b}{c\bar{z} + d} \qquad (a, b, c, d \in \mathbf{R}, \quad \begin{vmatrix} a & b \\ c & d \end{vmatrix} = 1).$$

§10. Immersion and Imbedding of Manifolds; Submanifolds

Let φ be a differentiable map from a manifold M into a manifold M', and let the dimensions of M and M' be n and m, respectively. If, at each point p of M, $(\varphi_*)_p$ is a 1:1 map, i.e., if $\text{rank}_p(\varphi) = n$, then φ is called an *immersion* of M into M'. If φ is an immersion, and if moreover φ is a 1:1 map, i.e., if $\varphi(p) \neq \varphi(q)$ for $p \neq q$, then φ is called an *imbedding* of M into M'. Figure 2.5, (1) and (2), shows an immersion and an imbedding of $\mathbf{R}^1$ into $\mathbf{R}^2$.

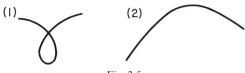

Fig. 2.5

DEFINITION.   If the manifolds M and M' satisfy the following two conditions, then M is called a *submanifold* of M':

(1) The set M is a subset of M'.
(2) The identity map i from M into M' is an imbedding of M into M'.

One should note, in this definition, that although M is a subset of M', it does not necessarily have the topology as a subspace of M'. However, by (2), i is a continuous map from M into M', so the topology of M is not weaker than the topology it would have as a subspace of M'. If the topology of M coincides with the topology of M as a subspace of M', then M is called a *regular submanifold* of M'. If M is a regular submanifold of M', and if M is a closed subset of M', then we call M a *closed submanifold* of M'.

If M is a submanifold of M', at each point p of M, $(i_*)_p$ is a $1:1$ linear map from $T_p(M)$ into $T_p(M')$, so we sometimes identify the subspace $i_* T_p(M)$ of $T_p(M')$ with $T_p(M)$, and call $i_* T_p(M)$ the tangent space of M at p. The tangent vectors of M' belonging to $i_* T_p(M)$ are called the tangent vectors of M' tangent to M. In particular, if M' is the affine space $\mathbf{R}^m$, then the n-dimensional plane passing through p and spanned by $i_* T_p(M)$ ($n = \dim M$), is called the tangent plane of M at p.

Problem 1. Let M be an n-dimensional submanifold of $\mathbf{R}^m$. Let $(x^1, \ldots, x^n)$ be a local coordinate system of M at the point p. Let $(y^1, \ldots, y^m)$ be the standard coordinate system of $\mathbf{R}^m$, and set $y^k \circ i = \varphi^k(x^1, \ldots, x^n)$. Show that the equations for the tangent plane of M at p can be written as

$$y^k = y^k(p) + \sum_{i=1}^{n} t^i \frac{\partial \varphi^k}{\partial x^i}(p) \qquad (k = 1, \ldots, m),$$

where $(t^1, \ldots, t^n)$ are parameters.

PROPOSITION.   *Let φ be a differentiable map from a manifold M to a manifold M'. Let N and N' be submanifolds of M and M', respectively, and suppose the image $\varphi(N)$ of N by φ is contained in N'. Let ψ be the map from N into N' obtained by restricting φ to N. If N' is a regular submanifold of M', then ψ is a differentiable map from N into N'.*

Proof. (1) We shall first show that ψ is a continuous map from N into N'. Let $p \in N$, $\psi(p) = p'$, and let $V_{N'}$ be a neighborhood of p' in N'.

By assumption, N' is a regular submanifold of M', so there is a neighborhood V of p' in M' such that $V_{N'} = V \cap N'$. Since φ is of course continuous, taking a neighborhood U of p in M small enough, we have $\varphi(U) \subset V$. The identity map i from N into M is continuous, and since $U \cap N = i^{-1}(U)$, $U \cap N$ is a neighborhood of p in N. By the definition of ψ, we have $\psi(U \cap N) \subset V_{N'}$. Hence ψ is continuous.

(2) We shall now prove the differentiability of ψ from its continuity. (We shall not use the hypothesis that N' is a regular submanifold of M' anymore.) Let i be the identity map from N into M, and i' the identity map from N' into M'. Since φ, ψ, i, i' are all continuous, we can choose neighborhoods U_N and U of $p \in N$ in N and M, respectively, and neighborhoods $V_{N'}$ and V of $\psi(p) \in N'$ in N' and M', respectively, such that

$$\varphi(U) \subset V, \qquad \psi(U_N) \subset V_{N'}$$
$$i(U_N) \subset U, \qquad i'(V_{N'}) \subset V$$

hold. Let dim $M = n$, dim $N = r$, dim $M' = m$, dim $N' = s$. By the theorem of §7, we can assume that we have a local coordinate system $(x^1, \ldots, x^r)$ of N on U_N and a local coordinate system $(y^1, \ldots, y^n)$ of M on U such that

$$y^k \circ i = x^k \quad (k = 1, \ldots, r), \qquad y^j \circ i = 0 \quad (j = r+1, \ldots, n)$$

hold. Similarly, we may assume that we have local coordinate systems $(\bar{x}^1, \ldots, \bar{x}^s)$ and $(\bar{y}^1, \ldots, \bar{y}^m)$, of N' on $V_{N'}$ and of M' on V, respectively, such that

$$\bar{y}^k \circ i' = \bar{x}^k \quad (k = 1, \ldots, s), \qquad \bar{y}^j \circ i' = 0 \quad (j = s+1, \ldots, m)$$

hold. Since φ is differentiable, we can write

$$\bar{y}^k \circ \varphi = \bar{\varphi}^k(y^1, \ldots, y^n) \qquad (k = 1, \ldots, m),$$

where $\bar{\varphi}^k$ is a C^∞ function in n variables. For $p \in U_N$, we have

$$\begin{aligned}
\bar{x}^k(\psi(p)) = \bar{y}^k(i'(\psi(p))) &= \bar{y}^k(\varphi(i(p))) \\
&= \bar{\varphi}^k(x^1(p), \ldots, x^r(p), 0, \ldots, 0).
\end{aligned}$$

Hence ψ is differentiable at p.

Problem 2.   Prove that the circle S^1 cannot be immersed in $\mathbf{R}^1$.

Example.   Let us give an example of a submanifold which is not regular. Let T^2 be the 2-dimensional torus. The torus T^2 can be considered as the set of all pairs $(e^{2\pi i\theta}, e^{2\pi i\eta})$ of complex numbers of absolute value 1. For an arbitrary real number t, let $\varphi(t) = (e^{2\pi it}, e^{2\pi i\alpha t})$, where α is a fixed irrational number. φ is an imbedding of $\mathbf{R}^1$ into T^2. By Kronecker's approximation theorem,* for arbitrary ε, θ, $\eta \in \mathbf{R}$, there exist $t \in \mathbf{R}$ and integers m, n such that

$$|\theta - t - m| < \varepsilon, \quad |\eta - \alpha t - n| < \varepsilon$$

hold. Hence, for an arbitrary point $(e^{2\pi i\theta}, e^{2\pi i\eta})$ of T^2, $\varphi(t)$ can be found arbitrarily close to it. Hence $\varphi(\mathbf{R}^1)$ is dense in T^2. On the other hand, since φ is an imbedding of $\mathbf{R}^1$ into T^2, we can put a topology and a differentiable structure on $\varphi(\mathbf{R}^1)$ so that φ becomes a diffeomorphism, and then $\varphi(\mathbf{R}^1)$ becomes a submanifold of T^2. The topology on $\varphi(\mathbf{R}^1)$ thus defined does not coincide with the topology as a subspace of T^2. Hence $\varphi(\mathbf{R}^1)$ is not a regular submanifold of T^2.

Problem 3. Let $\varphi: \mathbf{R}^1 \to T^2$ be defined by $\varphi(t) = (e^{2\pi it}, e^{2\pi i\alpha t})$, where α is a fixed rational number. Show that φ defines an imbedding of S^1 into T^2.

Problem 4. Let M be a submanifold of M'. Let $L(M)$ be the set of points q of M' with the following property: for q, there is a sequence $\{p_n\}$ of points in M such that $\{i(p_n)\}$ converges to q, and $\{p_n\}$ has no accumulation point in M. Prove the following: (1) M is regular if and only if $L(M) \cap M = \varnothing$, (2) M is a closed submanifold if and only if $L(M) = \varnothing$.

Now let M be a submanifold of M', and let the dimensions of M and M' be n and m, respectively. We can choose local coordinate systems $(x^1, \ldots, x^n)$ and $(y^1, \ldots, y^m)$ of M and M', respectively, at each point p of M, such that $x^k(p) = 0$ $(k = 1, \ldots, n)$, $y^j(p) = 0$ $(j = 1, \ldots, m)$, $y^k \circ i = x^k$ $(k = 1, \ldots, n)$ and $y^j \circ i = 0$ $(j = n + 1, \ldots, m)$. Let U_M be a sufficiently small neighborhood of p in M so that $(x^1, \ldots, x^n)$ are defined on U_M. If M is a regular submanifold of M', then a neighborhood U of p in M' exists such that $U_M = U \cap M$. Taking U sufficiently small so that $(y^1, \ldots, y^m)$ are defined on U, we can verify easily that $U \cap M = \{q \in U \mid y^j(q) = 0, j = n + 1, \ldots, m\}$. Conversely, we have the following theorem:

THEOREM 1. *Let M be a subset of an m-dimensional manifold M'. For each point p of M, assume that there are a neighborhood U of p in M' and C^∞ functions $f^1, \ldots, f^k$ on U (where k is independent of p), satisfying the two conditions* (1) $U \cap M = \{q \in U \mid f^i(q) = 0, i = 1, \ldots, k\}$ *and* (2) $(df^1)_p, \ldots, (df^k)_p$ *are linearly independent. Then M is a regular submanifold of M' of dimension $m - k$.*

*Cf. L. Pontryagin, *Topological Groups*, 2nd ed., Gordon and Breach, New York, 1966, p. 257, or Hardy-Wright, *An Introduction to the Theory of Numbers*, Oxford Univ. Press, London and New York, 1960, p. 373ff.

Proof. Put the topology on M as a subspace of M'. Then certainly M becomes a Hausdorff space. For each point p of M, to the functions $f^1, \ldots, f^k$ satisfying the hypotheses of the throrem, we can add $m - k$ C^∞ functions $y^1, \ldots, y^{m-k}$, such that $(dy^1)_p, \ldots, (dy^{m-k})_p, (df^1)_p, \ldots,$ $(df^k)_p$ are linearly independent.

Let $m - k = n$, and set $y^{n+i} = f^i$ $(i = 1, \ldots, k)$. Then $(dy^1)_p, \ldots,$ $(dy^m)_p$ are linearly independent. Hence $(y^1, \ldots, y^m)$ is a local coordinate system on a neighborhood V' of p. Choose $a > 0$ sufficiently small, and $V_a' = \{q \in V' \mid |y^i(q)| < a, i = 1, \ldots, m\}$. V_a' is diffeomorphic to the cube $Q_a^m = \{(u^i) \mid |u^i| < a, i = 1, \ldots, n\}$ in $\mathbf{R}^m$ by the correspondence $q \to (y^i(q))$. Futhermore, since $V_a = V_a' \cap M = \{q \in V' \mid y^{n+i}(q) = 0,$ $i = 1, \ldots, k\}$, the neighborhood V_a of p in M is homeomorphic to the cube $Q_a^n = \{(u^i) \mid |u^i| < a, i = 1, \ldots, n\}$ in $\mathbf{R}^n$ by the correspondence $q \to \psi_p(q) = (y^1(q), \ldots, y^n(q))$. Since this holds for an arbitrary point p of M, M becomes a topological manifold of dimension n, and (V_a, ψ_p) is a coordinate neighborhood containing p. Writing the restriction of y^i to V_a as x^i, $(x^1, \ldots, x^n)$ becomes a local coordinate system of M on V_a. Similarly, for another point q of M, take $b > 0$, a neighborhood W_b' of q in M', and a local coordinate system $(w^1, \ldots, w^m)$ of M' on W_b' satisfying the conditions as before. If we let z^i be the restriction of w^i to $W_b = W_b' \cap M$, then $(z^1, \ldots, z^n)$ is a local coordinate system of M on W_b. Now, if $V_a \cap W_b \neq \varnothing$, then of course $V_a' \cap W_b' \neq \varnothing$, and if $q \in V_a' \cap W_b'$, then $w^i(q) = F'^i(y^1(q), \ldots, y^m(q))$, where $F'^i(u^1, \ldots, u^m)$ is a C^∞ function of $(u^1, \ldots, u^m)$. Hence, if $q \in V_a \cap W_b$, then $z^i(q) = F^i(x^1(q), \ldots, x^n(q))$, where $F^i(u^1, \ldots, u^n) = F'^i(u^1, \ldots, u^n, 0, \ldots, 0)$. This shows that the transformation between the local coordinate systems $(x^1, \ldots, x^n)$ and $(z^1, \ldots, z^n)$ on $V_a \cap W_b$ is of class C^∞. Hence M is a C^∞ manifold. From the definitions of the topology and the differentiable structure of M, it is easy to check that M is a regular submanifold of M'.

COROLLARY.   *Let M' be an m-dimensional manifold and let $f^1, \ldots, f^k$ be C^∞ functions on M'. Set $M = \{p \in M' \mid f^i(p) = 0, i = 1, \ldots, k\}$. If at each point p of M, $(df^1)_p, \ldots, (df^k)_p$ are linearly independent, then M is an $(m - k)$-dimensional closed submanifold of M'.*

Remark. If, for a finite number of functions $f^1, \ldots, f^k$ on M', $M = \{p \in M' \mid f^i(p) = 0, i = 1, \ldots, k\}$ is a closed submanifold of M', then we call M a submanifold of M' defined by the equations $f^1 = 0, \ldots,$ $f^k = 0$. In this case, it is not necessarily true that the number of elements in a maximal independent subset of $\{(df^1)_p, \ldots, (df^k)_p\}$ $(p \in M)$ is equal to dim M' − dim M. For example, if M is defined by $f = 0$, then M is also defined by $(f)^2 = 0$, but for $p \in M$, we have $(d(f)^2)_p = 2f(p)$ $\cdot (df)_p = 0$, and $d(f)^2$ is equal to 0 on M.

Example 1. The n-dimensional sphere S^n is a closed submanifold of $\mathbf{R}^{n+1}$, defined by the equation $(x^1)^2 + \ldots + (x^{n+1})^2 = r^2$.

Example 2. Let (x, y) be the coordinates in the affine plane, and set $f = xy$. The equation $f = 0$ defines two straight lines $x = 0$ and $y = 0$, and $f = 0$ does not define a closed submanifold of the plane. In this case, df is 0 at $(0, 0)$. If a closed submanifold M of an m-dimensional manifold M' $(M \neq M')$ is defined by a single equation $f = 0$, it is not necessarily true that the dimension of M is $m - 1$. For example, a k-dimensional submanifold of $\mathbf{R}^m$, defined by $x^{k+1} = \ldots = x^m = 0$, can be defined by a single equation $(x^{k+1})^2 + \ldots + (x^n)^2 = 0$.

Problem 5. Let $f = \sum_{i,j=1}^{n} a_{ij} x^i x^j + \sum_{i=1}^{n} b_i x^i + c$, and $M = \{q \in \mathbf{R}^n | f(q) = 0\}$. Find a condition on f for df not to be 0 at each point of M, and find a condition for M to be a submanifold of $\mathbf{R}^n$.

THEOREM 2. *A compact manifold M is diffeomorphic to a closed submanifold of an affine space of a sufficiently high dimension, that is, M can be imbedded in an affine space.*

Proof. Choose a local coordinate system $(x^1, \ldots, x^n)$ around p in M, so that $x^i(p) = 0$ $(i = 1, \ldots, n)$. Let $a > b > 0$ be positive numbers, sufficiently small, and set $V_p = \{q | |x^i(q)| < a, i = 1, \ldots, n\}$ and $W_p = \{q | |x^i(q)| < b, i = 1, \ldots, n\}$. Then W_p and V_p are neighborhoods of p, and $\overline{W}_p \subset V_p$. Let $g(t)$ be a C^∞ function, defined for $-\infty < t < +\infty$, and satisfying the condition: $g(t) = 1$ for $|t| \leq b$, $0 < g(t) < 1$ for $b < |t| < a$, and $g(t) = 0$ for $|t| \geq a$.* We now define a function f_p on M by

$$f_p(q) = g(x^1(q)) \ldots g(x^n(q)), \quad q \in V_p$$
$$= 0 \qquad\qquad\qquad q \notin V_p.$$

The function f_p is of class C^∞ and we have: $f_p(q) = 1$ for $q \in \overline{W}_p$, $0 < f_p(q) < 1$ for $q \in V_p - \overline{W}_p$, and $f_p(q) = 0$ for $q \notin V_p$. Now $\{W_p, p \in M\}$ is an open cover of M, and since M is compact, we can choose a finite number of points $p_1, \ldots, p_k$ so that $\cup_{\alpha=1}^{k} W_{p_\alpha} = M$. Set $f_{p_\alpha} = f_\alpha$, $V_{p_\alpha} = V_\alpha$, $W_{p_\alpha} = W_\alpha$, and denote the local coordinate system on W_α by $x_\alpha^1, \ldots, x_\alpha^n$. Define $(n + 1)k$ functions f_α^i $(i = 0, 1, \ldots, n; \alpha = 1, \ldots,$

*We shall give an example of such a function after the proof of this theorem.

k) on M by

$$f_\alpha^0 = f_\alpha$$
$$f_\alpha^i(q) = f_\alpha(q)x_\alpha^i(q), \qquad q \in V_\alpha$$
$$= 0 \qquad q \notin V_\alpha \qquad (i = 1, \ldots, n).$$

The function f_α^i is of class C^∞, and for $i \geq 1$ and $q \in \overline{W}_\alpha$, we have $f_\alpha^i(q) = x_\alpha^i(q)$. Let $N = (n + 1)k$, and define a C^∞ map φ from M into $\mathbf{R}^N$ by $\varphi(q) = (f_\alpha^i(q))$. Since M is compact, $\varphi(M)$ is a compact subset of $\mathbf{R}^N$, and hence closed. The map φ is $1:1$. In fact, let $\varphi(p) = \varphi(q)$ and $p \in W_\alpha$. Then $f_\alpha^0(p) = 1$, but since $\varphi(p) = \varphi(q)$, we get $f_\alpha^0(q) = f_\alpha^0(p) = 1$, so that $q \in \overline{W}_\alpha$. Since p, $q \in \overline{W}_\alpha$, we have $x_\alpha^i(p) = f_\alpha^i(p) = f_\alpha^i(q) = x_\alpha^i(q)$, so that we get $p = q$, and hence φ is $1:1$. Now let p be an arbitrary point of M, and let $p \in W_\alpha$. On W_α we have $f_\alpha^i = x_\alpha^i$, and hence $\partial f_\alpha^i(p)/\partial x_\alpha^j = \delta_j^i$. Hence $\mathrm{rank}_p\varphi \geq n$, but $\mathrm{rank}_p\varphi \leq n$ is clear, so that we have $\mathrm{rank}_p\varphi = n$ at each point p. Hence φ is an imbedding of M into $\mathbf{R}^N$. If we put a differentiable structure on $\varphi(M)$ so that φ is a diffeomorphism, then $\varphi(M)$ is a closed submanifold of M.

Remark. It is known that a theorem more general than the one above holds (Whitney's theorem). Namely, if M is an n-dimensional para-compact manifold, then M is diffeomorphic to a closed submanifold of a $(2n + 1)$-dimensional affine space.* For paracompact manifolds, see §14.

The C^∞ function mentioned in the proof of Theorem 2 can be constructed, for example, as follows. Let $f(t)$ be defined by

$$f(t) = e^{-1/t^2} \qquad \text{for} \quad t > 0$$
$$= 0 \qquad \text{for} \quad t \leq 0.$$

The function $f(t)$ is of class C^∞. Let $h(t) = f(t)/(f(t) + f(1 - t))$. Then $h(t)$ is also of class C^∞, and we have: $h(t) = 0$ for $t \leq 0$, $0 < h(t) < 1$ for $0 < t < 1$, and $h(t) = 1$ for $1 \leq t$. Set $g(t) = h((t + a)/(a - b)) \cdot h((-t + a)/(a - b))$. Then we have: $0 \leq g(t) \leq 1$, $g(t) = 0$ for $|t| \geq a$, and $g(t) = 1$ for $|t| \leq b$. The function $g(t)$ is the desired function. (Fig. 2.6).

The following Lemma 1, which has important applications, has been proved essentially at the beginning of the proof of the theorem.

---

*For the proof of Whitney's theorem, see, for example, L. Auslander and R. MacKenzie, *Introduction to Differentiable Manifolds*, McGraw-Hill, New York, 1963.

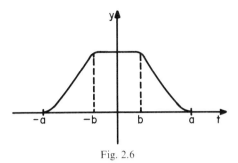

Fig. 2.6

LEMMA 1. *For an arbitrary neighborhood U of a point p of a manifold M, one can choose a sufficiently small neighborhood V of p and a C^∞ function f on M so that*

(1) $\overline{V} \subset U$,
(2) $0 \leq f \leq 1$, $f = 1$ on V, and $f = 0$ on the complement U^c of U.

LEMMA 2. *Let h be a C^∞ function defined on a neighborhood U of p on M. We can choose a neighborhood V of p sufficiently small so that $\overline{V} \subset U$, and a C^∞ function g on M such that $g = h$ on V and $g = 0$ on the complement U^c of U.*

Proof. First choose a neighborhood W of p so that $\overline{W} \subset U$, and then for W, choose a neighborhood V of p and a C^∞ function f on M, satisfying the conditions of Lemma 1. Set

$$g(q) = f(q)h(q) \qquad \text{for} \quad q \in U$$
$$= 0 \qquad \text{for} \quad q \in U^c.$$

Then g is the desired function.

Now if M is a submanifold of M', and if g is a Riemannian metric on M', then g induces a Riemannian metric h on M in a natural manner. Define h_p by

$$h_p(u, v) = g_{i(p)}(i_* u, i_* v)$$

for $u, v \in T_p(M)$. Then h_p is a positive inner product on $T_p(M)$. Let a local coordinate system around p in M be $(x^1, \ldots, x^n)$, let a local coordinate system around $p = i(p)$ in M' be $(y^1, \ldots, y^m)$, and set

$$y^k \circ i = a^k(x^1, \ldots, x^n) \qquad (k = 1, \ldots, m).$$

Then the components $h_{ab} = h(\partial/\partial x^a, \partial/\partial x^b)$ of h with respect to $(x^1, \ldots, x^n)$ are

$$h_{ab} = \sum_{i,j=1}^{m} \frac{\partial a^i}{\partial x^a} \frac{\partial a^j}{\partial x^b} g_{ij} \qquad (a, b = 1, \ldots, n).$$

Hence h is of class C^∞ and is a Riemannian metric on M.

In particular, if M' is the m-dimensional Euclidean space E^m, and $(y^1, \ldots, y^m)$ is the standard coordinate system, then since $g_{ij} = \delta_{ij}$, we have

$$h_{ab} = \sum_{i=1}^{m} \frac{\partial a^i}{\partial x^a} \frac{\partial a^i}{\partial x^b}.$$

The Riemannian metric $\sum_{a,b} h_{ab} \, dx^a \, dx^b$ of M is called the Riemannian metric of the submanifold M of M' determined from the Riemannian metric of M'.

Problem 6. Using cylindrical coordinates, find the Riemannian metric on the cylinder $x^2 + y^2 = r^2$ in the (x, y, z)-space.

Problem 7. Using spherical coordinates, find the Riemannian metric on the unit sphere $x^2 + y^2 + z^2 = 1$.

§11. Vector Fields and Derivations

For each point p of a manifold M, pick a tangent vector X_p at p. The correspondence $X : p \to X_p$ is called a *vector field* on M.

If $(x^1, \ldots, x^n)$ is a local coordinate system on an open set U of M, then at each point p of U, X_p can be expressed uniquely as

$$X_p = \sum_{i=1}^{n} \xi^i(p) \left(\frac{\partial}{\partial x^i} \right)_p. \tag{1}$$

Then n functions ξ^i $(i = 1, \ldots, n)$ on U are called the *components* of X with respect to the local coordinate system $(x^1, \ldots, x^n)$. If $(\bar{x}^1, \ldots, \bar{x}^n)$ is another local coordinate system on U, and $\bar{\xi}^i$ the components of X with respect to $(\bar{x}^1, \ldots, \bar{x}^n)$, then we have

$$\bar{\xi}^i(p) = \sum_{j=1}^{n} \frac{\partial \bar{x}^i}{\partial x^j}(p) \xi^j(p), \qquad p \in U. \tag{2}$$

Thus, the property that the components of X are continuous or of class C^r at one point does not depend on the choice of the local coordinate system. If all the components of X are continuous or of class C^r on U, then we say that X is continuous of class C^r on U. If X is continuous or of class C^r on a neighborhood of each point of M, then X is said to be continuous or of class C^r on M. From now on, for simplicity, we consider only C^∞ vector fields, and denote the set of all C^∞ vector fields on M by $\mathfrak{X}(M)$. We also agree that by a vector field, we always mean a C^∞ vector field, unless otherwise stated.

Let $C^\infty(M)$ denote the set of all C^∞ functions on M. For $X, Y \in \mathfrak{X}(M)$ and $f \in C^\infty(M)$, the correspondences

$$p \to f(p)X_p; \qquad p \to X_p + Y_p$$

again define vector fields on M. These vector fields are denoted by fX and $X + Y$, respectively, and are called the product of f and X, and the sum of X and Y. The following rules for products and sums hold: if $f, g \in C^\infty(M)$ and $X, Y \in \mathfrak{X}(M)$, then

$$f(gX) = (fg)X$$
$$f(X + Y) = fX + fY, \qquad (f + g)X = fX + gX.$$

Now, if $(x^1, \ldots, x^n)$ is a local coordinate system on U, we can define a vector field $\partial/\partial x^i$ on U by $p \to (\partial/\partial x^i)_p$. The components ξ^i of a vector field with respect to $(x^1, \ldots, x^n)$ are C^∞ functions on U, and X can be expressed on U as

$$X = \sum_{i=1}^{n} \xi^i \frac{\partial}{\partial x^i}.$$

For $X \in \mathfrak{X}(M)$ and $f \in C^\infty(M)$, we can define a function Xf on M by

$$(Xf)(p) = X_p f \qquad (p \in M).$$

At each point p of a coordinate neighborhood U, we have

$$(Xf)(p) = \sum_{i=1}^{n} \xi^i(p) \frac{\partial f}{\partial x^i}(p).$$

Hence Xf is of class C^∞. Xf is called the derivative of f by the vector field X. We have the following rules on differentiation by X:

$$X(f + g) = Xf + Xg,$$
$$X(fg) = Xf \cdot g + f \cdot Xg.$$

We shall now have a short interlude to introduce the definitions of an algebra and a derivation of it.

DEFINITION 1.   Let $\mathfrak{A}$ be a vector space over a field K.* If, for any two elementa a, b of $\mathfrak{A}$, there is a third elemnt $a \circ b$ of $\mathfrak{A}$, and if the conditions

$$\lambda(a \circ b) = (\lambda a) \circ b = a \circ \lambda b$$
$$(a + b) \circ c = a \circ c + b \circ c, \qquad (\lambda \in K; \, a, b, c \in \mathfrak{A})$$
$$a \circ (b + c) = a \circ b + a \circ c$$

are satisfied, then $\mathfrak{A}$ is called an *algebra*.

If $\mathfrak{A}$ further satisfies

$$a \circ (b \circ c) = (a \circ b) \circ c,$$

then $\mathfrak{A}$ is called an *associative algebra*.

If a mapping D from an algebra $\mathfrak{A}$ to itself has the property that

$$D(\lambda a) = \lambda Da, \quad D(a + b) = Da + Db,$$
$$D(a \circ b) = (Da) \circ b + a \circ (Db), \qquad (\lambda \in K; \, a, b \in \mathfrak{A})$$

then D is called a *derivation* of the algebra $\mathfrak{A}$.

Let $\mathfrak{A}$ be an associative algebra and let $\mathfrak{M}$ be a vector space over K. If, for any element a of $\mathfrak{A}$, and for any element x of $\mathfrak{M}$, there is defined an element ax of $\mathfrak{M}$, and if the conditions

$$a(\lambda x) = (\lambda a)x = \lambda(ax),$$
$$a(x + y) = ax + ay,$$
$$(a + b)x = ax + bx \qquad (\lambda \in K; \, a, b \in \mathfrak{A}; \, x, y \in \mathfrak{M})$$
$$a(bx) = (a \circ b)x$$

are satisfied, then $\mathfrak{M}$ is called an $\mathfrak{A}$-*module*.

Example.   If we define the sum and the product of functions in the usual manner, then $C^\infty(M)$ is an associative algebra. For $f \in C^\infty(M)$ and $X \in \mathfrak{X}(M)$, we have defined $fX \in \mathfrak{X}(M)$. Using this product, $\mathfrak{X}(M)$ becomes a $C^\infty(M)$-module.

For $X \in \mathfrak{X}(M)$ and $f \in C^\infty(M)$, if we set

$$D_X f = Xf, \tag{3}$$

then D_X is a derivation of the associative algebra $C^\infty(M)$. We shall

---

*We do not assume that $\mathfrak{A}$ is finite dimensional.

now prove that, conversely, any derivation of $C^\infty(M)$ is of the form D_X.

We shall first prove the following lemma.

LEMMA. *If for vector fields X and Y on M, $D_X = D_Y$ holds, then X and Y are equal.*

Proof. It suffices to show that if $D_X f = 0$ for all $f \in C^\infty(M)$, then $X_p = 0$ for an arbitrary point p of M. Let $(x^1, \ldots, x^n)$ be a local coordinate system around p. Let $X = \sum_i \xi^i \partial/\partial x^i$ on a neighborhood of p. By Lemma 2 of §10, there is a $f^i \in C^\infty(M)$, coinciding with x^i on a neighborhood of p. We have $(Xf^i)(p) = \sum_j \xi^j(p) \partial f^i(p)/\partial x^j$. Since $f^i = x^i$ on a neighborhood of p, we get $\partial f^i(p)/\partial x^j = \delta^i_j$. Hence $(Xf^i)(p) = \xi^i(p)$. However, by assumption, $Xf^i = 0$. Hence $\xi^i(p) = 0$. Since i was an arbitrary index, we have $X_p = 0$.

THEOREM 1. *If D is a derivation of the associative algebra $C^\infty(M)$, then there is a unique vector field X on M such that $D = D_X$.*

Proof. (1) We shall prove that if f is 0 at all points of an open set U, then so is Df. Let p be an arbitrary point of U. By Lemma 2 of §10, there are neighborhoods U_p and W_p of p, and $f_p \in C^\infty(M)$, such that $\overline{W}_p \subset U_p \subset U$, $f_p = 1$ on W_p, and $f_p = 0$ on the complement of U_p. Set $g = 1 - f_p$. Then we have $gf = f$. In fact, if $q \notin U_p$, then $f(q) = 0$, so $g(q)f(q) = f(q) = 0$, and if $q \in U_p$, then $g(q) = 1$, so $g(q)f(q) = f(q)$. Hence $Df = D(gf) = (Dg)f + g(Df)$. However, since $f(p) = g(p) = 0$, we have $(Df)(p) = 0$, that is, Df is 0 at each point of U.

(2) Let p be an arbitrary point of M, and let h be a C^∞ function defined on a neighborhood V of p. By Lemma 2 of §10, there are a neighborhood $U(\overline{U} \subset V)$ and an $f \in C^\infty(M)$, such that $f = h$ on U. If $f' \in C^\infty(M)$ is also equal to h on U, then by (1), Df is equal to Df' on U. Hence, for any $f \in C^\infty(M)$ agreeing with h on a neighborhood of p, the value of Df at p is independent of the choice of f. Hence we can define $X_p h \in \mathbf{R}$ by $X_p h = (Df)(p)$. That X_p is a tangent vector at p can be shown easily from the fact that D is a derivation. Hence a vector field $X : p \to X_p$ on M is defined, and if $f \in C^\infty(M)$, then by the definition of X_p, we have $X_p f = (Df)(p)$, so that $Xf = Df$.

(3) Let us prove that X is of class C^∞. Let $(x^1, \ldots, x^n)$ be a local coordinate system on V and let $p \in V$. If $f^i \in C^\infty(M)$ is a function coinciding with x^i on a neighborhood of p, then $X_q x^i = (Df^i)(q)$ for q in some neighborhood of p, and since $Df^i \in C^\infty(M)$, $q \to X_q x^i$ is of class C^∞. That there is only one $X \in \mathfrak{X}(M)$ such that $D = D_X$ is shown in the lemma.

Let D_1, D_2 be derivations of an algebra $\mathfrak{A}$. Define a linear map $[D_1, D_2]$ from $\mathfrak{A}$ to $\mathfrak{A}$ by

$$[D_1, D_2]a = D_1(D_2a) - D_2(D_1a).$$

Then $[D_1, D_2]$ is also a derivation of $\mathfrak{A}$. In fact, if $a, b \in \mathfrak{A}$, then we have

$$
\begin{aligned}
[D_1, & D_2](a \circ b) \\
&= D_1((D_2a) \circ b + a \circ (D_2b)) - D_2((D_1a) \circ b + a \circ (D_1b)) \\
&= (D_1(D_2a)) \circ b + (D_2a) \circ (D_1b) + (D_1a) \circ (D_2b) + a \circ (D_1(D_2b)) \\
&\quad - (D_2(D_1a)) \circ b - (D_1a) \circ (D_2b) - (D_2a) \circ (D_1b) - a \circ (D_2(D_1b)) \\
&= ([D_1, D_2]a) \circ b + a \circ ([D_1, D_2]b).
\end{aligned}
$$

$[D_1, D_2]$ is called the *commutator product* of the derivations D_1, D_2 of $\mathfrak{A}$. The commutator product has the following two properties:

$$[D_1, D_2] = -[D_2, D_1],$$

$$[D_1, [D_2, D_3]] + [D_2, [D_3, D_1]] + [D_3, [D_1, D_2]] = 0.$$

The latter is called the *Jacobi identity* for commutator products.

In particular, if we consider the commutator product $[D_X, D_Y]$ of the derivations D_X, D_Y of $C^\infty(M)$, then since this is also a derivation of $C^\infty(M)$, by the theorem above there is a vector field Z on M such that $[D_X, D_Y] = D_Z$. The vector field Z is called the *commutator product* or *bracket* of the vector fields X and Y, and denoted by $Z = [X, Y]$. That is, we have

$$[D_X, D_Y] = D_{[X, Y]}.$$

Hence, for $f \in C^\infty(M)$, we have

$$[X, Y]f = X(Yf)) - (Y(Xf)).$$

Commutator products of vector fields have the following properties:

$$
\begin{aligned}
&[X + Y, Z] = [X, Z] + [Y, Z], \quad [X, Y + Z] = [X, Y] + [X, Z], \\
&[X, Y] = -[Y, X], \\
&[fX, gY] = fg[X, Y] + f(Xg)Y - g(Yf)X, \qquad (f, g \in C^\infty(M)) \\
&[X, [Y, Z]] + [Y, [Z, X]] + [Z, [X, Y]] = 0.
\end{aligned}
\tag{4}
$$

Let the components of X, Y with respect to a local coordinate system $(x^1, \ldots, x^n)$ on U be ξ^i, η^i. Then, by (4), on U we have

$$[X, Y] = \left[\sum_i \xi^i \frac{\partial}{\partial x^i}, \sum_j \eta^j \frac{\partial}{\partial x^j} \right] = \sum_{i, j} \left[\xi^i \frac{\partial}{\partial x^i}, \eta^j \frac{\partial}{\partial x^j} \right]$$

$$= \sum_{i, j} \left\{ \xi^i \eta^j \left[\frac{\partial}{\partial x^i}, \frac{\partial}{\partial x^j} \right] + \xi^i \frac{\partial \eta^j}{\partial x^i} \frac{\partial}{\partial x^j} - \eta^j \frac{\partial \xi^i}{\partial x^j} \frac{\partial}{\partial x^i} \right\}.$$

However, since $[\partial/\partial x^i, \partial/\partial x^j] = 0$, we have

$$[X, Y] = \sum_{j=1}^{n} \left(\sum_{i=1}^{n} \left(\xi^i \frac{\partial \eta^j}{\partial x^i} - \eta^i \frac{\partial \xi^j}{\partial x^i} \right) \right) \frac{\partial}{\partial x^j} \tag{5}$$

on U.

Remark. We have used Theorem 1 to define $[X, Y]$ above. However, we can instead take a local coordinate system at each point p of M, and use

$$p \to [X, Y]_p = \sum_{j=1}^{n} \left(\sum_{i=1}^{n} \left(\xi^i(p) \frac{\partial \eta^j}{\partial x^i}(p) - \eta^i(p) \frac{\partial \xi^j}{\partial x^i}(p) \right) \right) \left(\frac{\partial}{\partial x^j} \right)_p$$

to define $[X, Y]$. In this case, we have to prove that the definition of $[X, Y]_p$ does not depend on the choice of the local coordinate system.

Problem 1. Show that the alternate definition of $[X, Y]_p$ given above does not depend on the choice of (x^i).

DEFINITION 2. Let $\mathfrak{G}$ be an algebra over K. For any two elements X, Y of $\mathfrak{G}$, we denote the product of X and Y by $[X, Y]$. If, for any X, Y, $Z \in \mathfrak{G}$, the following conditions (1) and (2) are satisfied, then we call $\mathfrak{G}$ a *Lie algebra* over K:

(1) $[X, Y] = -[Y, X]$,
(2) $[X, [Y, Z]] + [Y, [Z, X]] + [Z, [X, Y]] = 0$.

If a nonsingular linear transformation θ of $\mathfrak{G}$ satifies

$$\theta[X, Y] = [\theta(X), \theta(Y)]$$

for arbitrary X, $Y \in \mathfrak{G}$, then θ is called an *automorphism* of the Lie algebra $\mathfrak{G}$.

Examples. (1) Let $\mathfrak{A}$ be an associative algebra over K. For a, $b \in \mathfrak{A}$, set

$[a, b] = ab - ba$ (where ab is the product in the associative algebra $\mathfrak{A}$).

Then $\mathfrak{A}$ becomes a Lie algebra with respect to the new multiplication $[a, b]$.

(2) Let $\vartheta(\mathfrak{A})$ be the set of all derivations of an algebra $\mathfrak{A}$ over K. For $\lambda \in K$, D, $D' \in \vartheta(\mathfrak{A})$, we define λD, $D + D'$ by $(\lambda D)a = \lambda(Da)$, $(D + D')a = Da + D'a$, for $a \in \mathfrak{A}$. Then $\vartheta(\mathfrak{A})$ becomes a vector space over K. Furthermore, $\vartheta(\mathfrak{A})$ is a Lie algebra with respect to the commutator product.

From what was proved above, we get the following theorem:

THEOREM 2. *The set $\mathfrak{X}(M)$ of all C^∞ vector fields on a manifold M is a Lie algebra with respect to the commutator product, and this Lie algebra can be identified with the Lie algebra formed by the set of all derivations of the associative algebra $C^\infty(M)$ of all C^∞ functions on M.*

Let φ be a differentiable map from a manifold M to a manifold M'. As before, the linear map φ^* from $C^\infty(M')$ to $C^\infty(M)$ induced by φ is defined by

$$\varphi^*(f) = f \circ \varphi \qquad (f \in C^\infty(M')).$$

If a derivation D of $C^\infty(M)$ and a derivation D' of $C^\infty(M')$ satisfy

$$\varphi^* \circ D' = D \circ \varphi^*, \tag{6}$$

then D and D' are said to be *φ-related*. If D_1 and D_1', D_2 and D_2' are both φ-related, then $[D_1, D_2]$ and $[D_1', D_2']$ are φ-related. In fact, for an arbitrary $f \in C^\infty(M')$, we have $[D_1, D_2]\varphi^*(f) = D_1(D_2\varphi^*(f)) - D_2(D_1\varphi^*(f)) = \varphi^*(D_1'(D_2'f) - D_2'(D_1'f)) = \varphi^*[D_1', D_2']f$.

Now, let us assume that D and D' are φ-related. By Theorem 1, for D and D', there are vector fields X and X' on M and M', such that $D = D_X$ and $D' = D_{X'}$. Then for an arbitrary $f \in C^\infty(M')$, we have

$$((D \circ \varphi^*)f)(p) = X_p(f \circ \varphi) = ((\varphi^*)_p X_p)f,$$
$$((\varphi^* \circ D')f)(p) = (D_{X'}f)(\varphi(p)) = X'_{\varphi(p)}f.$$

Hence, at an arbitrary point p of M,

$$(\varphi_*)_p X_p = X'_{\varphi(p)} \tag{7}$$

holds. Hence X' is uniquely determined by X. Hence for a given D and φ, there is at most one derivation D' of $C^\infty(M)$ which is φ-related to D.

If (7) holds for a vector field X on M and a vector field X' on M', then we say that X and X' are φ-*related*. We have shown above that if D_X and $D_{X'}$ are φ-related, then X and X' are φ-related. The converse is also true. That is, if X and X' are φ-related, then it can be proved in a similar fashion that D_X and $D_{X'}$ are φ-related.

Now if X_1 and X_1', X_2 and X_2' are both φ-related, then D_{X_1} and $D_{X_1'}$, D_{X_2} and $D_{X_2'}$ are both φ-related, and hence $[D_{X_1}, D_{X_2}] = D_{[X_1, X_2]}$ and $[D_{X_1'}, D_{X_2'}] = D_{[X_1', X_2']}$ are φ-related. Hence $[X_1, X_2]$ and $[X_1', X_2']$ are φ-related.

If X and X' are φ-related, then since X' is uniquely determined by X, we write $X' = \varphi_* X$, and call X' the *projection* of X by φ. Then we have

$$\varphi_*[X_1, X_2] = [\varphi_* X_1, \varphi_* X_2]. \tag{8}$$

If φ is a diffeomorphism of M, then φ^* is a $1:1$ mapping from $C^\infty(M)$ onto $C^\infty(M)$, and for arbitrary $f, g \in C^\infty(M)$, we have $\varphi^*(fg) = \varphi^*(f)\varphi^*(g), \varphi^*(f + g) = \varphi^*(f) + \varphi^*(g), \varphi(\lambda f) = \lambda \varphi^*(f) (\lambda \in R)$. This shows that φ^* is an automorphism* of the associative algebra $C^\infty(M)$.

In general, if D is a derivation of an algebra $\mathfrak{A}$, and α an automorphism of $\mathfrak{A}$, then $\alpha^{-1} D\alpha$ is a derivation of $\mathfrak{A}$. In fact, for $a, b \in \mathfrak{A}$, we have

$$\begin{aligned}
(\alpha^{-1} D\alpha)(a \circ b) &= \alpha^{-1}(D(\alpha a \circ \alpha b)) \\
&= \alpha^{-1}((D\alpha a) \circ \alpha b + \alpha a \circ D\alpha b) \\
&= (\alpha^{-1} D\alpha a) \circ b + a \circ (\alpha^{-1} D\alpha b).
\end{aligned}$$

In particular, if $\mathfrak{A} = C^\infty(M)$ and φ is a diffeomorphism of M, then $\varphi^{*-1} D_X \varphi^*$ is a derivation of $C^\infty(M)$, so there is an $X' \in \mathfrak{X}(M)$ such that $\varphi^{*-1} D_X \varphi^* = D_{X'}$. Hence $\varphi^* D_{X'} = D_X \varphi^*$, and X and X' are φ-related. That is, $X' = \varphi_* X$ and, by (7),

$$(\varphi_* X)_{\varphi(p)} = (\varphi_*)_p X_p \tag{9}$$

holds at each point p of M. If we replace p by $\varphi^{-1}(p)$ in (9), then we have

$$(\varphi_* X)_p = (\varphi_*)_{\varphi^{-1}(p)} X_{\varphi^{-1}(p)}. \tag{9'}$$

From $\varphi^{*-1} D_X \varphi^* = D_{\varphi_* X}$, we have

$$\varphi^*((\varphi_* X)f) = X(\varphi^* f) \tag{10}$$

---

*In general, an automorphism of an algebra $\mathfrak{A}$ is a $1:1$ linear map α from $\mathfrak{A}$ onto $\mathfrak{A}$, satisfying $\alpha(a \circ b) = (\alpha a) \circ (\alpha b)$ for any $a, b \in \mathfrak{A}$.

for arbitrary $f \in C^\infty(M)$. From (9'), we get

$$\varphi_*(\lambda X + \mu Y) = \lambda(\varphi_* X) + \mu(\varphi_* Y) \qquad (\lambda, \mu \in \mathbf{R}).$$

Furthermore, from (8), we see that

$$\varphi_*[X, Y] = [\varphi_* X, \varphi_* Y]$$

holds for arbitrary $X, Y \in \mathfrak{X}(M)$. That is, a diffeomorphism φ of M defines an automorphism φ_* of the Lie algebra $\mathfrak{X}(M)$.

Remark. A diffeomorphism φ of M defines an automorphism φ^* of $C^\infty(M)$, but if M is paracompact (§14), conversely it is known that every automorphism of $C^\infty(M)$ is defined by a diffeomorphism of M. Similarly, one can conjecture that every automorphism of $\mathfrak{X}(M)$ is of the form φ_*, but no proof or counter example seems to be known.

§12. Vector Fields and One-Parameter Transformation Groups

If φ is a differentiable curve defined on an open interval (a, b), and if the tangent vector $\varphi'(t)$ to φ at $\varphi(t)$ is equal to the value of a vector field X at $\varphi(t)$, i.e., if $\varphi'(t) = X_{\varphi(t)}$ holds for all $t \in (a, b)$, then φ is called an *integral curve* of X (Fig. 2.7). Take a local coordinate system (x^i) and set $\varphi^i(t) = x^i(\varphi(t))$. Denote the components of X by ξ^i. Then $\varphi'(t) = X_{\varphi(t)}$ means $d\varphi^i(t)/dt = \xi^i(\varphi^1(t), \ldots, \varphi^n(t))$ $(i = 1, \ldots, n)$. Namely, φ^i is a solution of a system of differential equations

$$\frac{du^i}{dt} = \xi^i(u^1, \ldots, u^n) \qquad (i = 1, \ldots, n). \tag{1}$$

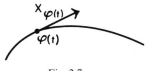

Fig. 2.7

LEMMA. *Let φ and ψ be integral curves of X defined on open intervals I and J, respectively, containing 0. If $\varphi(0) = \psi(0)$, then $\varphi = \psi$ at each point of $I \cap J$.*

Proof. $I \cap J$ is an open interval containing 0. Let $E = \{t \in I \cap J | \varphi(t) = \psi(t)\}$. Then, since $0 \in E$, E is not empty. By the continuity of φ and ψ, it is clear that E is a closed subset of $I \cap J$. Now let us show that E is an open set. For a $t_0 \in E$, set $p_0 = \varphi(t_0) = \psi(t_0)$. Let $(x^1, \ldots,$

x^n) be a local coordinate system around p_0, and set $\varphi^i(t) = x^i(\varphi(t))$, $\psi^i(t) = x^i(\psi(t))$. φ^i, ψ^i are both defined in a neighborhood of t_0, and both are solutions of the system of differential equations (1) satisfying the same initial condition: $\varphi^i(t_0) = x^i(p_0)$, $\psi^i(t_0) = x^i(p_0)$. Hence, by the uniqueness of the solution of a system of ordinary differential equations, φ^i and ψ^i coincide on a sufficiently small neighborhood of t_0. This means that a sufficiently small neighborhood of t_0 is contained in E. Hence E is an open set. Since $I \cap J$ is connected and E is a nonempty open and closed subset of $I \cap J$, we conclude that $E = I \cap J$. Thus, at each point of $I \cap J$, we have $\varphi = \psi$.

DEFINITION 1. Let φ_t be diffeomorphisms of M, defined for each real value t, and satisfying the following conditions:

(1) $\varphi_s \circ \varphi_t = \varphi_{s+t}$ $(s, t \in \mathbf{R})$
(2) $(t, p) \to \varphi_t(p)$ gives a differentiable map from $\mathbf{R} \times M$ to M.

Then the family $\{\varphi_t | t \in \mathbf{R}\}$ of diffeomorphisms φ_t is called a *one-parameter group of transformations of the manifold M*.

In what follows, we shall often call a diffeomorphism of M simply a transformation of M.

Problem 1. Let $\{\varphi_t\}$ be a one-parameter group of transformations of M. Show that
(a) φ_0 is the identity map of M and $\varphi_{-t} = \varphi_t^{-1}$,
(b) $\{\varphi_t\}$ form an abelian group.

$\{\varphi_t(p) | t \in \mathbf{R}\}$ is called the *orbit* of the one-parameter group of transformations through the point p of M. Define a curve θ_p by $\theta_p(t) = \varphi_t(p)$. By condition (2), θ_p is differentiable. The image of θ_p is the orbit of p.

Now, for a one-parameter group of transformations $\{\varphi_t\}$, we can define a vector field X on M by

$$X_p f = \left[\frac{df(\varphi_t(p))}{dt} \right]_{t=0}. \tag{2}$$

This vector field X is called the *infinitesimal transformation* of the one-parameter group of transformations $\{\varphi_t\}$. θ_p is an integral curve of X such that $\theta_p(0) = p$.

Problem 2. Show that X defined by (2) is a C^∞ vector field on M.

A vector field X, which is an infinitesimal transformation of a one-parameter group of transformations of M, is said to be *complete*.

If X is complete, then there is only one one-parameter group of transformations which has X as its infinitesimal transformation. In fact,
suppose that the infinitesimal transformations of $\{\varphi_t\}$ and $\{\psi_t\}$ are both
equal to X. Let $\theta_p(t) = \varphi_t(p)$ and $\eta_p(t) = \psi_t(p)$. Then θ_p and η_p are
both integral curves of X defined on $(-\infty, +\infty)$, and $\theta_p(0) = \eta_p(0) = p$.
Hence, by the lemma, we have $\theta_p(t) = \eta_p(t)$ for all t. Hence we have
$\varphi_t(p) = \psi_t(p)$ for all $t \in \mathbf{R}$ and all $p \in M$. Hence $\{\varphi_t\} = \{\psi_t\}$. Thus the
correspondence between the set of all one-parameter groups of trans-
formations on M and the set of all complete vector fields on M is $1:1$,
and we denote the one-parameter group of transformations, whose
infinitesimal transformation is the complete vector field X, by the symbol

$$\varphi_t = \operatorname{Exp} tX.$$

Problem 3.   Let M be the open submanifold of $\mathbf{R}^2$ obtained by deleting $(0, 0)$ from
$\mathbf{R}^2$. Show that the vector field $\partial/\partial x$ on M is not complete.

Example.   Let $(x^1, \ldots, x^n)$ be the standard coordinate system on
$\mathbf{R}^n$. Let $A = (a_j^i)$ be an $n \times n$ matrix. We call a vector field on $\mathbf{R}^n$ of
the form

$$X = \sum_{i, j = 1}^{n} a_j^i x^j \frac{\partial}{\partial x^i}$$

an infinitesimal linear transformation of $\mathbf{R}^n$. The vector field X is complete and we have

$$(\operatorname{Exp} tX)(p) = \exp tA \begin{bmatrix} x^1(p) \\ \cdot \\ \cdot \\ \cdot \\ x^n(p) \end{bmatrix}. \tag{3}$$

Here, $\exp tA$ is the exponential function of the matrix A.*

Proof.   Denote the right-hand side of Eq. (3) by $\varphi_t(p)$. Since $\exp tA$
is a nonsingular matrix, φ_t is a nonsingular linear transformation of
$\mathbf{R}^n$. Hence φ_t is a diffeomorphism. Clearly $(t, p) \to \varphi_t(p)$ is differentiable.
Also, since $\exp sA \cdot \exp tA = \exp(s + t)A$, we have $\varphi_s \circ \varphi_t = \varphi_{s+t}$, so

*Cf. C. Chevalley, *Theory of Lie Groups*, Princeton Univ. Press, Princeton, New Jersey,
1946, p. 5.

$\{\varphi_t\}$ is a one-parameter group of transformations of $\mathbf{R}^n$. However,

$$\frac{d}{dt}\begin{bmatrix} x^1(\varphi_t(p)) \\ \cdot \\ \cdot \\ \cdot \\ x^n(\varphi_t(p)) \end{bmatrix} = \left(\frac{d}{dt}\exp tA\right)\begin{bmatrix} x^1(p) \\ \cdot \\ \cdot \\ \cdot \\ x^n(p) \end{bmatrix} = A\exp tA\begin{bmatrix} x^1(p) \\ \cdot \\ \cdot \\ \cdot \\ x^n(p) \end{bmatrix}.$$

Hence if Y is the infinitesimal transformation of $\{\varphi_t\}$, then the ith component $(dx^i(\varphi_t(p))/dt)_{t=0}$ of Y_p is equal to $\sum_j a_j^i x^j(p)$. Hence the infinitesimal transformation Y of $\{\varphi_t\}$ is equal to X, so we have (3).

Problem 4. Show that a vector field in $\mathbf{R}^n$ of the form $X = \sum_{i=1}^n b^i\,\partial/\partial x^i\ (b^i \in R)$ is complete, and the Exp tX is a parallel translation $x^i \to x^i + tb^i\ (i = 1, \ldots, n)$.

If X is not complete, then X does not correspond to a one-parameter group of transformations, but will correspond to a "local one-parameter group of local transformations" in a 1:1 manner.

DEFINITION 2.   Let U and V be open subsets of a manifold M. A diffeomorphism φ from U onto V is called a *local transformation* of M. U is called the domain of definition of φ and V is called the range of φ.

Let φ and ψ be two local transformations and let U and U' be their respective domains of definition. If $U'' = \{q \in U' \mid \psi(q) \in U\} = \psi^{-1}(\psi(U') \cap U)$ is not empty, then we define the product $\varphi \circ \psi$ of φ and ψ as a local transformation with its domain of definition U'' and $(\varphi\psi)(q) = \varphi(\psi(q))$ for $q \in U''$. If U'' is empty, then $\varphi \circ \psi$ is not defined (Fig. 2.8).

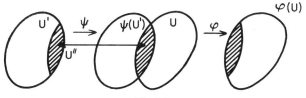

Fig. 2.8

DEFINITION 3.   A *local one-paramter group of local transformations* of M is a set $\{U_\alpha,\ \varepsilon_\alpha,\ \varphi_t^{(\alpha)}\}_{\alpha \in A}$, where U_α is an open set of M, ε_α a positive number, and $\varphi_t^{(\alpha)}$ a local transformation on M for each t, $|t| < \varepsilon_\alpha$, satisfying the following conditions:

(1) $\{U_\alpha\}_{\alpha \in A}$ is an open cover of M.
(2) The domain of definition of $\varphi_t^{(\alpha)}$ ($|t| < \varepsilon_\alpha$) contains U_α, and $\varphi_0^{(\alpha)}$ is

the identity transformation on U_α; the map $(t, p) \to \varphi_t^{(\alpha)}(p)$ is a differentiable map from $(-\varepsilon_\alpha, \varepsilon_\alpha) \times U_\alpha$ into M.

(3) If $|t|$, $|s|$, $|s + t| < \varepsilon_\alpha$, then $\varphi_t^{(\alpha)}\varphi_s^{(\alpha)}$ is defined and the domain of definition of it contains U_α and $(\varphi_t^{(\alpha)}\varphi_s^{(\alpha)})(q) = \varphi_{t+s}^{(\alpha)}(q)$ holds for $q \in U_\alpha$.

(4) If $U_\alpha \cap U_\beta \neq \varnothing$, then for each point $p \in U_\alpha \cap U_\beta$, one can choose $\varepsilon < \mathrm{Min}(\varepsilon_\alpha, \varepsilon_\beta)$, such that, for $|t| < \varepsilon$, $\varphi_t^{(\alpha)}$ and $\varphi_t^{(\beta)}$ agree on a sufficiently small neighborhood of p.

Now, let $G_l = \{U_\alpha, \varepsilon_\alpha, \varphi_t^{(\alpha)}\}_{\alpha \in A}$, $G_l' = \{V_i, \eta_i, \psi_t^{(i)}\}_{i \in I}$ be two local one-parameter groups of local transformations. Suppose the following condition is satisfied: if $U_\alpha \cap V_i \neq \varnothing$, then for each point $p \in U_\alpha \cap V_i$, there is a $\delta > 0$, $\delta < \mathrm{Min}(\varepsilon_\alpha, \eta_i)$, such that, for $|t| < \delta$, $\varphi_t^{(\alpha)}$ and $\psi_t^{(i)}$ agree on a sufficiently small neighborhood of p. Then we say that G_l and G_l' are *equivalent* and write $G_l \sim G_l'$.

Problem 4. Show that the equivalence defined above satisfies the equivalence relations.

To G_l we can associate a vector field X on M as follows. For $p \in M$, and f a C^∞ function in a neighborhood of p, we define the value X_p of the vector field X at p by

$$X_p f = \left(\frac{df(\varphi_t^{(\alpha)}(p))}{dt}\right)_{t=0} \qquad (p \in U_\alpha).$$

By the condition (4) of G_l, the definition of X_p does not depend on the choice of U_α for which $p \in U_\alpha$. If $G_l \sim G_l'$, then the vector fields corresponding to G_l and G_l' coincide, because of the definition of the equivalence. From condition (3) of G_l, one sees that for $p \in U_\alpha$, $\theta_p^{(\alpha)} : t \to \varphi_t^{(\alpha)}(p)$ is an integral curve of X, and $\theta_p^{(\alpha)}(0) = p$. The vector field X is called the *infinitesimal transformation of the local one-parameter group G_l of local transformations.*

THEOREM 1. *For any vector field X on a manifold M, there exists a local one-parameter group of local transformations, whose infinitesimal transformation is X. If two local one-parameter groups of local transformations have the same vector field as their infinitesimal tranformations, then they are equivalent.*

Proof. (1) The uniqueness up to equivalence. Suppose $G_l = \{U_\alpha, \varepsilon_\alpha, \varphi_t^{(\alpha)}\}_{\alpha \in A}$ and $G_l' = \{V_i, \eta_i, \psi_t^{(i)}\}_{i \in I}$ have, as their infinitesimal transformations, the same vector field X. We shall show that $G_l \sim G_l'$. If $U_\alpha \cap V_i \neq \varnothing$, then we set $\varepsilon_{\alpha i} = \mathrm{Min}(\varepsilon_\alpha, \eta_i)$. For $p \in U_\alpha \cap V_i$, $\theta_p^{(\alpha)} : t \to \varphi_t^{(\alpha)}(p)$ and $\eta_p^{(i)} : t \to \psi_t^{(i)}(p)$ are integral curves of X, defined for $|t| < \varepsilon_{\alpha i}$, and $\theta_p^{(\alpha)}(0) = \eta_p^{(i)}(0) = p$. Hence, by the lemma, for $|t| < \varepsilon_{\alpha i}$, we have $\theta_p^{(\alpha)}(t) = \eta_p^{(i)}(t)$, i.e., $\varphi_t^{(\alpha)}(p) = \psi_t^{(i)}(p)$. Hence $\varphi_t^{(\alpha)}$ and $\psi_t^{(i)}$ coincide on $U_\alpha \cap V_i$ for $|t| < \varepsilon_{\alpha i}$, and thus G_l and $G_l{}'$ are equivalent.

(2) The existence of G_l. Let $(x^1, \ldots, x^n)$ be a local coordinate system on a neighborhood V of p, such that $x^i(p) = 0$ $(i = 1, \ldots, n)$. Let $X = \sum_{i=1}^{n} \xi^i \, \partial/\partial x^i$ on V, and consider the system of differential equations in $\varphi^1, \ldots, \varphi^n$

$$\frac{d\varphi^i}{dt} = \xi^i(\varphi^1, \ldots, \varphi^n) \qquad (i = 1, \ldots, n). \tag{4}$$

For a sufficiently small $\delta_1 > 0$, set $U_1 = \{q \in V \mid |x^i(q)| < \delta_1, i = 1, \ldots, n\}$. By the existence theorem of solutions of a system of differential equations, for sufficiently small ε_1, $\varepsilon_1 > 0$, there is a unique set of n C^∞ functions $\varphi^i(t; q)$ $(i = 1, \ldots, n)$, such that each $\varphi^i(t; q)$ is defined on $(-\varepsilon_1, \varepsilon_1) \times U_1$, as a function of t, $\varphi^i(t; q)$ satisfies the initial condition

$$\varphi^i(0; q) = x^i(q) \qquad (i = 1, \ldots, n) \qquad q \in U_1, \tag{5}$$

and $\varphi^i(t; q)$ form a solution of (4). For $|t| < \varepsilon_1$, $q \in U_1$, let $\varphi_t^{(p)}(q)$ be the point of V whose coordinates are $(\varphi^1(t; q), \ldots, \varphi^n(t; q))$. By (5), $\varphi_0^{(p)}$ is the identity transformation on U_1. Hence, by taking $\varepsilon_p > 0$ and $\delta_p > 0$ sufficiently small, and setting $U_p = \{q \in V \mid |x^i(q)| < \delta_p\}$, we can have $\varphi_t^{(p)}(U_p) \subset U_1$ for all $|t| < \varepsilon_p$. Hence, if $|s|$, $|t|$, $|s + t| < \varepsilon_p$, then $\varphi^i(t + s; q)$, $\varphi^i(t; \varphi_s^{(p)}(q))$ $(q \in U_p)$ are defined. Now, if we set $\varphi^i(t + s; q) = \psi^i(t)$, it is easy to see that $\{\psi^i(t)\}$ is a solution of (4), and that $\psi^i(0) = \varphi^i(s; q)$. Hence, by the uniqueness of solutions of (4), we have $\varphi^i(t + s; q) = \varphi^i(t; \varphi_s^{(p)}(q))$, so $\varphi_{t+s}^{(p)}(q) = \varphi_t^{(p)}(\varphi_s^{(p)}(q))$. holds. If we let $s = -t$, then $\varphi_0^{(p)}(q) = q = \varphi_t^{(p)}(\varphi_{-t}^{(p)}(q))$. Similarly, we have $q = \varphi_{-t}^{(p)}(\varphi_t^{(p)}(q))$ $(q \in U_p)$. Thus $\varphi_t^{(p)}$ is a diffeomorphism from U_p onto $\varphi_t^{(p)}(U_p)$, i.e., $\varphi_t^{(p)}$ is a local transformation, and its domain of definition contains U_p. Hence $G_l = \{U_p, \varepsilon_p, \varphi_t^{(p)}\}_{p \in M}$ is a local one-parameter group of local transformations, and the infinitesimal transformation of G_l is the given vector field X.

By Theorem 1, we see that there is a one-to-one correspondence between the vector fields on M and the equivalence classes of local one-parameter groups of local transformations. A local one-parameter group G_l of local transformations corresponding to a vector field X is called a local one-parameter group of local transformations generated by X. G_l is determined uniquely by X up to equivalence.

THEOREM 2. *A vector field on M is complete if and only if among the equivalent local one-parameter group of local transformations generated by X, there is a $G_l = \{U_\alpha, \varepsilon_\alpha, \varphi_t^{(\alpha)}\}_{\alpha \in A}$ such that $\inf_\alpha \varepsilon_\alpha > 0$.*

Proof. Suppose G_l satisfying the condition in the theorem exists. Set $\inf_\alpha \varepsilon_\alpha = \varepsilon$, and for $|t| < \varepsilon$, define a map $\tilde{\varphi}_t$ from M to M by $\tilde{\varphi}_t(p) = \varphi_t^{(\alpha)}(p)$, for $p \in U_\alpha$. $\tilde{\varphi}_t$ is well defined because, if $p \in U_\alpha \cap U_\beta$, then $t \to \varphi_t^{(\alpha)}(p)$, $t \to \varphi_t^{(\beta)}(p)$ ($|t| < \varepsilon$) are both integral curves of X, starting from p at $t = 0$, so for $|t| < \varepsilon$, we have $\varphi_t^{(\alpha)}(p) = \varphi_t^{(\beta)}(p)$. Since $\tilde{\varphi}_t$ agrees with $\varphi_t^{(\alpha)}$ on U_α, it is differentiable. Now, suppose $p \in U_\beta$ and $\varphi_s^{(\beta)}(p) \in U_\alpha$ ($|s| < \varepsilon$). If $|t| < \varepsilon$, $|t + s| < \varepsilon$, then $\varphi_t^{(\alpha)}(\varphi_s^{(\beta)}(p))$ and $\varphi_{t+s}^{(\beta)}(p)$ are both defined, so by condition (3) and the lemma, $\varphi_t^{(\alpha)}(\varphi_s^{(\beta)}(p)) = \varphi_{t+s}^{(\beta)}(p)$ as long as $|t| < \varepsilon$ and $|t + s| < \varepsilon$ hold. In particular, we get $\varphi_{-s}^{(\alpha)}(\varphi_s^{(\beta)}(p)) = \varphi_0^{(\beta)}(p) = p$. Hence, by the definition of $\tilde{\varphi}_s$, we have $\tilde{\varphi}_{-s}(\tilde{\varphi}_s(p)) = p$ for any p and any s, $|s| < \varepsilon$. Hence $\tilde{\varphi}_s$ is a diffeomorphism of M, and we have $\tilde{\varphi}_s^{-1} = \tilde{\varphi}_{-s}$. Moreover, we have shown that if $|t|, |s|, |t + s| < \varepsilon$, then we have $\tilde{\varphi}_{t+s} = \tilde{\varphi}_t \tilde{\varphi}_s$. Now, we shall define a diffeomorphism $\tilde{\varphi}_t$ of M for an arbitrary real number t. For this, choose a positive integer m such that $|t/m| < \varepsilon$, and set $\tilde{\varphi}_t = (\tilde{\varphi}_{t/m})^m$. For this definition to make sense, if l is another positive integer such that $|t/l| < \varepsilon$, then we must be able to show that $(\varphi_{t/m})^m = (\varphi_{t/l})^l$. For this, it suffices to show that $\tilde{\varphi}_s = (\tilde{\varphi}_{s_1})^l$, $\tilde{\varphi}_{s'} = (\tilde{\varphi}_{s_1})^m$, where we have set $t/m = s$, $t/l = s'$, and $t/ml = s_1$.

Since $s = ls_1$, for $1 \leq k \leq l$, we have $|ks_1| < \varepsilon$, so that $\tilde{\varphi}_{ks_1} = \tilde{\varphi}_{s_1} \tilde{\varphi}_{(k-1)s_1}$, and by induction we have $\tilde{\varphi}_{ks_1} = (\tilde{\varphi}_{s_1})^k$. Hence we have $\tilde{\varphi}_s = (\tilde{\varphi}_{s_1})^l$, and similarly $\tilde{\varphi}_{s'} = (\tilde{\varphi}_{s_1})^m$. Thus we have justified the definition of $\tilde{\varphi}_t$ above. Again, from its definition, $\tilde{\varphi}_t$ is clearly a diffeomorphism of M. That $\{\tilde{\varphi}_t | t \in \mathbf{R}\}$ is a one-parameter group of transformations of M, and that its infinitesimal transformation is X, can be seen easily from the construction of $\tilde{\varphi}_t$.

(2) Conversely, let X be a complete vector field. Set $\varphi_t = \mathrm{Exp}\, tX$. Take an arbitrary $\varepsilon > 0$. Then $\{M, \varepsilon, \varphi_t\}$ is a one-parameter group of local transformations generated by X.

COROLLARY. *If M is compact, then every vector field on M is complete.*

In fact, let $G_l = \{U_\alpha, \varepsilon_\alpha, \varphi_t^{(\alpha)}\}_{\alpha \in A}$ be generated by X. Since M is compact, we can choose a finite number of $\alpha_1, \ldots, \alpha_k \in A$, so that $\{U_{\alpha_i}\}_{i=1,\ldots,k}$ is an open cover of M. Set $G_{l'} = \{U_{\alpha_i}, \varepsilon_{\alpha_i}, \varphi_t^{(\alpha_i)}\}_{i=1,\ldots,k}$. Then G_l and $G_{l'}$ are equivalent, and $G_{l'}$ satisfies the condition of Theorem 2. Hence X is complete.

Let θ be a transformation of M (i.e., a diffeomorphism of M). Let $G_l = \{U_\alpha, \varepsilon_\alpha, \varphi_t^{(\alpha)}\}_{\alpha \in A}$ be a local one-parameter group of local transformations. Then

$$G_l^\theta = \{\theta(U_\alpha),\ \varepsilon_\alpha,\ \theta\varphi_t^{(\alpha)}\theta^{-1}\}_{\alpha \in A}$$

is also a local one-parameter group of local transformations.

THEOREM 3.   *Let θ be a transformation of M and let X be a vector field on M. If G_l is a local one-parameter group of local transformations generated by X, then G_l^θ is a local one-parameter group of local transformations generated by $\theta_* X$.*

COROLLARY.   *If X is complete, then so is $\theta_* X$, and we have*

$$\text{Exp } t\theta_* X = \theta(\text{Exp } tX)\theta^{-1}.$$

Proof of Theorem 3.   It suffices to show that $\theta_* X$ is the infinitesimal transformation of G_l^θ. By the definition of $\theta_* X$, we have $(\theta_* X)_p = (\theta_*)_{\theta^{-1}(p)} X_{\theta^{-1}(p)}$. Hence, if f is a function in a neighborhood of p, then

$$(\theta_* X)_p f = X_{\theta^{-1}(p)}(f \circ \theta).$$

Suppose $\theta^{-1}(p) \in U_\alpha$. Then we have

$$X_{\theta^{-1}(p)}(f \circ \theta) = \left[\frac{df(\theta(\varphi_t^{(\alpha)}(\theta^{-1}(p))))}{dt}\right]_{t=0}.$$

Hence, if $p \in \theta(U_\alpha)$, then

$$(\theta_* X)_p f = \left[\frac{df((\theta\varphi_t^{(\alpha)}\theta^{-1})(p))}{dt}\right]_{t=0},$$

so $\theta_* X$ is the infinitesimal transformation of G_l^θ.

Using G_l, we can define the commutator product $[X, Y]$ of vector fields in a geometrical manner. That is, the following theorem holds.

THEOREM 4.   *Let X and Y be vector fields on M, and let $G_l = \{U_\alpha, \varepsilon_\alpha,\ \varphi_t^{(\alpha)}\}_{\alpha \in A}$ be a local one-parameter group of local transformations generated by X. Then we have*

$$[X, Y]_p = \lim_{t \to 0} \frac{1}{t}\{Y_p - (\varphi_t^{(\alpha)}{}_*)_q Y_q\} \qquad (p \in U_\alpha), \qquad (6)$$

where $q = \varphi_{-t}^{(\alpha)}(p)$ and the limit on the right side is taken with respect to the natural topology of the tangent space $T_p(M)$. We abbreviate (6) as

$$[X, Y] = \lim_{t \to 0} \frac{1}{t}\{Y - (\varphi_t)_* Y\}.$$

Proof.   Take a neighborhood U of p sufficiently small so that it is contained in U_α, and let $(x^1, \ldots, x^n)$ be a local coordinate system on U such that $x^i(p) = 0$ $(i = 1, \ldots, n)$. Omitting the superscript α, we shall write φ_t instead of $\varphi_t^{(\alpha)}$. We set $X = \sum_i \xi^i \partial/\partial x^i$ and $Y = \sum_i \eta^i \partial/\partial x^i$. Since $t \to \varphi_t(x)$ $(x \in U)$ is an integral curve of X, setting $x^i(\varphi_t(x)) = \varphi^i(t; x^1, \ldots, x^n)$, we have

$$\frac{d\varphi^i(t; x^1, \ldots, x^n)}{dt} = \xi^i(\varphi^1(t; x), \ldots, \varphi^n(t; x)) \qquad (i = 1, \ldots, n). \quad (7)$$

Let U_1 be a neighborhood of p and $\varepsilon > 0$ such that $U_1 \subset U_\alpha$, $\varphi_t(q)$ is defined for $-\varepsilon < t < \varepsilon$ and $q \in U_1$. Define $g^i(t, q) = g_t^i(q)$ by

$$g^i(t, q) = \frac{x^i \circ \varphi_t - x^i}{t}(q) \qquad \text{for} \quad t \neq 0$$

$$= \xi^i(q) \qquad \text{for} \quad t = 0.$$

Using (7), we see that $g^i(t, q)$ is a C^∞ function on $(-\varepsilon, \varepsilon) \times U_1$. We have $x^i \circ \varphi_t = x^i + tg^i(t, q)$. Hence, for $-\varepsilon < t < \varepsilon$ and $q \in U_1$, we have

$$Y_q(x^i \circ \varphi_t) = \sum_j \eta^j(q) \left(\delta^i_j + t \frac{\partial g_t^i}{\partial x^j}(q) \right)$$

and

$$\frac{1}{t}(Y_p x^i - Y_q(x^i \circ \varphi_t)) = \frac{\eta^i(q) - \eta^i(p)}{-t} - \sum_j \eta^j(q) \frac{\partial g_t^i}{\partial x^j}(q).$$

If $q = \varphi_{-t}(p)$, then

$$\lim_{t \to 0}(Y_p x^i - Y_q(x^i \circ \varphi_t)) = \sum_k \frac{\partial \eta^i}{\partial x^k}(p)\xi^k(p) - \sum_j \eta^j(p)\frac{\partial \xi^i}{\partial x^j}(p).$$

By (5) of §11, this is the ith component of $[X, Y]_p$, so that (6) is proved.

Now, let X be a complete vector field, and let, for any $f \in C^\infty(M)$, $((\text{Exp } tX)^*f)(p) = F(t, p)$. If we fix p, then $F(t, p)$ is a C^∞ function of t, so by Taylor's theorem, we have

$$F(t, p) = F(0, p) + t\frac{dF}{dt}(0, p) + \ldots + \frac{t^n}{n!}\frac{d^nF}{dt^n}(0, p)$$

$$+ \frac{t^{n+1}}{(n+1)!}\frac{d^{n+1}F}{dt^{n+1}}(\theta t, p) \qquad (0 < \theta < 1),$$

for any positive integer n. Here, we have

$$(D_X^m f)(p) = \frac{d^m F}{dt^m}(0, p).$$

We verify this equality by induction. For $m = 1$, this is nothing but the definition of $D_X f = Xf$. Suppose the equality is true up to and including m. Since $D_X^{m+1} f = D_X(D_X^m f)$, we have

$$(D_X^{m+1} f)(p) = X_p\left(\frac{d^m F}{dt^m}(0, p)\right).$$

To carry out the calculation further, we need the relation

$$\frac{d^m F}{dt^m}(0, (\text{Exp } sX)(p)) = \frac{d^m F}{dt^m}(s, p),$$

which can be proved easily by induction on m, the case for $m = 1$ being proved by using $F(t, (\text{Exp } sX)(p)) = F(s + t, p)$, i.e., from the fact that we have a one-parameter group of transformations. Continuing now on the calculation of $(D_X^{m+1} f)(p)$, we have

$$X_p\left(\frac{d^m F}{dt^m}(0, p)\right) = \lim_{s \to 0} \frac{1}{s}\left[\frac{d^m F}{dt^m}(0, (\text{Exp } sX)(p)) - \frac{d^m F}{dt^m}(0, p)\right]$$

$$= \lim_{s \to 0} \frac{1}{s}\left[\frac{d^m F}{dt^m}(s, p) - \frac{d^m F}{dt^m}(0, p)\right]$$

$$= \frac{d^{m+1} F}{dt^{m+1}}(0, p),$$

proving the desired result. Hence for an arbitrary integer n, we have

$$((\text{Exp } tX)^* f)(p) = \sum_{m=0}^{n} \frac{t^m}{m!}(D_X^m f)(p) + O(t^n).*$$

This is called *Taylor's formula*.

§13. The Infinitesimal Motion of a Riemannian Manifold

Let M be a Riemannian manifold and let g be its Riemannian metric. For vector fields X and Y on M, define a C^∞ function on M by $p \to g_p(X_p, Y_p) \in \mathbf{R}$. This function is denoted by $g(X, Y)$. Let $(x^1, \ldots, x^n)$

---

*$O(t^n)$ is the symbol for a quantity such that $O(t^n)/t^n \to 0$ as $t \to 0$.

be a local coordinate system on an open set U, and let g_{ij}, ξ^i, η^i be the components of g, X, Y, respectively, with respect to (x^i). Then $g(X, Y)$ is equal to

$$\sum_{i,j}^{n} g_{ij}\xi^i\eta^j$$

on U.

Let X be a vector field on M. If, for arbitrary vector fields Y, Z on M,

$$X(g(Y, Z)) = g([X, Y], Z) + g(Y, [X, Z]) \tag{1}$$

holds, then X is called an *infinitesimal motion* of the Riemannian manifold M or a *Killing vector field*.

Let $(x^1, \ldots, x^n)$ be a local coordinate system on U, and let ξ^i be the components of the infinitesimal motion X with respect to (x^i). Setting $Y = \partial/\partial x^j$, $Z = \partial/\partial x^k$ in (1), we obtain a system of differential equations for ξ^i:

$$\sum_{i=1}^{n}\left(g_{ki}\frac{\partial\xi^i}{\partial x^j} + g_{ji}\frac{\partial\xi^i}{\partial x^k} + \frac{\partial g_{jk}}{\partial x^i}\xi^i\right) = 0 \qquad (j, k = 1, \ldots, n). \tag{2}$$

This is called *Killing's differential equations*. X is an infinitesimal motion if and only if the set of components ξ^i of X is a solution of Killing's differential equations.

The reason why a vector field X satisfying (1) is called an infinitesimal motion is given by the following theorem.

THEOREM. *Let X be a complete vector field on a Riemannian manifold M. The vector field X is an infinitesimal motion if and only if $\operatorname{Exp} tX$ is a motion for each $t \in \mathbf{R}$.*

First we shall prove the following lemma. This lemma will be used often later.

LEMMA. *For any given tangent vector v at a point p on a manifold M, there is a C^∞ vector field X on M such that $X_p = v$.*

Proof. Taking a local coordinate system, it is easy to see that on a sufficiently small neighborhood U of p, there is a vector field Y on U, such that $Y_p = v$. Choose a neighborhood V of p sufficiently small so that $\overline{V} \subset U$ and so that we can apply Lemma 1 of §10 to get a C^∞ function f on M such that $f(p) = 1$ and $f(q) = 0$ for $q \notin U$. We define a vector field X on M by

$$X_q = f(q)Y_q, \qquad q \in U.$$
$$= 0 \qquad q \notin U.$$

This X will do the job.

Proof of the theorem. By the lemma, the map $\mathfrak{X}(M) \to T_p(M)$ given by $Y \to Y_p$ is an onto map. Hence, setting $\varphi_t = \mathrm{Exp}\, tX$, the condition for φ_t to be a motion is that

$$g_q(Y_q, Z_q) = g_p((\varphi_{t*})_q Y_q, (\varphi_{t*})_q Z_q) \quad (q = \varphi_t^{-1}(p))$$

holds for an arbitrary point p of M and arbitrary vector fields Y and Z. However, since $(\varphi_{t*})_q Y_q = (\varphi_{t*}Y)_p$ and $(\varphi_{t*})_q Z_q = (\varphi_{t*}Z)_p$, the equality above can be written as $g(Y, Z)(\varphi_t^{-1}(p)) = g(\varphi_{t*}Y, \varphi_{t*}Z)(p)$, which means

$$\varphi_{-t}{}^*(g(Y, Z)) - g(\varphi_{t*}Y, \varphi_{t*}Z) = 0. \tag{3}$$

Let us set $f(t: Y, Z) = -\varphi_{-t}{}^*(g(Y, Z)) + g(\varphi_{t*}Y, \varphi_{t*}Z)$, and compute $(df/dt)_{t=0}$. Differentiating the first term in $f(t; Y, Z)$ with respect to t and setting $t = 0$, we get $X(g(Y, Z))$. For the second term, we observe that

$$\lim_{t \to 0} \frac{1}{t}[g(\varphi_{t*}Y, \varphi_{t*}Z) - g(Y, Z)]$$
$$= \lim_{t \to 0} \frac{1}{t}[g(\varphi_{t*}Y - Y, \varphi_{t*}Z) + g(Y, \varphi_{t*}Z - Z)]$$
$$= -g([X, Y], Z) - g(Y, [X, Z]).$$

Hence we have

$$X(g(Y, Z)) - g([X, Y], Z) - g(Y, [X, Z]) \equiv \left[\frac{d}{dt}f(t; Y, Z)\right]_{t=0}. \tag{4}$$

If φ_t is a motion, then, by (3), $f(t; Y, Z) = 0$, so the right member of (4) is 0, and hence X is an infinitesimal motion. Conversely, if X is an infinitesimal motion, then, by (4), for arbitrary Y and Z, we have $[df(t; Y, Z)/dt]_{t=0} = 0$. By computation we can check that

$$f(s + t; Y, Z) = \varphi_{-s}^*(f(t; Y, Z)) = f(s; \varphi_{t*}Y, \varphi_{t*}Z). \tag{5}$$

If we differentiate this equality with respect to s and set $s = 0$, then, since $[df(s; \varphi_{t*}Y, \varphi_{t*}Z)/ds]_{s=0} = 0$, we have

$$\frac{df(t; Y, Z)}{dt} = -Xf(t; Y, Z).$$

On the other hand we have

$$X(f(t;Y, Z)) = -X(\varphi^*_{-t}(g(Y, Z)) + X(g(\varphi_{t*}Y, \varphi_{t*}Z)) \qquad (6)$$

and, since X is the infinitesimal transformation of φ_t, we have

$$X(\varphi^*_{-t}(g(Y, Z)) = \varphi^*_{-t}(X(g(Y, Z))), \qquad (7)$$

as we see easily from the definition of differentiation by X. On the other hand, as X is an infinitesimal motion, we have

$$X(g(\varphi_{t*}Y, \varphi_{t*}Z)) = g([X, \varphi_{t*}Y], \varphi_{t*}Z) + g(\varphi_{t*}Y, [X, \varphi_{t*}Z])$$

and we also have $\varphi_{t*}X = X$ by Theorem 3 of §12. Hence, by (8) of §11, we obtain

$$\begin{aligned} X(g(\varphi_{t*}Y, \varphi_{t*}Z)) &= g(\varphi_{t*}[X, Y], \varphi_{t*}Z) + g(\varphi_{t*}, \varphi_{t*}[X, Z]) \\ &= f(t;[X, Y], Z) + f(t;Y, [X, Z]) + \varphi^*_{-t}(g[X, Y], Z)) \\ &\quad + \varphi^*_{-t}(g[Y, [X, Z]). \end{aligned} \qquad (8)$$

It follows from (6), (7), (8) and the fact that X is an infinitesimal motion that

$$X(f(t;Y, Z)) = f(t;[X, Y], Z) + f(t;Y, [X, Z]).$$

Therefore the function $f(t;Y, Z)$ satisfies, as a function of t, the differential equation

$$\frac{df(t;Y, Z)}{dt} = -f(t;[X, Y], Z) - f(t;Y, [X, Z]). \qquad (9)$$

We see easily from the definition of $f(t;Y, Z)$ that we have the following properties:

$$f(t;Y, Z) = f(t;Z, Y), \qquad f(t;Y + Y', Z) = f(t;Y, Z) + f(t;Y', Z),$$
$$f(t;hY, Z) = (\varphi^*_{-t}h)f(t;Y, Z) \quad \text{for any function } h.$$

Let p be an arbitrary point of M and let $\{x^1, \ldots, x^n\}$ be a local coordinate system in a neighborhood U of p. Let $X_i = \partial/\partial x^i$ and $X = \sum_i \xi^i X_i$ on U. Then $[X, X_i] = \sum_j h^j_i X_j$ with $h^j_i = -\partial \xi^j/\partial x^i$. Since we can restrict our consideration for f to U, we put $f(t;X_i, X_k) = f_{ik}(t;x)(x \in U)$. Then $f(t;[X, X_i], X_k) + f(t;X_i, [X, X_k]) = \sum_j h^j_i(\varphi_{-t}(x)f_{jk}(t;x) + \sum_j h^j_k(\varphi_{-t}(x))f_{ij}(t;x)$, and it follows from (9) that

$$\frac{df_{ik}(t;x)}{dt} = -\sum_j h^j_i(\varphi_{-t}(x))f_{jk}(t;x) - \sum_j h^j_k(\varphi_{-t}(x))f_{ij}(t;x) \qquad (10)$$

for each $x \in V$ and $|t| < \varepsilon$, where V is a neighborhood of p such that $\varphi_{-t}(V) \subset U$ for each t such that $|t| < \varepsilon$. This shows that, for each $x \in V$, the functions $f_{ik}(t; x)$ of t are the solutions of (10) with the initial condition $f_{ik}(0; x) = 0$. Obviously, the functions of t which are identically equal to zero form a solution of (10) with the same initial conditions. Hence, by the uniqueness theorem of the solution of (10), we see that $f_{ik}(t; x) = 0$ for each $x \in V$ and $|t| < \varepsilon$. Let $Y = \sum \eta^i X_i$ and $Z = \sum \zeta^i X_i$ on U. Then $f(t; Y, Z) = \sum_{i,k} \eta^i(\varphi_{-t}(x)) \zeta^k(\varphi_{-t}(x)) f_{ik}(t; x)$ for $|t| < \varepsilon$ on V and hence $f(t; Y, Z) = 0$ on V for $|t| < \varepsilon$ and for any Y and Z. Then using (5) repeatedly, we see that $f(t; Y, Z) = 0$ for any t. In particular, $f(t; Y, Z) = 0$ at p for any t and any Y and Z. Since p is an arbitrary point of M, we get $f(t; Y, Z) = 0$ and this proves that φ_t is an isometry for all $t \in \mathbf{R}$.

Problem 1. Prove that a vector field X in an n-dimensional Euclidean space is an infinitesimal motion if and only if it is of the form

$$X = \sum_{i=1}^{n} \left(\sum_{j=1}^{n} a_j^i x^j + b^i \right) \frac{\partial}{\partial x^i}, \qquad a_j^i = -a_i^j \qquad (i, j = 1, \ldots, n).$$

Problem 2. Lex $X = \xi \, \partial/\partial x + \eta \, \partial/\partial y$ be a vector field in the hyperbolic plane. Find a condition for X to be an infinitesimal motion.

Problem 3. If X and Y are infinitesimal motions, then show that $\lambda X + \mu Y$ ($\lambda, \mu \in R$) and $[X, Y]$ are also infinitesimal motions, and hence that the set of all infinitesimal motions of a Riemannian manifold forms a Lie algebra.

§14. Paracompact Manifolds and the Partition of Unity

DEFINITION 1. A family $\{U_\alpha\}_{\alpha \in A}$ of subsets of a topological space X is said to be *locally finite* if, for each point p of X, there is a neighborhood V_p of p, such that $V_p \cap U_\alpha \neq \varnothing$ for only a finite number of $\alpha \in A$. (If M is a manifold, a coordinate neighborhood system $\{(U_\alpha, \varphi_\alpha)\}_{\alpha \in A}$ of M is said to be locally finite if $\{U_\alpha\}_{\alpha \in A}$ is locally finite.)

DEFINITION 2. Let $\{U_\alpha\}_{\alpha \in A}$, $\{V_i\}_{i \in I}$ be two coverings of X, i.e., $X = \cup_{\alpha \in A} U_\alpha = \cup_{i \in I} V_i$. If, for each index i in I, there is an index α in A such that $V_i \subset U_\alpha$, then the covering $\{V_i\}_{i \in I}$ is called a *refinement* of the covering $\{U_\alpha\}_{\alpha \in A}$.

DEFINITION 3. If a topological space X satisfies the following two conditions, then X is called a *paracompact topological space*: (1) X is a Hausdorff space, (2) for an arbitrary open covering of X, there is a locally finite open covering which is a refinement of the former.

A manifold M is called a *paracompact manifold* if it is paracompact as a topological space.

LEMMA 1. *Let $\{U_\alpha\}_{\alpha \in A}$ be an open covering of paracompact manifold M. There is an open covering $\{V_\alpha\}_{\alpha \in A}$ of M, having the same index set A, such that $\bar{V}_\alpha \subset U_\alpha$ for each $\alpha \in A$.*

Proof. For each point p of M, choose a neighborhood V_p of p such that (1) V_p is a coordinate neighborhood of p and $\bar{V}_p$ is compact, (2) $\bar{V}_p$ is contained in some U_α. Since M is paracompact, there is a locally finite refinement $\{V_i'\}_{i \in I}$ of the open covering $\{V_p\}_{p \in M}$. For each α, set $I_\alpha = \{i \in I | \bar{V}_i' \subset U_\alpha\}$, $V_\alpha = \cup_{i \in I_\alpha} V_i'$. Then $\{V_\alpha\}_{\alpha \in A}$ is an open covering of M. If we can show

$$\bar{V}_\alpha = \bigcup_{i \in I_\alpha} \bar{V}_i', \qquad (*)$$

then, since each $\bar{V}_i'$ is contained in U_α, we have $\bar{V}_\alpha \subset U_\alpha$, and hence $\{V_\alpha\}_{\alpha \in A}$ would be the desired open covering of M. First, $\bar{V}_\alpha \supset \cup_{i \in I_\alpha} \bar{V}_i'$ is clear. Conversely, if p is an arbitrary point of $\bar{V}_\alpha$, then since $\{V_i'\}_{i \in I}$ is locally finite, we can take a neighborhood W of p sufficiently small so that the number of $i \in I_\alpha$ such that $W \cap V_i' \neq \emptyset$ is finite. Let $i_1, \ldots, i_k$ be the finite number of indices i ($\in I_\alpha$) such that $W \cap V_i' \neq \emptyset$. It suffices to prove that $p \in \cup_{a=1}^{k} \bar{V}_{i_a}$. So suppose $p \notin \cup_{a=1}^{k} \bar{V}_{i_a}$. Then there is a neighborhood W' of p, contained in W, and such that $W' \cap \cup_{a=1}^{k} \bar{V}_{i_a} = \emptyset$. If i is an element of I_α different from i_a ($a = 1, \ldots, k$), then since $W \cap V_i' = \emptyset$, we have naturally $W' \cap V_i' = \emptyset$. Hence, for an arbitrary $i \in I_\alpha$, we have $W' \cap V_i' = \emptyset$, and thus $W' \cap V_\alpha = \emptyset$, contradicting $p \in \bar{V}_\alpha$. Hence, $p \in \cup_{a=1}^{k} \bar{V}_{i_a}$, so $(*)$ holds, and the lemma is proved.

LEMMA 2. *For a compact subset K of a manifold M, and an open set U of M containing K, there is a C^∞ function f on M satisfying the following conditions: (1) $0 \leq f \leq 1$, (2) $f > 0$ on K, and (3) $f = 0$ on the complement U^c of U.*

Proof. For each point p of K, there are a C^∞ function f_p on M and a neighborhood V_p of p ($\bar{V}_p \subset U$), such that $0 \leq f_p \leq 1$, $f_p = 1$ on V_p, and $f_p = 0$ on U^c (cf. Lemma 1 of §10). Since K is compact, we can choose a finite number of points $p_1, \ldots, p_k$ of K so that $K \subset V_{p_1} \cup \ldots \cup V_{p_k}$. Set

$$f = \frac{1}{k}(f_{p_1} + \ldots + f_{p_k}).$$

This is the desired function.

For a function f on M, the closure of the subset $\{p \in M \mid f(p) \neq 0\}$ of M is called the *support* of f and is denoted by supp f.

DEFINITION 4. For a locally finite open covering $\{U_\alpha\}_{\alpha \in A}$ of a manifold M, suppose there is a family $\{f_\alpha\}_{\alpha \in A}$ of C^∞ functions on M satisfying the following conditions:

(1) $0 \leq f_\alpha \leq 1$ for each α.
(2) The support of f_α is contained in U_α for each α.
(3) $\sum_{\alpha \in A} f_\alpha(p) = 1$ for each point p in M.

Then $\{f_\alpha\}_{\alpha \in A}$ is called a *partition of unity* subordinate to the open cover $\{U_\alpha\}_{\alpha \in A}$.

Remark. Since $\{U_\alpha\}_{\alpha \in A}$ is locally finite, the number of U_α's containing p is finite, hence, except for a finite number of α, $f_\alpha(p) = 0$. Hence the sum $\sum_{\alpha \in A} f_\alpha(p)$ is a finite sum.

THEOREM 1. *Let M be a paracompact manifold, and let $\{U_\alpha\}_{\alpha \in A}$ be a locally finite open covering of M. If, for each α, $\bar{U}_\alpha$ is compact, then there is a partition of unity $\{f_\alpha\}_{\alpha \in A}$ subordinate to $\{U_\alpha\}_{\alpha \in A}$.*

Proof. By Lemma 1, we can pick open coverings $\{V_\alpha\}_{\alpha \in A}$ and $\{W_\alpha\}_{\alpha \in A}$ of M such that $\bar{W}_\alpha \subset U_\alpha$ and $\bar{V}_\alpha \subset W_\alpha$ for each α. By Lemma 2, for each α there is a C^∞ function g_α on M such that (1) $0 \leq g_\alpha \leq 1$, (2) $g_\alpha > 0$ on $\bar{V}_\alpha$, and (3) $g_\alpha = 0$ on W_α^c. The support of g_α is contained in $\bar{W}_\alpha$, and hence in U_α. Since $\{U_\alpha\}_{\alpha \in A}$ is locally finite, for each point p of M we can choose a neighborhood U of p sufficiently small so that the number of U_α's intersecting U is finite, so that g_α is 0 on U except for a finite number of α's. Hence, at each point p of M, we set $g(p) = \sum_{\alpha \in A} g_\alpha(p)$. Then g is a C^∞ function on M. Furthermore, since $\{V_\alpha\}_{\alpha \in A}$ is an open covering of M, $g_\alpha \geq 0$, and $g_\alpha > 0$ on V_α, we get $g > 0$. For each α, now set

$$f_\alpha = g_\alpha/g.$$

Then $\{f_\alpha\}_{\alpha \in A}$ is a partition of unity subordinate to $\{U_\alpha\}_{\alpha \in A}$.

Applications of the partition of unity. Example 1. Let us prove the existence of a Riemannian metric on a paracompact manifold M. Take an open covering $\{V_i\}_{i \in I}$ of M such that each V_i is a coordinate neighborhood of M with compact closure $\bar{V}_i$, and let $\{U_\alpha\}_{\alpha \in A}$ be a locally finite refinement of $\{V_i\}_{i \in I}$. Clearly U_α is a coordinate neighborhood, and $\overline{U}_\alpha$ is compact. Hence, by Theorem 1, there is a partition of unity $\{f_\alpha\}_{\alpha \in A}$ subordinate to $\{U_\alpha\}_{\alpha \in A}$. On the other hand, since U_α is a coordinate neighborhood, there is a Riemannian metric $g^{(\alpha)}$ on U_α. For example, we can pick $g^{(\alpha)}$ as follows. Let $(x^1, \ldots, x^n)$ be the local coordinate system on U_α, and let u, v be tangent vectors at a point p of U_α. Then set $g_p^{(\alpha)}(u, v) = \sum_{i=1}^n \xi^i \eta^i$, where ξ^i, η^i are the components of u, v, respectively, with respect to (x^i). For an arbitrary point p of M, and tangent vectors u, v at p, we set

$$g_p(u, v) = \sum_{\alpha \in A} f_\alpha(p) g_p^{(\alpha)}(u, v).$$

Since $f_\alpha(p) \neq 0$ for only a finite number of indices α, the right member is a finite sum, and g_p is a positive inner product on $T_p(M)$. Taking a coordinate neighborhood U of p sufficiently small, we have $U \cap U_\alpha \neq \emptyset$ for only a finite number of α's, and since the support of f_α is contained in U_α, the components of g with respect to the coordinate system of U can be written as $g_{ij} = \sum_\alpha f_\alpha g_{ij}^{(\alpha)}$ (the sum on the right side is a finite sum), so g is of class C^∞.

Example 2. Let M be a paracompact manifold and h a C^∞ function on an open set U of M. Let V be an open set of M such that $\overline{V} \subset U$. Then there is a C^∞ function g on M such that $g = h$ on V and $g = 0$ on the complement U^c of U.

Proof. For each point p of $\overline{V}$, there are a neighborhood V_p of p and a C^∞ function g_p on M, such that $\overline{V}_p \subset U$, $\overline{V}_p$ is compact, $g_p = h$ on V_p, and $g_p = 0$ on U^c (§10, Lemma 2). For each point p not in $\overline{V}$, choose a neighborhood V_p of p so that $\overline{V}_p$ is compact and $V_p \cap \overline{V} = \emptyset$, and set $g_p = 0$. Let $\{U_\alpha\}_{\alpha \in A}$ be a locally finite refinement of the open covering $\{V_p\}_{p \in M}$ of M. Since $\overline{U}_\alpha$ is compact, there is a partition of unity $\{f_\alpha\}_{\alpha \in A}$ subordinate to $\{U_\alpha\}_{\alpha \in A}$. Define a C^∞ function g_α on U_α as follows. Choose a V_p such that $U_\alpha \subset V_p$, and let g_α be the restriction of g_p to U_α. For each point p of M, let

$$g(p) = \sum_{\alpha \in A} f_\alpha(p) g_\alpha(p)$$

to define the function g. Using the properties of a partition of unity, we can easily verify that this g is the desired one.

Problem 1. Let U and V be open sets of a paracompact manifold M, and suppose that $\overline{V} \subset U$. If Y is a C^∞ vector field defined on U, then there is a C^∞ vector field X on M such that $X = Y$ on V and $X = 0$ on U^c. Prove this. (Cf. the proof of Example 2.)

Problem 2. If K is a compact subset of a paracompact manifold M, show that there is a C^∞ function on M which is equal to 1 on K.

§15. Some Remarks on the Topology of Manifolds

In order to make it easy to decide whether a given manifold is paracompact or not, we shall investigate the equivalence between paracompactness and various conditions on the topology. First we shall give various definitions on topological spaces. In what follows, X is a topological space.

(1) If, for each point p of X, there is a neighborhood U of p such that $\bar{U}$ is compact, then X is said to be *locally compact*.

(2) A family $\{O_\alpha\}_{\alpha \in A}$ of open sets of X is called a base for the open sets of X, if, for an arbitrary open set O of X and an arbitrary point p of O, there is an index α such that $p \in O_\alpha \subset O$. That is, $\{O_\alpha\}_{\alpha \in A}$ is called a base for the open sets of X, if an arbitrary open set of X can be expressed as a union of some open sets belonging to $\{O_\alpha\}_{\alpha \in A}$. If there is a base for the open sets of X consisting of a countable number of open sets of X, then we say that X *has a countable base*.

(3) If X is the union of a countable number of compact subsets, then X is said to be *σ-compact*.

(4) If, for an arbitrary open covering $\{U_\alpha\}_{\alpha \in A}$ of X, we can pick a countable subset $\{\alpha_i\}_{i=1, 2, \ldots}$ of A such that $\{U_{\alpha_i}\}_{i=1, 2, \ldots}$ is still a covering of X, then X is called a *Lindelöf space*.

LEMMA 1. *Let X be a topological manifold. The following three conditions are equivalent*: (1) *X has a countable base*, (2) *X is a Lindelöf space*, (3) *X is σ-compact.*

Proof. Suppose X has a countable base $\{O_i\}_{i \in N}$.* Let $\{U_\alpha\}_{\alpha \in A}$ be an arbitrary open covering of X. Let N' be the set of natural numbers i such that, for each i, there is at least one $\alpha \in A$ such that $O_i \subset U_\alpha$. Then $\{O_i\}_{i \in N'}$ is an open covering of X. Hence, if we pick one $\alpha \in A$ for each $i \in N'$ so that $O_i \subset U_\alpha$, and call it α_i, then $\{U_{\alpha_i}\}_{i \in N'}$ is an open covering of X. Hence (1) → (2). Suppose now that X is a Lindelöf space. For each point p of X, choose a neighborhood U_p of p such that $\bar{U}_p$ is compact. Since X is a Lindelöf space, we can choose a countable number of points $\{p_i\}_{i \in N}$ so that $\{U_{p_i}\}_{i \in N}$ is a cover of X. Then X is the union of a countable number of compact sets $\{\bar{U}_{p_i}\}_{i \in N}$, so that X is σ-compact, and (2) → (3) is proved. Now suppose X is σ-compact, and let X be the union of a countable number of compact sets $\{K_i\}_{i \in N}$. For each point of K_i, choose a neighborhood homeomorphic to a cube

*We shall denote the set $\{1, 2, \ldots\}$ of natural numbers by N.

in $\mathbf{R}^n$. Since K_i is compact, K_i is covered by a finite number of such cubes, and since X is the union of the K_i's, X is covered by a countable number of open sets homeomorphic to a cube in $\mathbf{R}^n$. However, a cube in $\mathbf{R}^n$ has a countable base as a topological space. Hence X also has a countable base, and $(3) \rightarrow (1)$ is proved.

LEMMA 2.   *If X is locally compact and σ-compact, then X is paracompact.*

Proof.   (1) Let K be an arbitrary compact subset of X. Then there is an open set V, containing K, such that $\bar{V}$ is compact. In fact, for each point p of K, choose a neighborhood W_p such that $\bar{W}_p$ is compact. Since K is compact, we can pick a finite number of points $p_1, \ldots, p_k$ such that $V = \cup_{i=1}^k W_{p_i}$ contains K. This is the desired V.

Since X is σ-compact, it is the union of a countable number of compact sets $\{K_n\}_{n \in N}$. Here we can assume that $K_1 \subset K_2 \ldots$ (In fact, we have only to rename $\cup_{i=1}^n K_i$ as K_n.)

(2) Define a sequence $V_1 \subset V_2 \ldots$ of open sets of X as follows. First, by (1), take an open set V_1, containing K_1, such that $\bar{V}_1$ is compact. Since $K_2 \cup \bar{V}_1$ is compact, pick an open set V_2, containing $K_2 \cup \bar{V}_1$, such that $\bar{V}_2$ is compact. We pick open sets V_{n+1} inductively by letting it contain $K_{n+1} \cup \bar{V}_n$, and requiring that $\bar{V}_{n+1}$ be compact.

From the construction of $\{V_n\}$, $\bar{V}_n \subset V_{n+1}$, $K_n \subset V_n$, and hence $\cup_{n=1}^{\infty} V_n = X$.

(3) Set $O_1 = V_2$, $O_2 = V_3$, and $O_n = \bar{V}_{n+1} - \bar{V}_{n-2}$ for $n \geq 3$. Furthermore, let $F_1 = \bar{V}_1$ and $F_n = \bar{V}_n - V_{n-1}$ for $n \geq 2$. Then O_n is an open set, F_n is compact, and $F_n \subset O_n$. Moreover, we have $\cup F_n = X$, and $O_n \cap O_m = \varnothing$ for $|n - m| \geq 3$.

(4) Now, let $\{U_\alpha\}_{\alpha \in A}$ be an arbitrary open cover of X. Let us construct a locally finite refinement of $\{U_\alpha\}_{\alpha \in A}$. Let us fix n for a moment, and for each point p of F_n, pick a U_α containing p. Pick a neighborhood $U_p^{(n)}$ of p such that $U_p^{(n)} \subset U_\alpha \cap O_n$. Since F_n is compact, we can choose a finite number of points p_i ($i = 1, \ldots, a_n$) of F_n so that $\{U_{p_i}^{(n)}\}_{i=1, \ldots, a}$ covers F_n. For each n, construct this $\{U_{p_i}^{(n)}\}$. Then $\mathfrak{U} = \{U_{p_i}^{(n)}, i = 1, \ldots, a_n; n = 1, 2, \ldots\}$ is an open cover of X. Since each $U_{p_i}^{(n)}$ is contained in some U_α, $\mathfrak{U}$ is a refinement of $\{U_\alpha\}$. Moreover, if $|n - m| \geq 3$, then $O_n \cap O_m = \varnothing$, and since $U_p^{(n)} \subset O_n$ and $U_q^{(m)} \subset O_m$, we have $U_{p_i}^{(n)} \cap U_{q_j}^{(m)} = \varnothing$. Hence there are only a finite number of elements of $\mathfrak{U}$ that intersect $U_{p_i}^{(n)}$. Hence clearly $\mathfrak{U}$ is locally finite. Thus X is paracompact.

LEMMA 3.   *Let X be a paracompact topological manifold, whose number of connected components is at most countable. Then X is σ-compact.*

Proof. It suffices to prove that each connected component is σ-compact. However, a connected component of X is an open set of X, and hence it is clear that it is paracompact. Thus it suffices to prove that a connected paracompact topological manifold is σ-compact. For each point p of X, choose a neighborhood U_p of p such that $\overline{U}_p$ is compact. Let $\mathfrak{B} = \{V_\alpha\}_{\alpha \in A}$ be a locally finite open covering which is a refinement of the covering $\{U_p\}_{p \in X}$ of X. $\overline{V}_\alpha$ is, of course, compact. Fix an index α_0 of A and let $\alpha \in A$. We shall say that V_{α_0} and V_α can be connected with a chain $\{V_{\alpha_i}\}$ if there is a finite sequence $\alpha_0, \alpha_1, \ldots, \alpha_n = \alpha$ of elements of A, such that $V_{\alpha_i} \cap V_{\alpha_{i+1}} \neq \varnothing$ for $i = 0, 1, \ldots, n-1$. We shall call n the length of the chain. If V_{α_0} and V_α can be connected with a chain, possibly there is more than one connecting chain, and we call the minimum length of the connecting chains the order of α. An arbitrary V_α can be connected to V_{α_0} by a chain. In fact, let A' be the set of indices α such that V_α can be connected to V_{α_0} by a chain, and let A'' be the set of indices α such that V_α cannot be connected to V_{α_0} by a chain. Set $X' = \cup_{\alpha \in A'} V_\alpha$, $X'' = \cup_{\alpha \in A''} V_\alpha$. Then X' and X'' are open sets of X, and we have $X' \cap X'' = \varnothing$ and $X' \cup X'' = X$. However $V_{\alpha_0} \subset X'$, so $X' \neq \varnothing$, and since we are assuming that X is connected, we conclude that $X'' = \varnothing$, so that $A'' = \varnothing$. Hence any V_α can be connected to V_{α_0} by a chain. Now let us show that there are only a finite number of α whose order is equal to a fixed n. For this, it suffices to show that, for any $\beta \in A$, there are only a finite number of $\gamma \in A$ such that $V_\beta \cap V_\gamma \neq \varnothing$. However, since $\mathfrak{B}$ is locally finite, for an arbitrary compact set K of X, there are only finite many γ such that $K \cap V_\gamma \neq \varnothing$. Since $\overline{V}_\beta$ is compact, the number of γ such that $\overline{V}_\beta \cap V_\gamma \neq \varnothing$ is finite.

Since the number of $\alpha \in A$ whose order is n is finite, the set A is countable. Since $X = \cup_{\alpha \in A} \overline{V}_\alpha$, and $\overline{V}_\alpha$ is compact, we conclude that X is σ-compact.

From Lemmas 1, 2, and 3, we obtain the following theorem.*

THEOREM 1. *If X is a topological manifold, then the following three conditions are equivalent:*

(1) *X has a countable base,*

(2) *X is σ-compact,*

(3) *X is paracompact and the number of connected components of X is at most countable.*

---

*For an example of a connected manifold which is not paracompact, see, e.g., J. G. Hocking and G. S. Young, *Topology*, Addison-Wesley, Reading, Massachusetts, 1961, p. 55–56.

Proof. We have (1) $\leftrightarrow$ (2) by Lemma 1, and (3) $\rightarrow$ (2) by Lemma 3. Assume that X is σ-compact. Then X is paracompact by Lemma 2, and X has a countable base by Lemma 1. Therefore, to show that (2) $\rightarrow$ (3), it suffices to show that if X has a countable base $\{O_n\}_{n \in N}$, then the number of connected components of X is at most countable. Let the connected components of X be $\{X_i\}_{i \in I}$, and set $N_i = \{n \in N | O_n \subset X_i\}$. Since O_n is a base of X and X_i is an open set, $N_i \neq \varnothing$, and since $X_i \cap X_j = \varnothing$ for $i \neq j$, we get $N_i \cap N_j = \varnothing$ for $i \neq j$. Hence, if we denote the smallest positive integer in N_i by n_i, then $n_i \neq n_j$ for $i \neq j$, so the correspondence from i to n_i is $1:1$, and thus I is at most countable.

Problem 1. A connected 1-dimensional paracompact manifold is homeomorphic to S^1 or $\mathbf{R}^1$. Prove this.

Let M' be a manifold with a countable base, and let M be a regular submanifold of M'. Then it is clear that M also has a countable base. However, even if M is not regular, we shall now show that M has a countable base if M is connected.*

LEMMA 4. *Let X be a connected topological manifold and let $\mathfrak{B} = \{V_\alpha\}_{\alpha \in A}$ be an open covering of X, satisfying the following two conditions: (a) Each V_α has a countable base as a subspace of X; (b) for each $\alpha \in A$, the number of $\beta \in A$ such that $V_\alpha \cap V_\beta \neq \varnothing$, is at most countable. Then X has a countable base.*

Proof. As in the proof of Lemma 3, we fix an $\alpha_0 \in A$. Then an arbitary V_α can be connected to V_{α_0} by a chain in $\mathfrak{B}$. By condition (b), there are at most only a countable number of $\alpha \in A$ whose order is equal to n. Hence A is countable. By condition (a), each V_α has a countable base, and since X is the union of a countable number of V_α, X itself has a countable base.

LEMMA 5. *Let X be a connected topological manifold and $\{V_m\}_{m \in N}$ a countable open covering of X. If each connected component of each V_m has a countable base as a subspace of X, then X has a countable base.*

Proof. Let $\{V_{m, \alpha}\}_{\alpha \in A_m}$ be the connected components of V_m. Then $\mathfrak{B} = \{V_{m, \alpha} | m \in N, \alpha \in A_m\}$ is an open covering of X. If we can show that there are at most only a countable number of elements of $\mathfrak{B}$ intersecting each $V_{m, \alpha}$, then, by Lemma 4, X has a countable base. For

---

*The proof below follows C. Chevalley, *Theory of Lie Groups*, pp. 96–98. For a proof using the Riemannian metric, see S. Kobayashi and K. Nomizu, *Foundations of Differential Geometry*, Vol. 1, Appendix 2.

this, it suffices to show, for a fixed k, that the number of $\beta \in A_k$, such that $V_{m,\alpha} \cap V_{k,\beta} \neq \varnothing$, is at most countable. The set $V_{m,\alpha} \cap V_k$ is an open subset of $V_{m,\alpha}$, so, by the hypothesis on $V_{m,\alpha}$, $V_{m,\alpha} \cap V_k$ has a countable base. Hence the number of connected components of $V_{m,\alpha} \cap V_k$ is at most countable. We let these connected components be $\{C_s\}_{s=1,2,\ldots}$. Since C_s is a connected subset of V_k, it is contained in a connected component of V_k. Let $V_{k,\beta(s)}$ be the connected component of V_k that contains C_s. Now suppose $V_{m,\alpha} \cap V_{k,\beta} \neq \varnothing$, and let $p \in V_{m,\alpha} \cap V_{k,\beta}$. The point p is contained in some C_s. For this C_s, we have $C_s \cap V_{k,\beta} \neq \varnothing$, so that $\beta = \beta(s)$. That is, if $V_{m,\alpha} \cap V_{k,\beta} \neq \varnothing$, then $\beta = \beta(s)$ for some s, so that such $V_{k,\beta}$ are countable in number.

LEMMA 6. *Let M be a connected n-dimensional manifold. If there is a differentiable map φ from M to $\mathbf{R}^n$, such that $\mathrm{rank}_p\varphi = n$ at each point p of M, then M has a countable base.*

Proof. By the hypothesis on φ, for each point p of M, we can take a neighborhood U_p of p sufficiently small so that φ is a homeomorphism from U_p onto a neighborhood V_p of $\varphi(p)$. Hence $D = \varphi(M)$ is a connected open subset of $\mathbf{R}^n$, and $\{V_p\}$ is an open covering of D. Since D has a countable base, we can find a countable open covering $\{V_i\}_{i \in I}$ which is a refinement of $\{V_p\}$. Furthermore, by Theorem 1, we can choose each V_i to be connected. Set $\varphi^{-1}(V_i) = U_i$. Then $\{U_i\}_{i \in N}$ is an open covering of M, and each connected component of U_i is mapped homeomorphically onto V_i by φ. Hence each connected component of U_i has a countable base. Thus, by Lemma 5, M has a countable base.

LEMMA 7. *A connected submanifold M of the affine space $\mathbf{R}^m$ has a countable base.*

Proof. Let $(x^1, \ldots, x^m)$ be the standard coordinate system on $\mathbf{R}^m$. Let y^i be the C^∞ function on M obtained by restricting x^i to M. Let n be the dimension of M, and let $I = (i_1, \ldots, i_n)$, where $i_1 < \ldots < i_n$ is a subsequence of the sequence $1, \ldots, m$. Set $M_I = \{p \in M \,|\, (dy^{i_1})_p, \ldots, (dy^{i_n})_p$ are linearly independent$\}$. Then M_I is an open submanifold of M, and we have $M = \cup_I M_I$. If we define $\varphi : M_I \to \mathbf{R}^n$ by $\varphi(p) = (y^{i_1}(p), \ldots, y^{i_n}(p))$, then φ is differentiable, and $\mathrm{rank}_p\varphi = n$ holds at each point p of M_I. Hence, by Lemma 6, each connected component of M_I has a countable base. However, $\{M_I\}$ is an open covering of M, so that, by Lemma 5, M also has a countable base.

THEOREM 2. *Let M' be a manifold with a countable base, and let M be a connected submanifold of M'. Then M also has a countable base.*

Proof.   Since M' has a countable base, we can choose an open covering $\{U_i\}_{i \in N}$ of M' such that each U_i is diffeomorphic to a cube in $\mathbf{R}^m$. Let $V_i = M \cap U_i$. Then $\{V_i\}_{i \in N}$ is an open covering of M. Each connected component V_i' of V_i is a connected submanifold of U_i, and since U_i can be considered as a cube in $\mathbf{R}^m$, we can consider V_i' as a submanifold of $\mathbf{R}^m$. Hence, by Lemma 7, V_i' has a countable base. V_i' is an arbitrary connected component of V_i and V_i is an open covering of M, so, by Lemma 5, M has a countable base.

Problem 2.   Under the hypothesis of Lemma 6, prove that for each point q in $\varphi(M)$, the inverse image $\varphi^{-1}(q)$ is at most a countable set.

§16. Complex Manifolds

If, for each point in a Hausdorff space M, there is a neighborhood on which a "complex local coordinate system" is defined, and if the changes between these complex local coordinate systems are "holomorphic," then M is called a complex manifold. We can define holomorphic functions on complex manifolds. In order to define these concepts precisely, we shall first discuss the complex n-dimensional number space $\mathbf{C}^n$, the complexification of a real vector space and the holomorphic functions in n complex variables.

A   Let $\mathbf{C}^n$ denote the set of all n-tuples $(z^1, \ldots, z^n)$ of complex numbers z^i $(i = 1, \ldots, n)$. We shall sometimes write (z^i) for $(z^1, \ldots, z^n)$. If we define the distance between two points $z = (z^i)$ and $w = (w^i)$ in $\mathbf{C}^n$ by

$$d(z, w) = \underset{i = 1, \ldots, n}{\text{Max}} |z^i - w^i|,$$

then $\mathbf{C}^n$ becomes a metric space and thus is endowed with the topology induced from the metric. This topological space is called the complex n-dimensional number space and is denoted by $\mathbf{C}^n$ again.

If we define a map from $\mathbf{C}^n$ to $\mathbf{R}^{2n}$ by $(z^1, \ldots, z^n) \to (x^1, y^1, \ldots, x^n, y^n)$ (where $z^k = x^k + iy^k$, $i = \sqrt{-1}$), then this map is a homeomorphism from $\mathbf{C}^n$ onto $\mathbf{R}^{2n}$.

B   Let F be a vector space over $\mathbf{R}$. From F we can construct a vector space F^C over $\mathbf{C}$, just as we construct the complex numbers from the real numbers.

Namely, we let F^C be the set of all symbols $u + iv$ $(u, v \in F)$. Here we agree that $u + iv = u' + iv'$ if and only if $u = u'$ and $v = v'$. We define the sum of $u + iv$ and $u' + iv'$, the product of $\alpha = a + ib \in \mathbf{C}$ and $u + iv$ as follows:

$$(u + iv) + (u' + iv') = (u + u') + i(v + v')$$
$$(a + ib)(u + iv) = (au - bv) + i(bu + av).$$

Then it is easy to check that F^C becomes a vector space over $\mathbf{C}$. By identifying the elements $u + i0$ (where 0 is the zero vector of F) of F^C with the elements u of F we can consider $F \subset F^C$. We call F^C the *complexification* of the real vector space F. For an element $a = u + iv$ of F^C we set $\bar{a} = u - iv$ and call $\bar{a}$ the *conjugate* of a with respect to F. The map $a \to \bar{a}$ is a *conjugate linear map* from F^C to F^C, i.e., $\overline{a + b} = \bar{a} + \bar{b}$ and $\overline{\lambda a} = \bar{\lambda} \bar{a}$ ($\lambda \in \mathbf{C}$) hold.

Now let F be n-dimensional and let $\{e_1, \ldots, e_n\}$ be a basis of F over $\mathbf{R}$ Let u, v be elements of F and set $u = \sum_{k=1}^{n} a^k e_k$, $v = \sum_{k=1}^{n} b^k e_k$. By the definition of the sum and the scalar multiplication in F^C we have

$$u + iv = \sum_{k=1}^{n} (a^k e_k + i b^k e_k) = \sum_{k=1}^{n} \alpha^k e_k, \qquad \alpha^k = a^k + i b^k.$$

If we consider $\{e_1, \ldots, e_n\}$ as elements of F^C, they are linearly independent over $\mathbf{C}$. In fact if $\sum_{k=1}^{n} \gamma^k e_k = 0$, $\gamma^k = c^k + i d^k$, then $(\sum_{k=1}^{n} c^k e_k) + i(\sum_{k=1}^{n} d^k e_k) = 0 = 0 + i0$. Hence we have $\sum_k c^k e_k = 0$ and $\sum_k d^k e_k = 0$. As $\{e_1, \ldots, e_n\}$ is a basis of F, we have $c^k = d^k = 0$ ($k = 1, \ldots, n$). Hence we get $\gamma^k = 0$ ($k = 1, \ldots, n$) and thus $\{e_1, \ldots, e_n\}$ is linearly independent over $\mathbf{C}$. We have shown that a basis $\{e_1, \ldots, e_n\}$ of F is also a basis of F^C over $\mathbf{C}$. Hence F^C is an n-dimensional vector space.

Let us consider the dual space $(F^C)^*$ of F^C. $(F^C)^*$ is by definition the complex vector space formed by all the linear functions on the complex vector space F^C. If f is an element of the dual space F^* of F, then for an arbitrary $u + iv \in F^*$, by defining

$$\tilde{f}(u + iv) = f(u) + if(v)$$

we can extend f to a linear function $\tilde{f}$ on F^C. If $\{f^1, \ldots, f^n\}$ is the basis of F^* dual to a basis $\{e_1, \ldots, e_n\}$ of F, then $\{\tilde{f}^1, \ldots, \tilde{f}^n\}$, as a set of elements in $(F^C)^*$, is linearly independent over $\mathbf{C}$. In fact, if $\sum_{k=1}^{n} \alpha_k \tilde{f}^k = 0$, $\alpha_k = a_k + i b_k$, then of course $(\sum_{k=1}^{n} \alpha_k \tilde{f}^k)(e_j) = 0$ ($j = 1, \ldots, n$). But the left-hand member is equal to α_j, so we have $\alpha_j = 0$ ($j = 1, \ldots, n$). Now, as F^C is n-dimensional, $(F^C)^*$ is also n-dimensional, and hence $\{\tilde{f}^1, \ldots, \tilde{f}^n\}$ is a basis of $(F^C)^*$.

On the other hand, for an element $h = f + ig$ ($f, g \in F^*$) of the complexification $(F^*)^C$ of F^*, if we let $\tilde{h} \in (F^C)^*$ be determined by

$$\tilde{h}(z) = \tilde{f}(z) + i\tilde{g}(z), \qquad z \in F^C, \tag{1}$$

then $h \to \tilde{h}$ is a linear map from $(F^*)^C$ to $(F^C)^*$ as vector spaces over $\mathbf{C}$. For $f \, \varepsilon \, F^*(\subset (F^*)^C)$, we note that $\tilde{f}$ coincides with the $\tilde{f}$ defined before. As the basis $\{f^1, \ldots, f^n\}$ of $(F^*)^C$ is mapped to the basis $\{\tilde{f}^1, \ldots, \tilde{f}^n\}$ of $(F^C)^*$ by $h \to \tilde{h}$, $(F^*)^C$ and $(F^C)^*$ are isomorphic. Thus we can identify the complexification $(F^*)^C$ of F^* and the dual space $(F^C)^*$ of the complexification F^C of F via (1).

C By the correspondence $(z^1, \ldots, z^n) \to (x^1, y^1, \ldots, x^n, y^n)$ we can identify $\mathbf{C}^n$ with $\mathbf{R}^{2n}$. Hence we can consider $\mathbf{C}^n$ as a $2n$-dimensional affine space. Let us denote the tangent vector space to $\mathbf{C}^n$ at p by T_p and its dual by T_p^*. Set $z^k = x^k + iy^k$. $(x^1, y^1, \ldots, x^n, y^n)$ is a standard coordinate system, so that $\{(\partial/\partial x^1)_p, (\partial/\partial y^1)_p, \ldots, (\partial/\partial x^n)_p, (\partial/\partial y^n)_p\}$ and $\{(dx^1)_p, (dy^1)_p, \ldots, (dx^n)_p, (dy^n)_p\}$ are bases of T_p and T_p^* over $\mathbf{R}$, respectively. Now let us consider the complexifications T_p^C and T_p^{*C} and let

$$(dz^k)_p = (dx^k)_p + i(dy^k)_p, \quad (d\bar{z}^k)_p = (dx^k)_p - i(dy^k)_p$$

$$\left(\frac{\partial}{\partial z^k}\right)_p = \frac{1}{2}\left\{\left(\frac{\partial}{\partial x^k}\right)_p - i\left(\frac{\partial}{\partial y^k}\right)_p\right\}, \quad \left(\frac{\partial}{\partial \bar{z}^k}\right)_p = \frac{1}{2}\left\{\left(\frac{\partial}{\partial x^k}\right)_p + i\left(\frac{\partial}{\partial y^k}\right)_p\right\}$$

$$(k = 1, \ldots, n) \tag{2}$$

Then as we get $(dx^k)_p = \{(dz^k)_p + (d\bar{z}^k)_p\}/2$, $(dy^k)_p = \{(dz^k)_p - (d\bar{z}^k)_p\}/2i$, $(\partial/\partial x^k)_p = (\partial/\partial z^k)_p + (\partial/\partial \bar{z}^k)_p$, and $(\partial/\partial y^k)_p = i\{(\partial/\partial z^k)_p - (\partial/\partial \bar{z}^k)_p\}$, we see that $\{(\partial/\partial z^1)_p, (\partial/\partial \bar{z}^1)_p, \ldots, (\partial/\partial z^n)_p, (\partial/\partial \bar{z}^n)_p\}$ and $\{(dz^1)_p, (d\bar{z}^1)_p, \ldots, (dz^n)_p, (d\bar{z}^n)_p\}$ are bases of T_p^C and T_p^{*C}, respectively.

Let f be a complex-valued function defined on an open set D of $\mathbf{C}^n$ and let u and v be the real and imaginary parts of f, respectively, i.e., let $f(p) = u(p) + iv(p)$. If u and v are both of class C^1, then we say that f is of class C^1. At each point p of D, define $(df)_p$, $(\partial/\partial x^k)_p f$, and $(\partial/\partial y^k)_p f$ by $(df)_p = (du)_p + i(dv)_p \in T_p^{*C}$, $(\partial/\partial x^k)_p f = (\partial/\partial x^k)_p u + i(\partial/\partial x^k)_p v$, and $(\partial/\partial y^k)_p f = (\partial/\partial y^k)_p u + i(\partial/\partial y^k)_p v$. Similarly, we define $(\partial/\partial z^k)_p f$ and $(\partial/\partial \bar{z}^k)_p f$ by

$$\left(\frac{\partial}{\partial z^k}\right)_p f = \frac{1}{2}\left\{\left(\frac{\partial}{\partial x^k}\right)_p f - i\left(\frac{\partial}{\partial y^k}\right)_p f\right\},$$

$$\left(\frac{\partial}{\partial \bar{z}^k}\right)_p f = \frac{1}{2}\left\{\left(\frac{\partial}{\partial x^k}\right)_p f + i\left(\frac{\partial}{\partial y^k}\right)_p f\right\}. \tag{3}$$

Then by an easy computation we see that

$$(df)_p = (du)_p + i(dv)_p = \sum_{k=1}^{n} \left\{ \left(\frac{\partial}{\partial z^k} \right)_p f \cdot (dz^k)_p + \left(\frac{\partial}{\partial \bar{z}^k} \right)_p f \cdot (d\bar{z}^k)_p \right\}. \quad (4)$$

Problem 1.   Verify Eq. (4).

D. Now let $n = 1$, and let D be an open set in the complex plane $C = C^1$. A complex-valued function f defined on D is said to be holomorphic at a point z_0 of D if

$$\lim_{h \to 0} \frac{f(z_0 + h) - f(z_0)}{h} \qquad (h \in C) \qquad (5)$$

exists. If f is holomorphic at all points of D, we say that f is holomorphic on D. Condition (5) means that

$$f(z_0 + h) - f(z_0) = c \cdot h + a(h) \cdot |h|, \qquad c \in C \qquad (6)$$

holds, where $a(h)$ approaches 0 as h approaches 0. Letting $z_0 = x_0 + iy_0$, $h = x + iy$, we can write (6) as

$$f(x_0 + x, y_0 + y) - f(x_0, y_0) = c(x + iy) + a(x, y)(x^2 + y^2)^{1/2}.$$

Hence if we separate f into its real part u and imaginary part v and set $c = a + ib$, then $f(z)$ is holomorphic at z_0 if and only if u and v are both totally differentiable at z_0 and the following equations are satisfied:

$$\frac{\partial u}{\partial x}(x_0, y_0) = a, \qquad \frac{\partial u}{\partial y}(x_0, y_0) = -b$$

$$\frac{\partial v}{\partial x}(x_0, y_0) = b, \qquad \frac{\partial v}{\partial y}(x_0, y_0) = a.$$

From these equations we get Cauchy-Riemann equations at z_0:

$$\left(\frac{\partial u}{\partial x} \right)_{z_0} = \left(\frac{\partial v}{\partial y} \right)_{z_0}, \qquad \left(\frac{\partial u}{\partial y} \right)_{z_0} = -\left(\frac{\partial v}{\partial x} \right)_{z_0},$$

and furthermore we get

$$f'(z_0) = c = \frac{1}{2} \left\{ \left(\frac{\partial u}{\partial x}(z_0) + i \frac{\partial v}{\partial x}(z_0) \right) - i \left(\frac{\partial u}{\partial y}(z_0) + i \frac{\partial v}{\partial y}(z_0) \right) \right\}$$

$$= \frac{1}{2} \left\{ \left(\frac{\partial}{\partial x} \right)_{z_0} f - i \left(\frac{\partial}{\partial y} \right)_{z_0} f \right\},$$

and hence we obtain

$$f'(z_0) = \left(\frac{\partial}{\partial z}\right)_{z_0} f.$$

We also find, by an easy computation, that the Cauchy-Riemann equations at z_0 are equivalent to $(\partial/\partial \bar{z})_{z_0} f = 0$. Hence f is holomorphic at z_0 if and only if u and v are totally differentiable at z_0 and

$$\left(\frac{\partial}{\partial \bar{z}}\right)_{z_0} f = 0.$$

If this is the case, then we have

$$f'(z_0) = \left(\frac{\partial}{\partial z}\right)_{z_0} f.$$

Let now D be an open set in $\mathbf{C}^n$, let f be a complex-valued function defined on D, and suppose that the real part u and the imaginary part v of f are both of class C^1. Furthermore, if at each point p of D we have

$$\left(\frac{\partial}{\partial \bar{z}^k}\right)_p f = 0 \qquad (k = 1, \ldots, n), \tag{7}$$

then we call f a holomorphic function defined on D.

That is, we say that f is holomorphic if, considered as a function of $2n$ real variables $x^1, y^1, \ldots, x^n, y^n$, f is of class C^1, and if for each k, when all the complex variables except z^k are fixed, f is holomorphic as a function of z^k.

If f is holomorphic on D, then f is analytic on D, i.e., f can be expanded into a power series in some neighborhood of each point of D. The real part and imaginary part of f are analytic functions of $x^1, y^1, \ldots, x^n$, y^n.*

Remark. It is known that if f is continuous and holomorphic in each variable z^k, i.e., if (7) holds, then f is holomorphic.

If ϕ is a map from an open set D of $\mathbf{C}^n$ into $\mathbf{C}^m$, we can write $\phi(z) = (\phi^1(z), \ldots, \phi^m(z))$, and each ϕ^i is a complex-valued function defined on D. If each ϕ^i is holomorphic on D, then ϕ is called a *holomorphic map* from D into $\mathbf{C}^m$.

---

*For properties of holomorphic functions of several complex variables see: H. Cartan, *Elementary Theory of Analytic Functions of One or Several Complex Variables*, Addison-Wesley, Reading, Massachusetts, 1963, Chapter IV.

E Let M be a Hausdorff space. Let $\{U_\alpha\}_{\alpha \in A}$ be an open cover of M and suppose that for each U_α there is a homeomorphism ψ_α from U_α onto an open set D_α of $\mathbf{C}^n$ satisfying the following property: if $U_\alpha \cap U_\beta \neq \varnothing$, then the map $f_{\beta\alpha} = \psi_\beta \circ \psi_\alpha^{-1}$ from the open set $\psi_\alpha(U_\alpha \cap U_\beta)$ of $\mathbf{C}^n$ onto the open set $\psi_\beta(U_\alpha \cap U_\beta)$ of $\mathbf{C}^n$ and the map $f_{\alpha\beta} = \psi_\alpha \circ \psi_\beta^{-1}$ from $\psi_\beta(U_\alpha \cap U_\beta)$ onto $\psi_\alpha(U_\alpha \cap U_\beta)$ are both holomorphic.

If M has an open cover $\{U_\alpha\}_{\alpha \in A}$ and a set of maps $\{\psi_\alpha\}_{\alpha \in A}$ with this property, then M is called a *complex manifold* of *complex dimension n*, and $\{(U_\alpha, \psi_\alpha)\}_{\alpha \in A}$ is called a *holomorphic coordinate neighborhood system* of M.

If we identify $\mathbf{C}^n$ with $\mathbf{R}^{2n}$, then a holomorphic map of an open set of $\mathbf{C}^n$ to an open set of $\mathbf{C}^n$, considered as a map between open sets in $\mathbf{R}^{2n}$, is analytic (because the real part and imaginary part of a holomorphic function are analytic). Hence, of course, a complex manifold of complex dimension n is a $2n$-dimensional (real) analytic manifold.

Let M be a complex manifold of complex dimension n and let $\{(U_\alpha, \psi_\alpha)\}_{\alpha \in A}$ be a holomorphic coordinate neighborhood system. Let U be an open set of M, ψ a homeomorphism from U onto an open set D of $\mathbf{C}^n$, and suppose they satisfy the following property:

If $U \cap U_\alpha \neq \varnothing$ $(\alpha \in A)$, then the map $\psi_\alpha \circ \psi^{-1}$ from $\psi(U \cap U_\alpha)$ to $\psi_\alpha(U \cap U_\alpha)$ and the map $\psi \circ \psi_\alpha^{-1}$ from $\psi_\alpha(U \cap U_\alpha)$ to $\psi(U \cap U_\alpha)$ are both holomorphic.

If this is the case, (U, ψ) is called a *holomorphic coordinate neighborhood* of M. For $q \in U$, set

$$\psi(q) = (z^1(q), \ldots, z^n(q)).$$

Then z^k $(k = 1, \ldots, n)$ is a complex-valued function defined on U, and we call $(z^1, \ldots, z^n)$ the *complex local coordinate system* on (U, ψ).

If M is considered a $2n$-dimensional analytic manifold, then a holomorphic coordinate neighborhood (U, ψ) of M is naturally a C^ω coordinate neighborhood, and if x^k and y^k are the real and imaginary parts of z^k, respectively, then $(x^1, y^1, \ldots, x^n, y^n)$ is a C^ω local coordinate system of M.

If $(z^1, \ldots, z^n)$ and $(w^1, \ldots, w^n)$ are two complex local coordinate systems defined in a neighborhood U of a point p, then for each point q in U we have a relation

$$w^k(q) = F^k(z^1(q), \ldots, z^n(q)).$$

Here $F^k(\zeta^1, \ldots, \zeta^n)$ is a holomorphic function in n complex variables

$\zeta, \ldots, \zeta^n$. As usual, we also write simply

$$w^k = F^k(z^1, \ldots, z^n).$$

We set

$$\frac{\partial w^k}{\partial z^j}(q) = \frac{\partial F^k}{\partial \zeta^j}(z^1(q), \ldots, z^n(q)).$$

F Let f be a complex-valued function defined on an open set E of a complex manifold M. For each point p of E, we can choose a holomorphic coordinate neighborhood (U, ψ) such that $p \in U \subset E$. If the function $f \circ \psi^{-1}$ defined on the open set $\psi(U)$ of $\mathbf{C}^n$ is holomorphic, then f is said to be *holomorphic* in a neighborhood of p. This definition does not depend on the choice of the holomorphic coordinate neighborhood (U, ψ). If f is holomorphic at all points of E, then f is said to be holomorphic on E. Suppose f is holomorphic on E, and let a complex local coordinate system in a neighborhood of p be $(z^1, \ldots, z^n)$. Then we can write

$$f(q) = F(z^1(q), \ldots, z^n(q)),$$

and the right-hand member is a holomorphic function on n variables. We write this simply as

$$f = F(z^1, \ldots, z^n)$$

and set

$$\frac{\partial f}{\partial z^k}(q) = \frac{\partial F}{\partial \zeta^k}(z^1(q), \ldots, z^n(q)).$$

G An n-dimensional complex manifold M is a $2n$-dimensional manifold, so that at each point p of M, the tangent vector space $T_p(M)$ and its dual $T_p^*(M)$ are defined. Let $(z^1, \ldots, z^n)$ be a complex local coordinate system, and let x^k and y^k be the real and imaginary parts of z^k, respectively. Then $\{(\partial/\partial x^1)_p, (\partial/\partial y^1)_p, \ldots, (\partial/\partial x^n)_p, (\partial/\partial y^n)_p\}$ is a basis of $T_p(M)$ and $\{(dx^1)_p, (dy^1)_p, \ldots, (dx^n)_p, (dy^n)_p\}$ is a basis of $T_p^*(M)$ dual to the former. As in the case of $\mathbf{C}^n$, set

$$\left(\frac{\partial}{\partial z^k}\right)_p = \frac{1}{2}\left\{\left(\frac{\partial}{\partial x^k}\right)_p - i\left(\frac{\partial}{\partial y^k}\right)_p\right\}, \quad \left(\frac{\partial}{\partial \bar{z}^k}\right)_p = \frac{1}{2}\left\{\left(\frac{\partial}{\partial x^k}\right)_p + i\left(\frac{\partial}{\partial y^k}\right)_p\right\}$$

$$(k = 1, \ldots, n), \quad (8)$$

$$(dz^k)_p = (dx^k)_p + i(dy^k)_p, \quad (d\bar{z}^k)_p = (dx^k)_p - i(dy^k)_p$$

so that $\{(\partial/\partial z^1)_p, (\partial/\partial \bar{z}^1)_p, \ldots, (\partial/\partial z^n)_p, (\partial/\partial \bar{z}^n)_p\}$ is a basis of $T_p^C(M)$

and $\{(dz^1)_p, (d\bar{z}^1)_p, \ldots, (dz^n)_p, (d\bar{z}^n)_p\}$ is a basis of $T_p^{*\,C}(M)$. Furthermore, we have

$$(dz^k)_p((\partial/\partial z^j)_p) = (d\bar{z}^k)_p((\partial/\partial \bar{z}^j)_p) = \delta_j^k$$
$$(dz^k)_p((\partial/\partial \bar{z}^j)_p) = (d\bar{z}^k)_p((\partial/\partial z^j)_p) = 0$$
$$(k, j = 1, \ldots, n).$$

As in the case of $\mathbf{C}^n$, if a complex-valued function f is of class C^1, then we can define $(\partial/\partial z^k)_p f$, $(\partial/\partial \bar{z}^k)_p f$, and $(df)_p$. For a complex-valued C^1 function f defined on E to be holomorphic, it is necessary and sufficient that

$$(\partial/\partial \bar{z}^k)_p f = 0 \qquad (k = 1, \ldots, n)$$

holds at each point p of E. If f is holomorphic, then we have

$$\left(\frac{\partial}{\partial z^k}\right)_p f = \frac{\partial f}{\partial z^k}(p) \qquad (k = 1, \ldots, n).$$

Similarly, if f is of class C^1, then we have

$$(df)_p = \sum_{k=1}^{n} \left\{\left(\frac{\partial}{\partial z^k}\right)_p f \cdot (dz^k)_p + \left(\frac{\partial}{\partial \bar{z}^k}\right)_p f \cdot (d\bar{z}^k)_p\right\}.$$

H   Let M and M' be complex manifolds of complex dimensions n and m, respectively, and let ϕ be a continuous map from M to M'. If, for each point p of M and each holomorphic function f on M' defined on a neighborhood of $\phi(p)$, $\phi^* f$ is also holomorphic in a neighborhood of p, then ϕ is called a *holomorphic map* from M to M'. Holomorphic maps are naturally differentiable. If ϕ is a one-to-one holomorphic map from M onto M', and if the inverse map ϕ^{-1} is also a holomorphic map from M' to M, then ϕ is called a *holomorphic isomorphism* (or *holomorphism*) from M onto M'.*

I   Let $(z^1, \ldots, z^n)$ be a complex local coordinate system on a neighborhood U of a point p of M. Define a linear transformation J_p of $T_p(M)$ by

$$J_p(\partial/\partial x^k)_p = (\partial/\partial y^k)_p$$
$$J_p(\partial/\partial y^k)_p = -(\partial/\partial x^k)_p$$
$$(k = 1, \ldots, n). \qquad (9)$$

Let us prove that the definition of J_p does not depend on the choice of the complex local coordinate system $(z^1, \ldots, z^n)$. To see this extend

---

*One can prove that if ϕ is one-to-one and holomorphic, then ϕ^{-1} is also holomorphic. For the case of one variable there is a proof on p. 175 in Cartan, *loc. cit.*

J_p to a linear transformation of the complex vector space $T_p^C(M)$, by setting $J_p(u + iv) = J_p u + i J_p v$ $(u, v \in T_p(M))$. Then by (9) we have

$$J_p\left(\frac{\partial}{\partial z^k}\right)_p = i\left(\frac{\partial}{\partial z^k}\right)_p, \quad J_p\left(\frac{\partial}{\partial \bar{z}^k}\right)_p = -i\left(\frac{\partial}{\partial \bar{z}^k}\right)_p \qquad (k = 1, \ldots, n). \qquad (10)$$

Hence if an element a of $T_p^C(M)$ is a linear combination of $(\partial/\partial z^k)_p$ $(k = 1, \ldots, n)$ only, then we have $J_p a = ia$, and if a is a linear combination of $(\partial/\partial \bar{z}^k)_p$ $(k = 1, \ldots, n)$ only, then we have $J_p a = -ia$.

Now if $(w^1, \ldots, w^n)$ is also a complex local coordinate system on the neighborhood U of p, and if $w^k = u^k + iv^k$, then define a linear transformation I_p of $T_p(M)$ by $I_p(\partial/\partial u^k)_p = (\partial/\partial v^k)_p$, $I_p(\partial/\partial v^k) = -(\partial/\partial u^k)_p$ $(k = 1, \ldots, n)$. If we extend I_p also to a linear transformation of $T_p^C(M)$, then as in (10) we have $I_p(\partial/\partial w^k)_p = i(\partial/\partial w^k)_p$, $I_p(\partial/\partial \bar{w}^k)_p = -i(\partial/\partial \bar{w}^k)_p$. But if we let

$$z^k = F^k(w^1, \ldots, w^n) \qquad (k = 1, \ldots, n),$$

then we have

$$\left(\frac{\partial}{\partial w^k}\right)_p = \sum_j \frac{\partial F^j}{\partial w^k}(p)\left(\frac{\partial}{\partial z^j}\right)_p, \quad \left(\frac{\partial}{\partial \bar{w}^k}\right)_p = \sum_j \frac{\overline{\partial F^j}}{\partial w^k}(p)\left(\frac{\partial}{\partial \bar{z}^j}\right)_p$$

$$(k = 1, \ldots, n),$$

where the bar ‾ indicates complex conjugation. Hence $(\partial/\partial w^k)_p$ and $(\partial/\partial \bar{w}^k)_p$ are linear combinations of $(\partial/\partial z^j)_p$ and $(\partial/\partial \bar{z}^j)_p$, respectively, and hence we have $J_p(\partial/\partial w^k)_p = i(\partial/\partial w^k)_p$ and $J_p(\partial/\partial \bar{w}^k)_p = -i(\partial/\partial \bar{w}^k)_p$. Hence J_p and I_p coincide, and this shows that the definition of J_p does not depend on the choice of the complex local coordinate system in the neighborhood of p.

From (9), it is clear that J_p satisfies

$$J_p^2 = -1, \qquad (11)$$

where 1 denotes the identity transformation of $T_p(M)$.

The correspondence J, that assigns to each point p of M the linear transformation J_p of $T_p(M)$, is called the *almost complex structure* attached to M.

Now let M and M' be complex manifolds and J and J' be the almost complex structures attached to M and M', respectively. Let ϕ be a differentiable map from M to M'. Then ϕ is holomorphic if and only if, for each point p of M, the following holds:

$$(\phi_*)_p \circ J_p = J'_{\phi(p)} \circ (\phi_*)_p. \qquad (12)$$

In fact, fix a p and let $(z^1, \ldots, z^n)$ and $(w^1, \ldots, w^m)$ be complex local coordinate systems on neighborhoods of p and $\phi(p)$, respectively, and set $z^k = x^k + iy^k$ and $w^k = u^k + iv^k$. If we set

$$\phi^* u^j = \alpha^j(x^1, y^1, \ldots, x^n, y^n), \qquad \phi^* v^j = \beta^j(x^1, y^1, \ldots, x^n, y^n),$$

then we have

$$\phi_*(\partial/\partial x^k)_p = \sum_{j=1}^{n} \{(\partial\alpha^j/\partial x^k)(p) \cdot (\partial/\partial u^j)_p + (\partial\beta^j/\partial x^k)(p) \cdot (\partial/\partial v^j)_p\},$$

$$\phi_*(\partial/\partial y^k)_p = \sum_{j=1}^{n} \{(\partial\alpha^j/\partial y^k)(p) \cdot (\partial/\partial u^j)_p + (\partial\beta^j/\partial y^k)(p) \cdot (\partial/\partial v^j)_p\},$$

$$(k = 1, \ldots, n).$$

If we compare $\phi_*(J_p(\partial/\partial x^k)_p)$ with $J'_{\phi(p)}(\phi_*(\partial/\partial x^k)_p)$, and $\phi_*(J_p(\partial/\partial y^k)_p)$ with $J'_{\phi(p)}(\phi_*(\partial/\partial y^k)_p)$ $(j = 1, \ldots, n)$, then we see that (12) holds at p if and only if

$$(\partial\alpha^j/\partial x^k)(p) = (\partial\beta^j/\partial y^k)(p), \qquad (\partial\alpha^j/\partial y^k)(p) = -(\partial\beta^j/\partial x^k)(p)$$
$$(j, k = 1, \ldots, n).$$

But this is the Cauchy-Riemann equation of $\phi^* w^j = \phi^* u^j + i\phi^* v^j = \alpha^j + i\beta^j$. Hence (12) holds at each point of M if and only if ϕ is holomorphic.

J Let us consider a complex manifold M as a differentiable manifold and let g be a Riemannian metric on M. The value g_p of g at $p \in M$ is a symmetric bilinear form on $T_p(M)$, and we can extend it to a symmetric bilinear form on $T_p^C(M)$ by defining, for $u + iv, u' + iv' \in T_p^C(M)$,

$$g_p(u + iv, u' + iv') = (g_p(u, u') - g_p(v, v')) + i(g_p(u, v') + g_p(v, u')).$$

If $(z^1, \ldots, z^n)$ is a complex local coordinate system, then set

$$g_{\alpha\beta}(p) = g_p((\partial/\partial z^\alpha)_p, (\partial/\partial z^\beta)_p), \qquad g_{\alpha\bar\beta}(p) = g_p((\partial/\partial z^\alpha)_p, (\partial/\partial \bar z^\beta)_p)$$
$$g_{\bar\alpha\beta}(p) = g_p((\partial/\partial \bar z^\alpha)_p, (\partial/\partial z^\beta)_p), \qquad g_{\bar\alpha\bar\beta}(p) = g_p((\partial/\partial \bar z^\alpha)_p, (\partial/\partial \bar z^\beta)_p)$$
$$(\alpha, \beta = 1, \ldots, n),$$

and call $g_{\alpha\beta}, g_{\alpha\bar\beta}, g_{\bar\alpha\beta}, g_{\bar\alpha\bar\beta}$ the components of g with respect to $(z^1, \ldots, z^n)$. It is easy to see that the following holds:

$$g_{\alpha\beta} = g_{\beta\alpha}, \qquad g_{\bar\alpha\bar\beta} = g_{\bar\beta\bar\alpha}, \qquad g_{\alpha\bar\beta} = g_{\bar\beta\alpha}, \qquad \bar g_{\alpha\beta} = g_{\bar\alpha\bar\beta}, \qquad \bar g_{\alpha\bar\beta} = g_{\bar\alpha\beta}.$$

If a Riemannian metric g on M satisfies

$$g_p(J_p u, J_p v) = g_p(u, v) \tag{13}$$

at each point p of M and for all u, $v \in T_p(M)$, then g is called a *Hermitian metric* of the complex manifold M. In terms of the components of g with respect to a complex local coordinate system, the condition that a Riemannian metric g be Hermitian is expressed as

$$g_{\alpha\beta} = g_{\bar{\alpha}\bar{\beta}} = 0 \qquad (\alpha, \beta = 1, \ldots, n).$$

Problem 2. Verify the last assertion above.

Now at each point p of M, set

$$T_p^+(M) = \{a \in T_p^C(M)| J_p a = ia\},$$
$$T_p^-(M) = \{a \in T_p^C(M)| J_p a = -ia\}.$$

$T_p^+(M)$ and $T_p^-(M)$ are subspaces of $T_p^C(M)$, and $T_p^+(M) \cap T_p^-(M) = (0)$. If $(z^1, \ldots, z^n)$ is a complex local coordinate system in a neighborhood of p, then $(\partial/\partial z^\alpha)_p$ $(\alpha = 1, \ldots, n)$ belongs to $T_p^+(M)$, while $(\partial/\partial \bar{z}^\alpha)_p$ $(\alpha = 1, \ldots, n)$ belongs to $T_p^-(M)$. Hence $T_p^C(M)$ is the direct sum of $T_p^+(M)$ and $T_p^-(M)$. For an element $a = u + iv$ $(u, v \in T_p(M))$ of $T_p(M)$, let $\bar{a} = u - iv$ be the conjugate of a.

Since $J_p\bar{a} = J_p u - iJ_p v$ and $J_p u, J_p v \in T_p(M)$, we have that $J_p\bar{a} = \overline{J_p a}$. From this we see easily that

$$\bar{T}_p^+(M) = T_p^-(M), \qquad \bar{T}_p^-(M) = T_p^+(M).$$

Furthermore, we have $\overline{(\partial/\partial z^\alpha)}_p = (\partial/\partial \bar{z}^\alpha)_p$. Using $T_p^+(M)$ and $T_p^-(M)$, we can rephrase the condition that the Riemannian metric be Hermitian as: $g_p(a, b) = 0$ for a, b both in $T_p^+(M)$ or both in $T_p^-(M)$.

If g is a Hermitian metric, for a, $b \in T_p^+(M)$, set

$$h_p(a, b) = g_p(a, \bar{b}).$$

Then h_p is a positive Hermitian form on the complex vector space $T_p^+(M)$.

Problem 3. Verify that h_p is a positive Hermitian form.

Moreover, if a and b both belong to $T_p^+(M)$, or if they both belong to $T_p^-(M)$, then $g_p(a, b) = 0$, and hence g_p is completely determined by h_p. (See the problems below.) This is the reason why a Riemannian metric satisfying (13) is called a Hermitian metric.

Problem 4. Show that an element $a = \sum_\alpha \{a^\alpha(\partial/\partial z^\alpha)_p + b^\alpha(\partial/\partial \bar{z}^\alpha)_p\}$ of T_p^C belongs to $T_p(M)$ if and only if $b^\alpha = \bar{a}^\alpha$ $(\alpha = 1, \ldots, n)$. Show also that if $a \in T_p(M)$, then $g_p(a, a) = 2\sum_{\alpha, \beta} g_{\alpha, \bar{\beta}} a^\alpha \bar{a}^\beta$ and $g_{\alpha\bar{\beta}} = h_p((\partial/\partial z^\alpha)_p, (\partial/\partial z^\beta)_p)$.

Problem 5. For an arbitrary element a_+ of $T_p^+(M)$, let $\bar{a}_+ = a_-$ and $a = a_+ + a_-$. Then show that a belongs to $T_p(M)$, and that for a Hermitian metric g we have

$$h_p(a_+, a_+) = g_p(a, a)/2$$

Let g be a Hermitian metric. For an arbitrary point p of M and for $u, v \in T_p(M)$, set

$$\omega_p(u, v) = g_p(J_p u, v).$$

Since g is Hermitian, we have $\omega_p(u, v) = g_p(J_p u, v) = g_p(J_p(J_p u), J_p v)$, but since $J_p^2 = -1$ we get $g_p(J_p(J_p u), J_p v) = -g_p(u, J_p v) = -g_p(J_p v, u) = -\omega_p(v, u)$. Hence

$$\omega_p(u, v) = -\omega_p(v, u)$$

holds for all $u, v \in T_p(M)$, and the correspondence ω which assigns ω_p to each point p of M is a differential form on M of degree 2 (cf. Chapter III). ω is called the differential form of degree 2 attached to the Hermitian metric g. If the exterior derivative $d\omega$ (cf. Chapter III) of ω is 0, then the Hermitian metric g is called a *Kählerian metric*.

Remark. Let g' be an arbitrary Riemannian metric on a complex manifold M. For an arbitrary point p of M, set

$$g_p(u, v) = \tfrac{1}{2}[g_p{}'(u, v) + g_p{}'(J_p u, J_p v)] \qquad (u, v \in T_p(M)).$$

Then $g : p \to g_p$ is a Hermitian metric. Hence if M is paracompact, M always admits a Hermitian metric. However an arbitrary paracompact complex manifold does not necessarily admit a Kählerian metric. In fact, examples of paracompact complex manifolds which do not admit a Kählerian metric are known. A complex manifold which admits a Kählerian metric is called a *Kählerian manifold*.

K Examples of complex manifolds

1. $\mathbf{C}^n$ and its open sets are complex manifolds. The Riemannian metric $(dx^1)^2 + (dy^1)^2 + \cdots + (dx^n)^2 + (dy^n)^2 = dz^1 d\bar{z}^1 + \cdots + dz^n d\bar{z}^n$ is Kählerian.

2. *Complex torus.* Consider $\mathbf{C}^n$ as a $2n$-dimensional vector space over $\mathbf{R}$, pick $2n$ linearly independent elements $a_1, \ldots, a_{2n}$ and set $\Gamma = \{\sum_{i=1}^{2n} m_i a_i | m_i$ an integer$\}$. $\mathbf{C}^n$ is an abelian group with respect to addition, and Γ is a subgroup of it. Let the quotient group $\mathbf{C}^n/\Gamma$ of $\mathbf{C}^n$ by Γ be denoted by T, and for an element a of $\mathbf{C}^n$, let $\pi(a)$ be the coset with respect to Γ containing a. π is a map from $\mathbf{C}^n$ to T. Let U be a subset of T. Defining U to be an open set of T if the inverse image $\pi^{-1}(U)$ of U is an open set of $\mathbf{C}^n$, we introduce a topology on T. Then each point $\pi(a)$ of T has a neighborhood homeomorphic to a neighborhood of $a \in \mathbf{C}^n$, and from

this fact we can prove that T is a complex manifold of complex dimension n. We call T a *complex torus*.*

Problem 6. Prove the statement above, and show that $\pi: \mathbf{C}^n \to T$ is a holomorphic map. Show that, as differentiable manifolds, T is diffeomorphic to the $2n$-dimensional torus T^{2n}. Show that the Kählerian metric on $\mathbf{C}^n$, as defined above, determines a Kählerian metric on T.

3. The *Riemann sphere* is a complex manifold of complex dimension 1.†

Problem 7. Show that a meromorphic function on the Riemann sphere is nothing but a holomorphic map from the Riemann sphere to itself.

4. *Complex projective space.* In the set $\mathbf{C}^{n+1} - \{0\}$ obtained by removing the origin 0 from $\mathbf{C}^{n+1}$, define an equivalence relation as follows: two points $z = (z^k)$ and $w = (w^k)$ are equivalent if there is a nonzero complex number λ such that $z^k = \lambda w^k$ $(k = 1, \ldots, n+1)$. The set of equivalence classes obtained by this equivalence relation is denoted by $P^n(\mathbf{C})$. As in the case of the real projective space in II, §2, a complex manifold structure of complex dimension n is defined on $P^n(\mathbf{C})$. This complex manifold $P^n(\mathbf{C})$ is called the n-dimensional complex projective space. $P^n(\mathbf{C})$ is a Kählerian manifold.

Problem 8. Show that $P^1(\mathbf{C})$ and the Riemann sphere are holomorphically isomorphic.

5. The hyperbolic plane H in II, §5, 9 is an open set of the complex plane, and hence a complex manifold. The Riemannian metric $(dx^2 + dy^2)/y^2 = dz \cdot dz/I(z)^2$ on H is a Kählerian metric on H.

L   As in the case of differentiable manifolds, for complex manifolds we define a complex submanifold as follows:

DEFINITION. A complex manifold M' is a complex submanifold of a complex manifold M if, when M and M' are considered as differentiable manifolds, M' is a submanifold of M and the injection map from M into M' is holomorphic.

For complex submanifolds, results similar to those in Theorem 1 and its corollaries in II, §10, hold, replacing C^∞ functions by holomorphic functions. We leave the proofs to the readers.

In II, §10, we have shown that a compact manifold is diffeomorphic to a closed submanifold of $\mathbf{R}^N$, and we have remarked that this is true for an arbitrary paracompact manifold (Whitney's theorem).

*Cf. Chapter IV, §5.

†One can find out about the Riemann sphere in any book on complex function theory. For example, cf. Cartan, *loc. cit.*, p. 89.

However, it is not true in general that "an arbitrary paracompact complex manifold is holomorphically isomorphic to a closed complex submanifold of $\mathbf{C}^N$ for some N."* In fact as we shall see below, a compact connected complex manifold in $\mathbf{C}^N$ consists of a single point and is 0-dimensional.

M LEMMA. *Let f be a holomorphic function on a connected complex manifold M. If the absolute value $|f|$ of f attains a local maximum at a point p_0 of M, then f is a constant.*

Proof. Let $(z^1, \ldots, z^n)$ be a complex local coordinate system in a neighborhood of p_0, and let $z^k(p_0) = 0$ $(k = 1, \ldots, n)$. Take $\varepsilon > 0$ sufficiently small so that when we set $U = \{p \mid |z^k(p)| < \varepsilon, k = 1, \ldots, n\}$, we have $|f(p)| \leqq |f(p_0)|$. We consider f as a function which is holomorphic in a neighborhood $U = \{(z^k) \mid |z^k| < \varepsilon, k = 1, \ldots, n\}$ of 0 in $\mathbf{C}^n$ and maximal at 0.

Let $(b^k) \in U$, and let $f(b^1 z, \ldots, b^n z) = F(z)$. Then $F(z)$ is holomorphic for $|z| < 1$, and $F(z)$ attains its maximum at $z = 0$. Hence F is constant for $|z| < 1$. Hence $f(b^1, \ldots, b^n) = f(0, \ldots, 0)$, and thus f is a constant on a neighborhood of p_0. Now using the connectivity of M, we see easily that f is constant on M.

COROLLARY 1. *The only holomorphic functions on a compact connected complex manifold M are the constant functions.*

In fact, if f is holomorphic, then $|f|$ is continuous on M. Since M is compact, $|f|$ attains its maximum at a point p_0 of M. Hence, by the lemma, f is a constant function.

COROLLARY 2. *A compact connected complex manifold in $\mathbf{C}^n$ consists of one point.*

In fact, let $(z^1, \ldots, z^n)$ be the standard coordinate system of $\mathbf{C}^n$. The restriction of each z^k to M is a holomorphic function on M. Hence z^k is a constant a^k on M. Hence M consists of one point $(a^1, \ldots, a^n)$.

§17. Almost Complex Structures

A Let V be an n-dimensional complex vector space. As is easily seen from the definition of a vector space, V can also be regarded as a vector space over $\mathbf{R}$. We shall write V_R for V when V is regarded as a vector

---

*The family of complex manifolds that are holomorphically isomorphic to closed complex submanifolds of C^N coincide with the family of complex manifolds called Stein manifolds. (This result is called the Remmert imbedding theorem of Stein manifolds.)

space over $\mathbf{R}$. As V is a vector space over $\mathbf{C}$, we can multiply an arbitrary element v of V by i, and then $v \rightarrow iv$ is a linear transformation of the real vector space V_R. We denote this linear transformation by I. Then clearly $I^2 = -1$ (1 is the identity transformation of V_R) holds. It is also easy to see that if $\{e_1, \ldots, e_n\}$ is a base of V over $\mathbf{C}$, then $\{e_1, Ie_1, \ldots, e_n, Ie_n\}$ is a base of V_R.

Conversely, let W be a vector space over R and let I be a linear transmation of W satisfying $I^2 = -1$. We can define the product $\alpha \cdot u$ of a complex number $\alpha = a + ib$ and an element u of W by

$$\alpha \cdot u = a \cdot u + b \cdot Iu.$$

If we define the product of a complex number and an element of W this way, we can consider W as a vector space over $\mathbf{C}$.

Problem 1. Show that if a real vector space W admits a linear transformation I satisfying $I^2 = -1$, then the dimension of W is even. Show also that a base $\{e_1, e_1', \ldots, e_n, e_n'\}$ can be so chosen that $Ie_k = e_k'$ and $Ie_k' = -e_k$ are satisfied.

A linear transformation I of a real vector space W satisfying $I^2 = -1$ is called a *complex structure* of the real vector space W.

Now let I be a complex structure of W, and consider the complexification W^C of W.* We extend I to a linear transformation on W^C by setting

$$I(u + iv) = Iu + iIv.$$

Take a base $\{e_1, e_1', \ldots, e_n, e_n'\}$ of W satisfying the condition of Problem 1 and set

$$u_k = \tfrac{1}{2}(e_k - ie_k'), \qquad \bar{u}_k = \tfrac{1}{2}(e_k + ie_k').$$

$\{u_1, \ldots, u_n, \bar{u}_1, \ldots, \bar{u}_n\}$ forms a base of W^C, and we have

$$Iu_k = iu_k, \quad I\bar{u}_k = -i\bar{u}_k \qquad (k = 1, \ldots, n).$$

Hence if we set

$$W^+ = \{a \in W^C | Ia = -ia\}, \qquad W^- = \{a \in W^C | Ia = -ia\},$$

then W^C decomposes into the direct sum of W^+ and W^-, and we have $\overline{W}^+ = W^-$. The elements of W^C belonging to W^+ are said to be of *holomorphic type* and those belonging to W^- are said to be of *antiholomorphic type*.

*Using I, we can consider W as a complex vector space, but we note here that we are not considering W as a complex vector space when we consider W^C.

Problem 2. Suppose the complexification W^C of a real vector space W decomposes into the direct sum of two subspaces W_1 and W_2, and suppose W_1 and W_2 are conjugate to each other (i.e., $\overline{W}_1 = W_2$). Define a linear transformation I of W^C by $Ia = ia$ for $a \in W_1$ and $Ia = -ia$ for $a \in W_2$. Then show that I leaves W invariant and defines a complex structure on W.

B Let M be a $2n$-dimensional manifold. At each point p of M, suppose a complex structure J_p is given on the tangent space $T_p(M)$, satisfying the following condition: Let $(x^1, \ldots, x^{2n})$ be an arbitrary local coordinate system on an open set U of M. For $p \in U$, let the matrix representation of the linear transformation J_p on $T_p(M)$ with respect to the base $\{(\partial/\partial x^1)_p, \ldots, (\partial/\partial x^{2n})_p\}$ be

$$J_p\left(\frac{\partial}{\partial x^i}\right)_p = \sum_{k=1}^{2n} J_i^k(p)\left(\frac{\partial}{\partial x^k}\right)_p. \tag{1}$$

The $(2n)^2$ functions J_i^k on U are of class C^∞. (As usual, it can be proved that this condition does not depend on the choice of the local coordinate system.)

When the condition above is satisfied, the correspondence J, that assigns J_p to each point p, is called a C^∞ *almost complex structure*, or simply just an almost complex structure. A manifold with an almost complex structure J is called an *almost complex manifold*.* The J_i^k's $(k, i = 1, \ldots, 2n)$ in (1) are called the components of J with respect to (x^i).

If M is a complex manifold and J is the almost complex structure attached to M, then we can choose $(x^1, \ldots, x^{2n})$ so that the J_i^k's are constants. Hence, naturally, J is an almost complex structure in the sense defined above. Conversely, we have the following:

LEMMA. *Let M be a $2n$-dimensional manifold with an almost complex structure J. Suppose there is an open cover $\mathfrak{U} = \{U\}$ of M, which satisfies the following condition: There is a local coordinate system $(x^1, y^1, \ldots, x^n, y^n)$ on each open set U of $\mathfrak{U}$, such that for each point q of U,*

$$J_q(\partial/\partial x^k)_q = (\partial/\partial y^k)_q$$
$$J_q(\partial/\partial y^k)_q = -(\partial/\partial x^k)_q \qquad (k = 1, \ldots, n)$$

---

*An even-dimensional manifold does not always admit an almost complex structure. For example, there is no almost complex structure on the 4-dimensional sphere S^4. Cf. N. E. Steenrod, *The Topology of Fibre Bundles*, Princeton Univ. Press, Princeton, New Jersey, 1951, §41.

are satisfied. Then M is a complex manifold, and J is the almost complex structure attached to the complex structure.

Proof. Let U, $V \in \mathfrak{U}$ and $U \cap V \neq \varnothing$. Let $(x^1, y^1, \ldots, x^n, y^n)$ and $(u^1, v^1, \ldots, u^n, v^n)$ be local coordinate systems on U and V, respectively, satisfying the condition in the lemma. On $U \cap V$ we let

$$u^i = \phi^i(x^1, \ldots, x^n, y^1, \ldots, y^n),$$
$$v^i = \psi^i(x^1, \ldots, x^n, y^1, \ldots, y^n).$$

Then on $U \cap V$ we have

$$\frac{\partial}{\partial x^i} = \sum_j \frac{\partial \phi^j}{\partial x^i} \frac{\partial}{\partial u^j} + \sum_j \frac{\partial \phi^j}{\partial x^i} \frac{\partial}{\partial v^j},$$

$$\frac{\partial}{\partial y^i} = \sum_j \frac{\partial \phi^j}{\partial y^i} \frac{\partial}{\partial u^j} + \sum_j \frac{\partial \phi^j}{\partial y^i} \frac{\partial}{\partial v^j}.$$

Applying J on both sides of these equations, we have

$$\frac{\partial}{\partial y^i} = \sum_j \frac{\partial \phi^j}{\partial x^i} \frac{\partial}{\partial v^j} - \sum_j \frac{\partial \phi^j}{\partial x^i} \frac{\partial}{\partial u^j},$$

$$\frac{\partial}{\partial x^i} = - \sum_j \frac{\partial \phi^j}{\partial y^i} \frac{\partial}{\partial v^j} + \sum_j \frac{\partial \phi^j}{\partial y^i} \frac{\partial}{\partial u^j}.$$

Comparing these equations, we get

$$\frac{\partial \phi^j}{\partial x^i} = \frac{\partial \psi^j}{\partial y^i}, \quad \frac{\partial \phi^j}{\partial y^i} = - \frac{\partial \psi^j}{\partial x^i} \qquad (i, j = 1, \ldots, n). \qquad (*)$$

Now if we let $z^i = x^i + iy^i$, $w^i = u^i + iv^i$, then $(z^1, \ldots, z^n)$ and $(w^1, \ldots, w^n)$ are complex local coordinate systems of U and V, respectively, and on $U \cap V$ we have

$$w^k = f^k(z^1, \ldots, z^n), \qquad f^k = \phi^k + \sqrt{-1}\psi^k.$$

But then, by (*), f^k is holomorphic in $(z^1, \ldots, z^n)$, and hence M is a complex manifold.

C   In what follows, we let M be an almost complex manifold and J the almost complex structure of M. If X is a vector field on M, then we can define a vector field JX by setting

$$(JX)_p = J_p X_p$$

at each point $p \in M$. $X \to JX$ is a linear transformation of the vector space $\mathfrak{X}(M)$ formed by the vector fields on M, and $J^2(X) = J(JX)) = -X$ holds for all $X \in \mathfrak{X}(M)$.

Now let $\mathfrak{X}^C(M)$ be the complexification of the real vector space $\mathfrak{X}(M)$. An element A of $\mathfrak{X}^C(M)$ is expressed uniquely as $A = X + iY$ ($X, Y \in \mathfrak{X}(M)$). An element A of $\mathfrak{X}^C(M)$ is called a *complex vector field** on M, and the value of A at each point p of M is defined by

$$A_p = X_p + iY_p \in T_p^C(M).$$

Extend the linear transformation $J: X \to JX$ of $\mathfrak{X}(M)$ to a linear transformation of $\mathfrak{X}^C(M)$ by

$$J(X + iY) = JX + iJY.$$

We also define the bracket $[A, B]$ of $A = X + iY$, $B = X' + iY'$ by

$$[A, B] = ([X, X'] - [Y, Y']) + i([X, Y'] + [Y, X']).$$

Then $\mathfrak{X}^C(M)$ becomes a Lie algebra over $\mathbf{C}$.

If $A_p \in T_p^+(M)$, i.e., if A_p is of holomorphic type for each point p of M, then we say that A is of *holomorphic type*. We also say that A is of *antiholomorphic type* if A_p is antiholomorphic for each point p. If A is of holomorphic type, then $JA = iA$, and conversely, if $JA = iA$, then A is of holomorphic type. Similarly for antiholomorphic type. For an arbitrary vector field A, set

$$A^+ = (A - iJA)/2, \qquad A^- = (A + iJA)/2,$$

so that A^+ is of holomorphic type and A^- is of antiholomorphic type, and we have

$$A = A^+ + A^-.$$

Problem 3. If $A = A_1 + A_2$, where A_1 is holomorphic and A_2 is antiholomorphic, then show that $A_1 = A^+$ and $A_2 = A^-$.

Problem 4. Show that A is a real vector field (i.e., $A \in \mathfrak{X}(M)$) if and only if $\overline{A}^+ = A^-$. Here, $\overline{A}^+$ is the conjugate of the element A^+ in $\mathfrak{X}^C(M)$ with respect to $\mathfrak{X}(M)$.

Denote the set of vector fields of holomorphic type and of antiholomorphic type by $\mathfrak{X}^+(M)$ and $\mathfrak{X}^-(M)$, respectively. Then these are subspaces of $\mathfrak{X}^C(M)$, and $\mathfrak{X}^C(M)$ is the direct sum of these subspaces.

---

*We have assumed here that M is an almost complex manifold, but a complex vector field can of course be defined for an arbitrary manifold.

DEFINITION. If, for any $A, B \in \mathfrak{X}^+(M)$, the bracket $[A, B]$ of A and B also belongs to $\mathfrak{X}^+(M)$, then we say that the complex structure J is *integrable*.*

Since $[\overline{A, B}] = [\overline{A}, \overline{B}]$ and $\overline{\mathfrak{X}^+}(M) = \mathfrak{X}^-(M)$, if J is integrable, then for $A, B \in \mathfrak{X}^-(M)$, we have $[A, B] \in \mathfrak{X}^-(M)$.

Now for arbitrary $X, Y \in \mathfrak{X}(M)$, set

$$S(X, Y) = [X, Y] + J[JX, Y] + J[X, JY] - [JX, JY]. \qquad (2)$$

Then S is a skew symmetric bilinear form from the product $\mathfrak{X}(M) \times \mathfrak{X}(M)$ of the $C^\infty(M)$-module $\mathfrak{X}(M)$ to $\mathfrak{X}(M)$. That is,

$$S(X + X', Y) = S(X, Y) + S(X', Y),$$
$$S(X, Y + Y') = S(X, Y) + S(X, Y'),$$
$$S(fX, Y) = S(X, fY) = fS(X, Y), \qquad S(X, Y) = -S(Y, X).$$

Problem 5. Prove the identities above.

THEOREM 1. *For an almost complex structure J to be completely integrable, it is necessary and sufficient that*

$$S(X, Y) = [X, Y] + J[JX, Y] + J[X, JY] - [JX, JY] = 0 \qquad (3)$$

holds for all $X, Y \in \mathfrak{X}(M)$.

Proof. For elements $A = X + iY$, $B = X' + iY'$ of $\mathfrak{X}^C(M)$, set

$$S(A, B) = [A, B] + J[JA, B] + J[A, JB] - [JA, JB].$$

Then we have $S(A, B) = (S(X, X') - S(Y, Y')) + i(S(X, Y') + S(Y, X'))$. Hence if we assume $S(X, Y) = 0$ for all $X, Y \in \mathfrak{X}(M)$, then we have $S(A, B) = 0$ for all $A, B \in \mathfrak{X}^C(M)$. Now suppose (3) holds and let $A, B \in \mathfrak{X}^+(M)$. Since $JA = iA$ and $JB = iB$, we have $S(A, B) = 2([A, B] + iJ[A, B])$. But from our assumption we have $S(A, B) = 0$, so that $J[A, B] = i[A, B]$. This shows $[A, B] \in \mathfrak{X}^+(M)$. Hence if (3) holds, then J is integrable.

Conversely, suppose J is integrable. We shall show that $S(A, B) = 0$ holds for all $A, B \in \mathfrak{X}^C(M)$. Since $\mathfrak{X}^C(M)$ is the direct sum of $\mathfrak{X}^+(M)$ and $\mathfrak{X}^-(M)$, and since $S(A, B)$ is bilinear in A and B, it suffices to show $S(A, B) = 0$ for A and B in $\mathfrak{X}^+(M)$ or $\mathfrak{X}^-(M)$.

(i) If $A, B \in \mathfrak{X}^+(M)$, then $JA = iA$ and $JB = iB$, and since $[A, B] \in$

---

*One can define an almost complex structure on the 6-dimensional sphere S^6 which is not integrable. For this example, see: A. Fröhlicher, "Zur Differential-geometrie der komplexen Strukturen," *Math. Annalen*, **129** (1955), 50–95.

$\mathfrak{X}^{+}(M)$, we have $J[A, B] = i[A, B]$. Hence $S(A, B) = [A, B] + iJ[A, B] + iJ[A, B] + [A, B] = 2[A, B] - 2[A, B] = 0$. (ii) If $A, B \in \mathfrak{X}^{-}(M)$, then just as in (i), we get $S(A, B) = 0$. (iii) If $A \in \mathfrak{X}^{+}(M)$ and $B \in \mathfrak{X}^{-}(M)$, then $S(A, B) = [A, B] + iJ[A, B] - iJ[A, B] - [A, B] = 0$. (iv) If $A \in \mathfrak{X}^{-}(M)$ and $B \in \mathfrak{X}^{+}(M)$, similarly we get $S(A, B) = 0$.

The proof of Theorem 1 is complete.

D Let M be a complex manifold, and let J be the almost complex structure attached to M. Let $(z^1, \ldots, z^n)$ be a complex local coordinate system on an open set U. Then $A \in \mathfrak{X}^{C}(M)$ can be expressed as

$$A = \sum_{\alpha=1}^{n} \xi^{\alpha} \frac{\partial}{\partial z^{\alpha}} + \sum_{\alpha=1}^{n} \eta^{\bar{\alpha}} \frac{\partial}{\partial \bar{z}^{\alpha}},$$

where ξ^{α}, $\eta^{\bar{\alpha}}$ are complex-valued functions defined on U. $\{\xi^{\alpha}, \eta^{\bar{\alpha}}\}$ are called the components of the complex vector field A. A is of holomorphic type if and only if $\eta^{\bar{\alpha}} = 0$ ($\alpha = 1, \ldots, n$), and is of antiholomorphic type if and only if $\xi^{\alpha} = 0$ ($\alpha = 1, \ldots, n$). Also we have

$$A^{+} = \sum_{\alpha} \xi^{\alpha} \frac{\partial}{\partial z^{\alpha}}, \qquad A^{-} = \sum_{\alpha} \eta^{\bar{\alpha}} \frac{\partial}{\partial \bar{z}^{\alpha}}$$

on U. Also we have $A \in \mathfrak{X}(M)$ if and only if $\eta^{\bar{\alpha}} = \overline{\xi^{\alpha}}$ ($\alpha = 1, \ldots, n$).

Let A, B be of holomorphic type and set $A = \sum_{\alpha} \xi^{\alpha}(\partial/\partial z^{\alpha})$, $B = \sum_{\beta} \eta^{\beta}(\partial/\partial z^{\beta})$. Then we have $[A, B] = \sum_{\alpha} (\sum_{\beta} \xi^{\beta}(\partial \eta^{\alpha}/\partial z^{\beta}) - \eta^{\beta}(\partial \xi^{\alpha}/\partial z^{\beta})) \cdot (\partial/\partial z^{\alpha})$, and $[A, B]$ is also of holomorphic type, and hence J is integrable. Conversely, it is known that if there is an integrable almost complex structure J, then M is a complex manifold, and J is the almost complex structure attached to M.* In this book we shall prove this fact in III, §9, under the assumption that M and J are both of class C^{ω}.

Now let A be a vector field of holomorphic type on a complex manifold M. A is called a *holomorphic vector field* on M if all components ξ^{α} of A are holomorphic functions. If A is of antiholomorphic type, and if $\bar{A}$ is a holomorphic vector field, then A is called an antiholomorphic vector field.

If A is of holomorphic type, for A to be a holomorphic vector field it is necessary and sufficient that $[A, B] \in \mathfrak{X}^{-}(M)$ for arbitrary $B \in \mathfrak{X}^{-}(M)$. In fact, if $A = \sum_{\alpha} \xi^{\alpha}(\partial/\partial z^{\alpha})$ and $B = \sum_{\beta} \eta^{\bar{\beta}}(\partial/\partial \bar{z}^{\beta})$, then

$$[A, B] = -\sum_{\alpha} (\sum_{\beta} \eta^{\bar{\beta}}(\partial \xi^{\alpha}/\partial \bar{z}^{\beta}))(\partial/\partial z^{\alpha}) + \sum_{\beta} (\sum_{\alpha} \xi^{\alpha}(\partial \eta^{\bar{\beta}}/\partial z^{\alpha}))(\partial/\partial \bar{z}^{\beta}).$$

---

*A. Newlander and L. Nirenberg, Complex analytic coordinates in almost complex manifolds, *Ann. of Math*, **65** (1957), 391–404.

Hence if ξ^α are all holomorphic, then from, $\partial\xi^\alpha/\partial\bar{z}^\beta = 0$, we get $[A, B] \in$ $\mathfrak{X}^-(M)$. Conversely, if $[A, B] \in \mathfrak{X}^-(M)$, then $\sum_\beta \eta^\beta(\partial\xi^\alpha/\partial\bar{z}^\beta) = B\xi^\alpha = 0$ holds. But in a sufficiently small neighborhood around each point, we can find a B of antiholomorphic type coinciding with $\partial/\partial\bar{z}^\beta$. Hence ξ^α is holomorphic, and hence A is holomorphic. Similarly we can prove that A is a antiholomorphic if and only if $[A, B] \in \mathfrak{X}^+(M)$ for arbitrary $B \in \mathfrak{X}^+(M)$.

For a real vector field $X \in \mathfrak{X}(M)$ on M, if X^+ is a holomorphic vector field, then X is called an *analytic vector field*.* Since X is a real vector field, we have $X^- = \bar{X}^+$, so if X is an analytic vector field, then X is antiholomorphic. If X is an analytic vector field then X can be written as

$$X = \sum_\alpha \xi^\alpha \frac{\partial}{\partial z^\alpha} + \sum_\alpha \bar{\xi}^\alpha \frac{\partial}{\partial \bar{z}^\alpha}, \qquad \xi^\alpha \text{ holomorphic.}$$

Let us prove that $X \in \mathfrak{X}(M)$ is an analytic vector field if and only if

$$[X, JY] = J[X, Y], \qquad \text{for arbitrary } Y \in \mathfrak{X}(M), \tag{4}$$

holds.

Fix X, and for arbitrary $A \in \mathfrak{X}^C(M)$ set $F(A) = [X, JA] - J[X, A]$. Then $F(A + B) = F(A) + F(B)$. If we let $A = Y + iY'$, then $F(A) = F(Y) + iF(Y')$. Hence (4) is equivalent to $F(A) = 0$ for all $A \in \mathfrak{X}^+(M)$ and all $A \in \mathfrak{X}^-(M)$. Now, if $A \in \mathfrak{X}^+(M)$, from $JA = iA$ we have

$$F(A) = i[X, A] - J[X, A].$$

Similarly for $A \in \mathfrak{X}^-(M)$, we have

$$F(A) = -i[X, A] - J[X, A].$$

Hence (4) is equivalent to $[X, A] \in \mathfrak{X}^+(M)$ for $A \in \mathfrak{X}^+(M)$ and $[X, A] \in$ $\mathfrak{X}^-(M)$ for $A \in \mathfrak{X}^-(M)$. But X is a real vector field, so $[\overline{X, A}] = [X, \bar{A}]$, and $\bar{\mathfrak{X}}^-(M) = \mathfrak{X}^+(M)$. Hence if $[X, A] \in \mathfrak{X}^-(M)$ for $A \in \mathfrak{X}^-(M)$, then $[X, A] \in \mathfrak{X}^+(M)$ $A \in \mathfrak{X}^+(M)$ follows. Hence (4) is equivalent to $[X, A] \in$ $\mathfrak{X}^-(M)$ for $A \in \mathfrak{X}^-(M)$. Now let $A \in \mathfrak{X}^-(M)$. We have $[X, A] = [X^+, A] +$ $[X^-, A]$, but since J is integrable we have $[X^-, A] \in \mathfrak{X}^-(M)$, so that $[X, A] \in \mathfrak{X}^-(M)$ is equivalent to $[X^+, A] \in \mathfrak{X}^-(M)$. However, as we have already shown, $[X^+, A] \in \mathfrak{X}^-(M)$ for all $A \in \mathfrak{X}^-(M)$ if and only if X^+ is holomorphic, i.e., if and only if X is an analytic vector field. Hence X is an analytic vector field if and only if (4) holds.

*Although the terminology is confusing, one should not confuse this with a vector field of class C^ω.

THEOREM 2. *Let ϕ_t be a one-parameter transformation group on a complex manifold M and let X be the infinitesimal transformation of ϕ_t. Then X is an analytic vector field on M if and only if ϕ_t is a holomorphic isomorphism of M for each t.*

Proof. For an arbitrary vector field Y on M, set

$$\Phi(t; Y) = J((\phi_t)_* Y) - (\phi_t)_* JY.$$

Each ϕ_t is a holomorphic isomorphism of M if and only if $\Phi(t; Y) = 0$ for arbitrary Y and t ((12) of §16, I). Now at each point p of M we have

$$\lim_{t \to 0} \frac{1}{t}(\Phi(t; Y)_p - \Phi(0, Y)_p) = [X, JY]_p - J_p[X, Y]_p. \tag{5}$$

In fact, by §12, we have

$$\text{left member} = \lim_{t \to 0} \frac{1}{t}[J_p((\phi_t)_* Y_{\phi_t^{-1}(p)}) - (\phi_t)_*(JY)_{\phi_t^{-1}(p)}]$$

$$= \lim_{t \to 0} \frac{1}{t}[(JY)_p - (\phi_t)_*(JY)_{\phi_t^{-1}(p)} - J_p(Y_p - (\phi_t)_* Y_{\phi_t^{-1}(p)})]$$

$$= [X, JY]_p - J_p[X, Y]_p.$$

Hence if each ϕ_t is a holomorphic isomorphism, then $\Phi(t; Y) = 0$, so by (5) we have $[X, JY] - J[X, Y] = 0$ for arbitrary Y. Hence X is an analytic vector field. Conversely, if X is an analytic vector field, then by (5), for an arbitrary Y and p we have $[d\Phi(t; Y)_p/dt]_{t=0} = 0$. But by an easy computation we see that $\Phi(t + s; Y) = \Phi(s;(\phi_t)_* Y) + (\phi_s)_* \Phi(t; Y)$. Hence

$$\lim_{s \to 0} \frac{1}{s}(\Phi(t + s; Y)_p - \Phi(t; Y)_p)$$

$$= [\frac{d}{ds}\Phi(s;(\phi_t)_* Y)_p]_{s=0} - \lim_{s \to 0} \frac{1}{s}[\Phi(t; Y)_p - ((\phi_s)_* \Phi(t, Y))_p]$$

$$= -[X, \Phi(t; Y)]_p.$$

That is, $\Phi(t; Y)_p$, as a function of t, is a solution of the ordinary differential equation

$$\frac{d\Phi(t; Y)_p}{dt} = -[X, \Phi(t; Y)]_p. \tag{6}$$

By the definition of Φ, we have $\Phi(0; Y) = 0$. On the other hand, the vector field which is constantly zero is also a solution of (6). By the uniqueness of solutions we get $\Phi(t; Y)_p = 0$. Since this holds for arbitrary Y and p, we conclude that each ϕ_t is a holomorphic isomorphism.

CHAPTER

III

Differential Forms
and Tensor Fields

§1. p-Linear Functions

Let K be the field $\mathbf{R}$ of real numbers or the field $\mathbf{C}$ of complex numbers, and let $V_1, \ldots, V_p$ be vector spaces over K. A map f from the product set† $V_1 \times \ldots \times V_p$ into K is called a p-linear function on $V_1 \times \ldots \times V_p$ if f satisfies the condition

$$f(u_1, \ldots, u_{i-1}, \lambda u_i + \mu v_i, u_{i+1}, \ldots, u_p)$$
$$= \lambda f(u_1, \ldots, u_i, \ldots, u_p) + \mu f(u_1, \ldots, v_i, \ldots, u_p)$$
$$(\lambda, \mu \in K, \ i = 1, \ldots, n).$$

That is, if $f(u_1, \ldots, u_p)$ is linear in each variable u_i, then we say that f is p-linear. The linear function and the bilinear functions which we have studied in Chapter I are the cases where $p = 1$ and $p = 2$.

Let f_1, f_2 be p-linear functions on $V_1 \times \ldots \times V_p$, and $\lambda_1, \lambda_2 \in K$. Then the map $\lambda_1 f_1 + \lambda_2 f_2$ that assigns the element $\lambda_1 f_1(u_1, \ldots, u_p) + \lambda_2 f_2(u_1, \ldots, u_p)$ of K to $(u_1, \ldots, u_p)$ is also a p-linear function. The set of all p-linear functions on $V_1 \times \ldots \times V_p$ becomes a vector space over K by this definition of sums and scalar multiplications. The vector space is written as $V_1^* \otimes \ldots \otimes V_p^*$, and is called the *tensor product* of $V_1^*, \ldots, V_p^*$.

†In general, if $A_1, \ldots, A_p$ are sets, then the set of all p-tuples $(a_1, \ldots, a_p)$ $(a_i \in A_i)$ is called the product set of $A_1, \ldots, A_p$, and is denoted by $A_1 \times \ldots \times A_p$.

123

We now assume that each V_i is of finite dimension, and denote the dimension of V_i by n_i. Let $\{e_{i,1}, \ldots, e_{i,n_i}\}$ be a basis of V_i. If $u_i = \sum_{j=1}^{n_i} x_i^j e_{i,j}$ $(i = 1, \ldots, p; x_i^j \in K)$, then we have

$$f(u_1, \ldots, u_p) = \sum_{j_1, \ldots, j_p} x_1^{j_1} \ldots x_p^{j_p} f(e_{1, j_1}, \ldots, e_{p, j_p}).$$

Hence f is completely determined by $a_{j_1 \ldots j_p} = f(e_{1, j_1}, \ldots, e_{p, j_p})$. Conversely, for any given set of $a_{j_1 \ldots j_p} \in K$, by setting $f(u_1, \ldots, u_p) = \sum_{j_1, \ldots, j_p} a_{j_1 \ldots j_p} x_1^{j_1} \ldots x_p^{j_p}$, we get a p-linear function f. Hence $V_1^* \otimes \ldots \otimes V_p^*$ is an $n_1 \ldots n_p$-dimensional vector space over K.

Let $W_1, \ldots, W_q$ also be vector spaces over K, and let g be a q-linear function on $W_1 \times \ldots \times W_q$, f a p-linear function on $V_1 \times \ldots \times V_p$. If we let the element $f(v_1, \ldots, v_p)g(w_1, \ldots, w_q)$ of K correspond to element $(v_1, \ldots, v_p, w_1, \ldots, w_q)$ of $V_1 \times \ldots \times V_p \times W_1 \times \ldots \times W_q$ then we get a $(p + q)$-linear function on $V_1 \times \ldots \times V_p \times W_1 \times \ldots \times W_q$. We denote it by $f \otimes g$, and call it the *tensor product* of f and g. The following holds for tensor products:

(a) $(\lambda_1 f_1 + \lambda_2 f_2) \otimes g = \lambda_1(f_1 \otimes g) + \lambda_2(f_2 \otimes g),$

$\qquad\quad f \otimes (\lambda_1 g_1 + \lambda_2 g_2) = \lambda_1(f \otimes g_1) + \lambda_2(f \otimes g_2).$

(b) If $Z_1, \ldots, Z_r$ are vector spaces over K, and h is an r-linear function on $Z_1 \times \ldots \times Z_r$, then

$$(f \otimes g) \otimes h - f \otimes (g \otimes h).$$

From now on, we shall consider p-linear functions on the product $V^p = V \times \ldots \times V$ of p copies of the same vector space V. We write $\overset{p}{\otimes} V^*$ for $V^* \otimes \ldots \otimes V^*$. For $p = 0$, we set $\overset{0}{\otimes} V^* = K$. Let $f_p \in \overset{p}{\otimes} V^*$ $(p \geq 0)$, and assume that $f_p = 0$ except for a finite number of p. We denote by $T^* = \sum_{p=0}^{\infty} \overset{p}{\otimes} V^*$ the set of all such infinite-tuples $(f_0, f_1, \ldots, f_p, \ldots)$. The infinite-tuple $(f_0, f_1, \ldots, f_p, \ldots)$ is also denoted by $\sum_{p=0}^{\infty} f_p$. Here, as above, f_p are all 0 except for a finite number of p's. We define the sum of two elements and the scalar multiplication in T^* by

$$\sum_p f_p + \sum_p g_p = \sum_p (f_p + g_p), \qquad \lambda \sum_p f_p = \sum_p \lambda f_p.$$

Then T^* becomes a vector space over K. If we identify the element f_p of $\overset{p}{\otimes} V^*$ with the element $(0, \ldots, 0, f_p, 0, \ldots)$ of T^*, then $\overset{p}{\otimes} V^*$ becomes a subspace of T^*, and T^* is the direct sum of these subspaces. Furthermore, if we define the product of two elements in T^* by

$$(\sum_p f_p) \otimes (\sum_q g_q) = \sum_r h_r,$$
$$h_r = \sum_{p+q=r} f_p \otimes g_q,$$

then T^* becomes an associative algebra over K. (For the definition of an associative algebra cf. II, §11.) Here, for $f_0 \in K = \overset{0}{\otimes} V^*$, we set $f_0 \otimes g_q = f_0 g_q$. The associative algebra T^* is called the *covariant tensor algebra* of V. The unit element of K is the unit element of the associative algebra T^*. The elements of $\overset{p}{\otimes} V^*$, i.e., the p-linear functions on $V \times \ldots \times V$, †are also called the *covariant tensors of order p*. In particular, the elements of V^* are called *covariant vectors*.

Now let V be n-dimensional, let $\{e_1, \ldots, e_n\}$ be a basis of V, and let $\{\varphi^1, \ldots, \varphi^n\}$ be the dual basis of V^*. Then

$$\varphi^{j_1} \otimes \ldots \otimes \varphi^{j_p} \qquad (1 \leq j_k \leq n, k = 1, \ldots, p)$$

is a basis of $\overset{p}{\otimes} V^*$. In fact, for $f \in \overset{p}{\otimes} V^*$, set

$$f_{i_1, \ldots, i_p} = f(e_{i_1}, \ldots, e_{i_p}),$$

and let f' denote the element $\sum_{j_1, \ldots, j_p = 1}^n f_{j_1, \ldots, j_p} \varphi^{j_1} \otimes \ldots \otimes \varphi^{j_p}$ of $\overset{p}{\otimes} V^*$. Then, we have

$$f'(e_{i_1}, \ldots, e_{i_p}) = \sum_{j_1, \ldots, j_p} f_{j_1, \ldots, j_p} \varphi^{j_1}(e_{i_1}) \ldots \varphi^{j_p}(e_{i_p}) = f_{i_1, \ldots, i_p}.$$

Hence $f = f'$, that is, we have

$$f = \sum_{j_1, \ldots, j_p} f_{j_1, \ldots, j_p} \varphi^{j_1} \otimes \ldots \otimes \varphi^{j_p}.$$

On the other hand, the dimension of $\overset{p}{\otimes} V^*$ is exactly n^p. Hence the n^p elements $\varphi^{j_1} \otimes \ldots \otimes \varphi^{j_p}$ form a basis of $\overset{p}{\otimes} V^*$. For an element f of $\overset{p}{\otimes} V^*$, $\{f_{i_1, \ldots, i_p}\}$ is called the set of *components* of f with respect to the basis $\{e_1, \ldots, e_n\}$ of V.

§2. Symmetric Tensors and Alternating Tensors; the Exterior Product

We shall denote the permutation group formed by the permutations‡ of p letters $(1, \ldots, p)$ by $\mathfrak{S}_p$. If $\sigma \in \mathfrak{S}_p$, the image of k $(1 \leq k \leq p)$ by σ is denoted by $\sigma(k)$. If φ is a p-linear function on a vector space

---

†Since it is awkward to say a p-linear function on $V \times \cdots \times V$, we shall say instead, from now on, a p-linear function on V.

‡A permutation of the set $\{1, \ldots, p\}$ is a 1:1 mapping of the set onto itself.

V, for $u_1, \ldots, u_p \in V$ and $\sigma \in \mathfrak{S}_p$, let

$$(\sigma\varphi)(u_1, \ldots, u_p) = \varphi(u_{\sigma(1)}, \ldots, u_{\sigma(p)}). \tag{1}$$

Then $\sigma\varphi$ is also a p-linear function on V. If, for an arbitrary $\sigma \in \mathfrak{S}_p$, $\sigma\varphi = \varphi$, that is, $\varphi(u_{\sigma(1)}, \ldots, u_{\sigma(p)}) = \varphi(u_1, \ldots, u_p)$ holds, then φ is called a *symmetric p-linear function* or a *symmetric covariant tensor of order p*. If $\sigma\varphi = \varepsilon_\sigma\varphi (\varepsilon_\sigma$ is the signature† of the permutation σ) holds for all $\sigma \in \mathfrak{S}_p$, then φ is called an *alternating p-linear function* (or a *skew-symmetric p-linear* function) or an *alternating covariant tensor of order p*.

The set of symmetric p-linear functions and the set of alternating p-linear functions form subspaces of $\overset{p}{\otimes}V^*$. These subspaces are denoted by $\overset{p}{S}V^*$ and $\overset{p}{\wedge}V^*$, respectively. We set $\overset{0}{S}V^* = \overset{0}{\wedge}V^* = K$. For $p = 1$, we have $\overset{1}{S}V^* = \overset{1}{\wedge}V^* = V^*$.

Problem 1. φ is symmetric [alternating] if and only if, for any $u_1, \ldots, u_p \in V$ and $1 \leqq i < k \leqq p$, $\varphi(\ldots, u_i, \ldots, u_k, \ldots) = \varphi(\ldots, u_k, \ldots, u_i, \ldots)$ [$\varphi(\ldots, u_i, \ldots, u_k, \ldots) = -\varphi(\ldots, u_k, \ldots, u_i, \ldots)$] holds. φ is alternating if and only if, for any $(u_1, \ldots, u_p)$ such that $u_i = u_j$ for some $i \neq j$, $\varphi(u_1, \ldots, u_p) = 0$. Prove these statements. Show also that $\overset{p}{\wedge}V^* = (0)$ if p is larger than the dimension n of V, and that $\overset{n}{\wedge}V^*$ is 1-dimensional.

We define two linear transformations S_p and A_p of the vector space $\overset{p}{\otimes}V^*$ by

$$S_p\varphi = \sum_{\sigma \in \mathfrak{S}_p} \sigma\varphi, \qquad A_p\varphi = \sum_{\sigma \in \mathfrak{S}_p} \varepsilon_\sigma\sigma\varphi.$$

S_p and A_p are called, respectively, the *symmetrization operator* and the *alternation operator*.

For an arbitrary φ, $S_p\varphi$ is symmetric and $A_p\varphi$ is alternating. In fact, for $\tau \in \mathfrak{S}_p$, we have $\tau(S_p\varphi) = \sum_{\sigma \in \mathfrak{S}_p} \tau\sigma\varphi$. However, since $\mathfrak{S}_p$ is a group, $\sigma \to \tau\sigma$ is a 1:1 map from $\mathfrak{S}_p$ onto $\mathfrak{S}_p$, and hence $\sum_{\sigma \in \mathfrak{S}_p} \tau\sigma\varphi = \sum_{\sigma \in \mathfrak{S}_p} \sigma\varphi$. So $\tau(S_p\varphi) = S_p\varphi$, and thus $S_p\varphi$ is symmetric. Using $\varepsilon_{\sigma\tau} = \varepsilon_\sigma\varepsilon_\tau$, we can prove in a similar manner that $A_p\varphi$ is alternating. We also get

$$S_p^2\varphi = p!S_p\varphi, \qquad A_p^2\varphi = p!A_p\varphi. \tag{2}$$

Problem 2. Show that φ is symmetric [alternating] if and only if $S_p\varphi = p!\varphi$ [$A_p\varphi = p!\varphi$].

Let N^p denote the subspace of $\overset{p}{\otimes}V^*$ formed by all the φ such that $A_p\varphi = 0$. We shall prove that

$$\overset{p}{\otimes}V^* = \overset{p}{\wedge}V^* + N^p \quad \text{(direct sum).} \tag{3}$$

---

†That is, if σ is an even permutation, then $\varepsilon_\sigma = +1$, and if σ is an odd permutation, then $\varepsilon_\sigma = -1$.

For an arbitrary $f \in \overset{p}{\otimes} V^*$, we can write

$$f = \frac{1}{p!} A_p f + (f - \frac{1}{p!} A_p f).$$

From (2), we get $f - (1/p!) A_p f \in N^p$. Since $(1/p!) A_p f \in \overset{p}{\wedge} V^*$, we see that $\overset{p}{\otimes} V^*$ is the sum of the subspaces $\overset{p}{\wedge} V^*$ and N^p. If $f \in (\overset{p}{\wedge} V^*) \cap N^p$, then, since $f \in N^p$, $A_p f = 0$. On the other hand, since f is alternating, $A_p f = p! f$. Hence $f = 0$, that is, $(\overset{p}{\wedge} V^*) \cap N^p = (0)$, and so we get (3).

Now let $f \in N^p$ and $g \in \overset{q}{\otimes} V^*$. We shall prove that $f \otimes g \in N^{p+q}$. Let $u_1, \ldots, u_{p+q} \in V$. Then we have

$$(A_{p+q}(f \otimes g))(u_1, \ldots, u_{p+q})$$
$$= \sum_{\sigma \in \mathfrak{S}_{p+q}} \varepsilon_\sigma f(u_{\sigma(1)}, \ldots, u_{\sigma(p)}) g(u_{\sigma(p+1)}, \ldots, u_{\sigma(p+q)}).$$

The permutations $\sigma \in \mathfrak{S}_{p+q}$ which fix each of the letters $p+1, \ldots, p+q$ form a subgroup $\mathfrak{Z}$ of $\mathfrak{S}_{p+q}$. Let

$$\mathfrak{S}_{p+q} = \sigma_1 \mathfrak{Z} + \ldots + \sigma_k \mathfrak{Z}$$

be the coset decomposition of $\mathfrak{S}_{p+q}$ with respect to $\mathfrak{Z}$. The sum on the right side of the equality above can be written as

$$\sum_{s=1}^{k} (\sum_{\tau \in \mathfrak{Z}} \varepsilon_{\sigma_s \tau} f(u_{\sigma_s(\tau(1))}, \ldots, u_{\sigma_s(\tau(p))}) g(u_{\sigma_s(p+1)}, \ldots, u_{\sigma_s(p+q)})).$$

Fix s and let $v_i = u_{\sigma_s(i)}$ $(i = 1, \ldots, p)$. Then, since $v_{\tau(i)} = u_{\sigma_s(\tau(i))}$, the expression in the parentheses above is equal to

$$\varepsilon_{\sigma_s} g(u_{\sigma_s(p+1)}, \ldots, u_{\sigma_s(p+q)})(\sum_{\tau \in \mathfrak{Z}} \varepsilon_\tau f(v_{\tau(1)}, \ldots, v_{\tau(p)})).$$

The elements of $\mathfrak{Z}$ permute the set $\{1, \ldots, p\}$, and actually run through all the permutations of $\{1, \ldots, p\}$. Hence the second factor above is equal to $(A_p f)(v_1, \ldots, v_p)$, and since $A_p f = 0$, it is equal to 0. Thus we get $A_{p+q}(f \otimes g) = 0$, so that $f \otimes g \in N^{p+q}$.

Similarly, we can prove that $f \otimes g \in N^{p+q}$ for $f \in \overset{p}{\otimes} V^*$ and $g \in N^q$. For $f \in \overset{p}{\wedge} V^*$ and $g \in \overset{q}{\wedge} V^*$, we define an element $f \wedge g$ of $\overset{p+q}{\wedge} V^*$ by

$$f \wedge g = \frac{1}{p! q!} A_{p+q}(f \otimes g), \dagger \tag{4}$$

and call it the *exterior product of* f and g. If $f, f_1, f_2 \in \overset{p}{\wedge} V^*$, $g, g_1, g_2 \in$

---

†In many books the exterior product is defined by $f \wedge g = [1/(p+q)!] A_{p+q}(f \otimes g)$. We adopt (4) for our definition in order to avoid some cumbersome constants that would otherwise appear in the formulas we shall prove later.

$\overset{p}{\wedge} V^*$, and λ_1, $\lambda_2 \in K$, then we have

$$(\lambda_1 f_1 + \lambda_2 f_2) \wedge g = \lambda_1(f \wedge g) + \lambda_2(f_2 \wedge g)$$

$$f \wedge (\lambda_1 g_1 + \lambda_2 g_2) = \lambda_1(f \wedge g_1) + \lambda_2(f \wedge g_2). \qquad (5)$$

This is clear from the definition (4) of the exterior product.

Let us prove that

$$(f \wedge g) \wedge h = f \wedge (g \wedge h) \qquad (6)$$

holds, for $f \in \overset{p}{\wedge} V^*, g \in \overset{q}{\wedge} V^*, h \in \overset{r}{\wedge} V^*$. For this, let $a(f)$ be the $\overset{p}{\wedge} V^*$-component, $n(f)$ the N^p-component, of an element f of $\overset{p}{\otimes} V^*$ with respect to the direct sum decomposition (3) of $\overset{p}{\otimes} V^*$. We have

$$a(f) = \frac{1}{p!} A_p(f), \qquad n(f) = f - a(f).$$

We shall also write $f \square g$ for $a(f \otimes g)$. Then

$$f \square g = \frac{1}{(p + q)!} A_{p+q}(f \otimes g).$$

Hence we have

$$f \wedge g = \frac{(p + q)!}{p!q!} f \square g. \qquad (7)$$

We have $f \otimes g = f \square g + n(f \otimes g)$, and since we have proved $n(f \otimes g) \otimes h \in N^{p+q+r}$, we get $a((f \otimes g) \otimes h) = a((f \square g) \otimes h) = (f \square g) \square h$. Similarly, we get $a(f \otimes (g \otimes h)) = f \square (g \square h)$. However, since $(f \otimes g) \otimes h = f \otimes (g \otimes h)$, we conclude that $(f \square g) \square h = f \square (g \square h)$. From this, using (7), we easily obtain (6).

If $p = 0$ or $q = 0$, then f or g is an element of K, and by the definition (4) of the exterior product, we have

$$f \wedge g = fg.$$

Let

$$\wedge V^* = \sum_{p=0}^{n} \overset{p}{\wedge} V^*.\dagger$$

If we define the exterior product of two elements $\sum_p f_p$ and $\sum_q g_q$ of $\wedge V^*$ by

$$\left(\sum_p f_p \right) \wedge \left(\sum_q g_q \right) = \sum_r \left(\sum_{p+q=r} f_p \wedge g_q \right),$$

$\dagger$We remind the reader that $\overset{p}{\wedge} V^* = (0)$ for $p > n$ (Problem 1).

then $\wedge V^*$ becomes an associative algebra over K with respect to the exterior product. This associative algebra $\wedge V^*$ is called the *exterior algebra* of the vector space V^*.

Let $\psi_1, \ldots, \psi_p$ be p elements of $V^* = \overset{1}{\wedge} V^*$, and let $u_1, \ldots, u_p \in V$. We shall prove, by induction on p, that

$$(\psi_1 \wedge \ldots \wedge \psi_p)(u_1, \ldots, u_p) = \begin{vmatrix} \psi_1(u_1) & \psi_1(u_2) & \ldots & \psi_1(u_p) \\ \psi_2(u_1) & \psi_2(u_2) & \ldots & \psi_2(u_p) \\ \cdot & & & \\ \cdot & & & \\ \cdot & & & \\ \psi_p(u_1) & \psi_p(u_2) & \ldots & \psi_p(u_p) \end{vmatrix}. \quad (8)$$

For $p = 1$, this is clear, so suppose (8) holds for exterior products of $p - 1$ elements of V^*. By (4), the left member of (8) is equal to

$$\frac{1}{(p-1)!} A_p((\psi_1 \wedge \ldots \wedge \psi_{p-1}) \otimes \psi_p)(u_1, \ldots, u_p)$$

$$= \frac{1}{(p-1)!} \sum_{\sigma \in \mathfrak{S}_p} \varepsilon_\sigma (\psi_1 \wedge \cdots \wedge \psi_{p-1})(u_{\sigma(1)}, \ldots, u_{\sigma(p-1)}) \psi_p(u_{\sigma(p)}).$$

$$(*)$$

From the inductive hypothesis we have

$$(\psi_1 \wedge \ldots \wedge \psi_{p-1})(u_{\sigma(1)}, \ldots, u_{\sigma(p-1)}) = \det(\psi_i(u_{\sigma(j)}))_{i, j = 1, \ldots, p-1}.$$

If we identify $\mathfrak{S}_{p-1}$ with the subgroup $\mathfrak{Z}$ of $\mathfrak{S}_p$ consisting of those permutations that leave p fixed, then, using the definition of the determinant, the above becomes

$$(\psi_1 \wedge \ldots \wedge \psi_{p-1})(u_{\sigma(1)}, \ldots, u_{\sigma(p-1)}) = \sum_{\tau \in \mathfrak{Z}} \varepsilon_\tau \psi_1(u_{\sigma(\tau(1))}) \cdots$$
$$\psi_{p-1}(u_{\sigma(\tau(p-1))}).$$

Hence the right member of (*) is equal to

$$\frac{1}{(p-1)!} \sum_{\tau \in \mathfrak{Z}} \{ \sum_{\sigma \in \mathfrak{S}_p} \varepsilon_{\sigma\tau} \psi_1(u_{\sigma\tau(1)}) \cdots \psi_{p-1}(u_{\sigma\tau(p-1)}) \psi_p(u_{\sigma\tau(p)}).$$

If we fix τ, then $\sigma \to \sigma\tau$ is a $1:1$ map from $\mathfrak{S}_p$ onto $\mathfrak{S}_p$. Hence the expression between the parentheses above is equal to $\det (\psi_i(u_j))_{i, j = 1, \ldots, n}$. However the order of $\mathfrak{Z}$ is $(p-1)!$, so that the expression above is equal to $\det(\psi_i(u_j))_{i, j = 1, \ldots, n}$. Hence we have proved that (8) holds for $\psi_1 \wedge \ldots \wedge \psi_p$.

If $\sigma \in \mathfrak{S}_p$, from (8) we get $(\psi_{\sigma(1)} \wedge \ldots \wedge \psi_{\sigma(p)})(u_1, \ldots, u_p) = \varepsilon_\sigma(\psi_1 \wedge \ldots \wedge \psi_p)(u_1, \ldots, u_p)$. Since this holds for arbitrary $u_1, \ldots, u_p$, we have

$$\psi_{\sigma(1)} \wedge \ldots \wedge \psi_{\sigma(p)} = \varepsilon_\sigma \psi_1 \wedge \ldots \wedge \psi_p. \tag{9}$$

We also have

$$\psi_1 \wedge \ldots \wedge \psi_p = \sum_{\sigma \in \mathfrak{S}_p} \varepsilon_\sigma \psi_{\sigma(1)} \otimes \ldots \otimes \psi_{\sigma(p)}. \tag{10}$$

In fact, we have

$$(\sum_{\sigma \in \mathfrak{S}_p} \varepsilon_\sigma \psi_{\sigma(1)} \otimes \ldots \otimes \psi_{\sigma(p)})(u_1, \ldots, u_p) = \sum_{\sigma \in \mathfrak{S}_p} \varepsilon_\sigma \psi_{\sigma(1)}(u_1) \ldots$$
$$\psi_{\sigma(p)}(u_p),$$

and this is equal to $\det(\psi_i(u_j))_{i,\,j=1,\ldots,n}$, by the definition of a determinant. Hence, by (8), the two members of (10) take the same values for any $(u_1, \ldots, u_p)$, so that they are equal, and (10) holds.

As before, let $\{\varphi^1, \ldots, \varphi^n\}$ be a base of V^*, and let $\{e_1, \ldots, e_n\}$ be the base of V dual to $\{\varphi^1, \ldots, \varphi^n\}$. If $f \in \overset{p}{\wedge} V^*$, then naturally $f \in \overset{p}{\otimes} V^*$, so that we can write

$$f = \sum_{i_1,\ldots,i_p=1}^{n} f_{i_1 \ldots i_p} \varphi^{i_1} \otimes \ldots \otimes \varphi^{i_p}.$$

Here we have $f_{i_1 \ldots i_p} = f(e_{i_1}, \ldots, e_{i_p})$. Since f is alternating, $f_{i_{\sigma(1)} \ldots i_{\sigma(p)}} = \varepsilon_\sigma f_{i_1 \ldots i_p}$ for $\sigma \in \mathfrak{S}_p$. Let $I = (i_1, \ldots, i_p)$ denote an ordered set of indices satisfying $i_1 < \ldots < i_p$ $(1 \leq i_k \leq n)$. We shall denote the set of such I's by $\mathfrak{I}_p$. The set $\mathfrak{I}_p$ has $\binom{n}{p}$ elements. Then f can be written as

$$f = \sum_{(i_1,\ldots,i_p) \in \mathfrak{I}_p} (\sum_{\sigma \in \mathfrak{S}_p} f_{i_{\sigma(1)} \ldots i_{\sigma(p)}} \varphi^{i_{\sigma(1)}} \otimes \cdots \otimes \varphi^{i_{\sigma(p)}}).$$

Since $f_{i_{\sigma(1)} \ldots i_{\sigma(p)}} = \varepsilon_\sigma f_{i_1 \ldots i_p}$, the part inside the parenthesis above is equal to $f_{i_1 \ldots i_p} \varphi^{i_1} \wedge \cdots \wedge \varphi^{i_p}$ by (10). Hence, an arbitrary $f \in \overset{p}{\wedge} V^*$ can be written as

$$f = \sum_{(i_1,\ldots,i_p) \in \mathfrak{I}_p} f_{i_1 \ldots i_p} \varphi^{i_1} \wedge \cdots \wedge \varphi^{i_p}. \tag{11}$$

The $\binom{n}{p}$ elements $\varphi^{i_1} \wedge \cdots \wedge \varphi^{i_p}$ $((i_1, \ldots, i_p) \in \mathfrak{I}_p)$ of $\overset{p}{\wedge} V^*$ are linearly independent. In fact, if we let

$$\sum_{(i_1,\ldots,i_p) \in \mathfrak{I}_p} \lambda_{i_1 \ldots i_p} \varphi^{i_1} \wedge \cdots \wedge \varphi^{i_p} = 0,$$

then, by (10), we have

$$\sum_{(i_1,\ldots,i_p)\in \mathfrak{I}_p} \sum_{\sigma\in\mathfrak{S}_p} \varepsilon_\sigma \lambda_{i_1\cdots i_p} \, \varphi^{i_{\sigma(1)}} \otimes \cdots \otimes \varphi^{i_{\sigma(p)}} = 0.$$

Since $\varphi^{j_1} \otimes \ldots \otimes \varphi^{j_p}$ $(1 \leq j_k \leq n)$ are linearly independent, we get $\lambda_{i_1\cdots i_p} = 0$.

By the argument above, we have shown that the $\binom{n}{p}$ elements of $\overset{p}{\wedge} V^*$

$$\varphi^{i_1} \wedge \cdots \wedge \varphi^{i_p} \qquad (i_1 < \cdots < i_p)$$

form a basis of $\overset{p}{\wedge} V^*$. In particular, the dimension of $\overset{p}{\wedge} V^*$ is $\binom{n}{p}$, and

hence the dimension of $\wedge V^* = \sum_{p=0}^{n} \overset{p}{\wedge} V^*$ is $\sum_{p=0}^{n} \binom{n}{p} = 2^n$. So the exterior algebra is a finite dimensional associative algebra.

Let $\psi_1, \ldots, \psi_p, \theta_1, \ldots, \theta_q \in V^*$. By (9), we have $\psi_k \wedge \theta_1 \wedge \cdots \wedge \theta_q = (-1)^q \theta_1 \wedge \cdots \wedge \theta_q \wedge \psi_k$. The repeated use of this equality yields $(\psi_1 \wedge \cdots \wedge \psi_p) \wedge (\theta_1 \wedge \cdots \wedge \theta_q) = (-1)^{pq}(\theta_1 \wedge \cdots \wedge \theta_q) \wedge (\psi_1 \wedge \cdots \wedge \psi_p)$. Now let $f \in \overset{p}{\wedge} V^*$, $g \in \overset{q}{\wedge} V^*$, and express f and g as linear combinations of bases $\{\varphi^{i_1} \wedge \cdots \wedge \varphi^{i_p}\}$ of $\overset{p}{\wedge} V^*$ and $\{\varphi^{j_1} \wedge \cdots \wedge \varphi^{j_q}\}$ of $\overset{q}{\wedge} V^*$, respectively. Then, by the equality we just proved, we get

$$f \wedge g = (-1)^{pq} g \wedge f \quad (f \in \overset{p}{\wedge} V^*, g \in \overset{q}{\wedge} V^*). \tag{12}$$

In particular, we see that

$$\varphi \wedge \psi = -\psi \wedge \varphi, \quad \varphi \wedge \varphi = 0 \quad (\varphi, \psi \in V^*).$$

Problem 3. Let $f \in \overset{p}{\wedge} V^*$. If p is even, then f commutes with an arbitrary element of $\wedge V^*$. If p is odd, then $f \wedge f = 0$. Show this.

Problem 4. Let $\psi_1, \ldots, \psi_n \in V^*$ and set $\theta_i = \sum_j a_{ij}\psi_j$ $(i = 1, \ldots, n)$. Show that $\theta_1 \wedge \cdots \wedge \theta_n = \det(a_{ij})\psi_1 \wedge \cdots \wedge \psi_n$.

Problem 5. For n elements $\psi_1, \ldots, \psi_n$ of V^* to be a base of V^*, show that a necessary and sufficient condition is that $\psi_1 \wedge \cdots \wedge \psi_n \neq 0$.

Now, for $v \in V$ and $\varphi \in \overset{p}{\otimes} V^*$, we define $i(v)\varphi \in \overset{p-1}{\otimes} V^*$ by

$$i(v)\varphi(v_1, \ldots, v_{p-1}) = \varphi(v, v_1, \ldots, v_{p-1}).$$

This $i(v)\varphi$ is called the *interior product* of v and φ. If φ is alternating, then so is $i(v)\varphi$.

Let $\psi_1, \ldots, \psi_p$ be elements of V^* and $u_1, \ldots, u_p$ elements of V.

Then we have

$$(i(u_1)(\psi_1 \wedge \cdots \wedge \psi_p))(u_2, \ldots, u_p) = (\psi_1 \wedge \cdots \wedge \psi_p)(u_1, \ldots, u_p)$$

$$= \begin{vmatrix} \psi_1(u_1) & \psi_1(u_2) & \cdots & \psi_1(u_p) \\ \psi_2(u_1) & \psi_2(u_2) & \cdots & \psi_2(u_p) \\ \cdot & & & \\ \cdot & & & \\ \cdot & & & \\ \psi_p(u_1) & \psi_p(u_2) & \cdots & \psi_p(u_p) \end{vmatrix}$$

If we expand this determinant with respect to the first column, then we get $\sum_{i=1}^{p} (-1)^{i+1} \psi_i(u_1)(\psi_1 \wedge \cdots \wedge \psi_{i-1} \wedge \psi_{i+1} \cdots \wedge \psi_p)(u_2, \ldots, u_p)$. Since $\psi_i(u_1) = i(u_1)\psi_i$, and since $u_2, \ldots, u_p$ are arbitrary elements of V, setting $v = u_1$ yields

$$i(v)(\psi_1 \wedge \cdots \wedge \psi_p) = \sum_{i=1}^{p} (-1)^{i+1} i(v)\psi_i(\psi_1 \wedge \cdots \wedge \psi_{i-1} \wedge \psi_{i+1} \cdots \wedge \psi_p). \quad (13)$$

Let $\theta_1, \ldots, \theta_q \in V^*$, and set $\theta_1 \wedge \cdots \wedge \theta_q = \theta$, $\psi_1 \wedge \cdots \wedge \psi_p = \psi$. From (13), we have

$$i(v)(\psi \wedge \theta)$$

$$= \sum_{i=1}^{p} (-1)^{i+1} i(v)\psi_i(\psi_1 \wedge \cdots \wedge \psi_{i-1} \wedge \psi_{i+1} \wedge \cdots \wedge \psi_p \wedge \theta)$$

$$+ \sum_{i=1}^{p} (-1)^{p+j+1} i(v)\theta_j(\psi \wedge \theta_1 \wedge \cdots \wedge \theta_{j-1} \wedge \theta_{j+1} \wedge \cdots \wedge \theta_q)$$

$$= (i(v)\psi) \wedge \theta + (-1)^p \psi \wedge (i(v)\theta).$$

Since the elements of $\overset{p}{\wedge} V^*$ and $\overset{q}{\wedge} V^*$ can be expressed as linear combinations of $\varphi^{i_1} \wedge \cdots \wedge \varphi^{i_p}$ and $\varphi^{j_1} \wedge \cdots \wedge \varphi^{j_q}$, respectively, it follows from the formula we just proved that for $\psi = \overset{p}{\wedge} V^*$, $\theta \in \overset{q}{\wedge} V^*$,

$$i(v)(\psi \wedge \theta) = (i(v)\psi) \wedge \theta + (-1)^p \psi \wedge (i(v)\theta). \quad (14)$$

[Addendum] *The symmetric product and the symmetric algebra.* Similar to the case of alternating tensors, for symmetric tensors we can also define products in the following manner. Let M^p be the subspace of $\overset{p}{\otimes} V^*$ formed by the elements φ of $\overset{p}{\otimes} V^*$ satisfying $S_p\varphi = 0$. If $f \in \overset{p}{\otimes} V^*$, $g \in \overset{q}{\otimes} V^*$, and if f belongs to M^p or g belongs to M^q, then we can prove, as before, that $f \otimes g \in M^{p+q}$. For $f \in \overset{p}{S} V^*$, $g \in \overset{q}{S} V^*$, we define an element

$f \cdot g$ of $\overset{p+q}{S}{}^q V^*$ by

$$f \cdot g = \frac{1}{(p+q)!} S_{p+q}(f \otimes g).$$

Then we have

$$f \cdot g = g \cdot f, \qquad (f \cdot g) \cdot h = f \cdot (g \cdot h)$$
$$(\lambda_1 f_1 + \lambda_2 f_2) \cdot g = \lambda_1 (f_1 \cdot g) + \lambda_2 (f_2 \cdot g),$$
$$f \cdot (\lambda_1 g_1 + \lambda_2 g_2) = \lambda_1 (f \cdot g_1) + \lambda_2 (f \cdot g_2).$$

The element $f \cdot g$ is called the *symmetric product* of the symmetric tensors f and g. Defining products in $SV^* = \sum_{p=0}^{\infty} \overset{p}{S} V^*$ as we did in the previous case, SV^* becomes a commutative associative algebra. This associative algebra is called the symmetric algebra of the vector space V^*. If $\{\varphi^1, \ldots, \varphi^n\}$ is a base of V^*, then $(\varphi^1)^{p_1} \ldots (\varphi^n)^{p_n}$ (where $(\varphi)^q$ is the symmetric product of q φ's) $(p_1 + \ldots + p_n = p)$ form a base of $\overset{p}{S} V^*$. Moreover, by letting an indeterminate x_i correspond to φ^i, the polynomial ring $K[x_1, \ldots, x_n]$ and SV^* become isomorphic. For an element $f = \sum a_{p_1 \cdots p_n}(\varphi^1)^{p_1} \ldots (\varphi^n)^{p_n}$ of SV^* and an element u of V, we let $f(u) = \sum a_{p_1 \cdots p_n}(\varphi^1(u))^{p_1} \ldots (\varphi^n(u))^{p_n}$. Thus we consider f as a function on V, and call it a polynomial function on V.

§3. Covariant Tensor Fields on a Manifold and Differential Forms

Let M be an n-dimensional manifold. To each point a of M, assign a covariant tensor t_a on $T_a(M)$ of order p, $t_a \in \overset{p}{\otimes} T_a^*(M)$. The map $t : a \to t_a$ is called a *covariant tensor field* of order p on M.†

If t, s are both covariant tensor fields of order p, then, for each point a, by setting

$$(t + s)_a = t_a + s_a,$$

a tensor field $t + s$ of order p is defined. We call $t + s$ the *sum* of t and s. If f is a scalar field,* by setting

$$(ft)_a = f(a)t_a,$$

a covariant tensor field ft of order p is defined. It is easy to show that with respect to these operations, the set $\mathfrak{T}_p$ of all covariant tensors of order p forms a $C^\infty(M)$-module(II, §11).

†In particular, if $p = 0$, then $\overset{0}{\otimes} T_a^* = \mathbf{R}$, so t is a function on M. In this case, t is sometimes called a *scalar field*. If $p = 1$, then t is called a *covariant vector field* or a *differential form* of order 1.

Let t and s be covariant tensor fields of order p and q, respectively. By defining at each point a

$$(t \otimes s)_a = t_a \otimes s_a,$$

we get a covariant tensor field $t \otimes s$ of order $p + q$. We call $t \otimes s$ the *tensor product* of the tensor fields t and s.

If $p = 0$, then $t \otimes s$ is nothing but the product ts of the scalar field t and s. We have the following relations between sums and tensor products of covariant tensor fields:

$$(ft) \otimes s = t \otimes fs = f(t \otimes s) \qquad (f \text{ is a scalar field})$$
$$(t_1 + t_2) \otimes s = t_1 \otimes s + t_2 \otimes s, \qquad t \otimes (s_1 + s_2) = t \otimes s_1 + t \otimes s_2.$$

If U is an open set of M, then the restriction of t to U is a tensor field on the open submanifold U. If $(x^1, \ldots, x^n)$ is a local coordinate system on the open set U, then $\{(dx^1)_a, \ldots, (dx^n)_a\}$ is a base of $T_a{}^*(M)$. Hence the covariant tensor field t of order p can be written uniquely on U as

$$t = \sum_{i_1, \ldots, i_p = 1}^{n} t_{i_1 \cdots i_p}\, dx^{i_1} \otimes \cdots \otimes dx^{i_p},$$

where $dx^{i_1} \otimes \cdots \otimes dx^{i_p}$ is the tensor field on U defined by $a \to (dx^{i_1})_a \otimes \cdots \otimes (dx^{i_p})_a$, and $t_{i_1 \cdots i_p}$ is a function on U. $\{t_{i_1 \cdots i_p}\}$ is called the set of *components* of t with respect to the local coordinate system $(x^1, \ldots, x^n)$.

Problem 1. If $\{t_{i_1 \cdots i_p}\}$ and $\{s_{j_1 \cdots j_q}\}$ are the components of t and s, repectively, show that the components of $t \otimes s$ are $(t \otimes s)_{i_1 \cdots i_p j_1 \cdots j_q} = t_{i_1 \cdots i_p} s_{j_1 \cdots j_q}$.

If $(\bar{x}^1, \ldots, \bar{x}^n)$ is another local coordinate system on U, and if the components of t with respect to $(\bar{x}^1, \ldots, \bar{x}^n)$ are $\{\bar{t}_{i_1 \cdots i_p}\}$, then we have

$$t_{i_1 \cdots i_p} = \sum_{j_1, \ldots, j_p = 1}^{n} \frac{\partial \bar{x}^{j_1}}{\partial x^{i_1}} \cdots \frac{\partial \bar{x}^{j_p}}{\partial x^{i_p}} \bar{t}_{j_1 \cdots j_p},$$

$$\bar{t}_{i_1 \cdots i_p} = \sum_{j_1, \ldots, j_p = 1}^{n} \frac{\partial x^{j_1}}{\partial \bar{x}^{i_1}} \cdots \frac{\partial x^{j_p}}{\partial \bar{x}^{i_p}} t_{j_1 \cdots j_p}.$$

$$(1)$$

Hence the property that the components of t are of class C^r (or continuous) does not depend on the choice of the local coordinate system. The tensor field t is said to be of class C^r (or continuous) if all of the components of t are of class C^r (or continuous). If t and s are both of class C^r, then $t + s$ and $t \otimes s$ are also of class C^r.

If, at each point a, t_a is symmetric, then we call t a *symmetric covariant tensor field*, and if, at each point a, t_a is alternating, then we call t an *alternating covariant tensor field*. An alternating covariant tensor field of order p is called a *differential form* of order p, or simply a *p-form*. In particular, a differential form of order 1, i.e., a covariant vector field, is sometimes called a *Pfaffian*. If ω is a differential form of order p, and if ω is of class C^r (or continuous) as a covariant tensor field of order p, then we say that ω is of class C^r (or continuous).

Example 1. A Riemannian metric is a symmetric covariant tensor field of order 2.

Example 2. If f is a C^r function on M, then $df : a \to (df)_a$ is a Pfaffian of class C^{r-1}.

Example 3. Let g be a C^∞ Riemannian metric on M, and let X be a vector field on M. For each point a, set $\omega_a(v) = g_a(X_a, v)$, where $v \in T_a(M)$. Then $\omega_a \in T_a^*(M)$, and $\omega : a \to \omega_a$ is a C^∞ Pfaffian on M. Conversely, if ω is a C^∞ Pfaffian on M, then, at each point a, there is a unique $X_a \in T_a(M)$ such that $\omega_a(v) = g_a(X_a, v)$ for all $v \in T_a(M)$, and the map $a \to X_a$ is a (C^∞) vector field on M. Hence the vector fields X and the Pfaffians ω are in $1:1$ correspondence on a Riemannian manifold M. If $\{\xi^i\}$ are the components of X and $\{\alpha_i\}$ are the components of ω, then we have $\alpha_i = \sum_j g_{ij}\xi^j$.

If f is a C^∞ function, the vector field corresponding to the Pfaffian df is called the *gradient* of the scalar field f and is denoted by grad f.

Now, if ω and θ are differential forms of order p and q, respectively, then by letting

$$(\omega \wedge \theta)_a = \omega_a \wedge \theta_a$$

at each point a of M, we can define a differential form $\omega \wedge \theta$ of order $p + q$. We call $\omega \wedge \theta$ the *exterior product* of ω and θ.

If $(x^1, \ldots, x^n)$ is a local coordinate system on U, then ω can be written on U uniquely as

$$
\begin{aligned}
\omega &= \sum_{i_1 < \cdots < i_p} \alpha_{i_1 \cdots i_p} \, dx^{i_1} \wedge \cdots \wedge dx^{i_p} \\
&= \frac{1}{p!} \sum_{i_1, \ldots, i_p = 1}^{n} \alpha_{i_1 \cdots i_p} \, dx^{i_1} \wedge \cdots \wedge dx^{i_p}
\end{aligned}
\tag{2}
$$

Here, $\{\alpha_{i_1 \cdots i_p}\}$ are the components of α.

If ω and θ are differential forms of order p and q, respectively, then, by (12) of §2, we have

$$\omega \wedge \theta = (-1)^{pq}\theta \wedge \omega. \tag{3}$$

Problem 2.   If ω and θ are both of class C^r, then show that $\omega \wedge \theta$ is also of class C^r.

From now on, we shall only consider C^∞ tensor fields and C^∞ differential forms unless otherwise stated. We shall call C^∞ tensor fields and C^∞ defferential forms simply tensor fields and differential forms.

Let X be a vector field on M, and let t be a covariant tensor field of order $p(\geqq 1)$. For each point a of M, set

$$(i(X)t)_a = i(X_a)t_a.$$

The map $a \to (i(X)t)_t$ defines a covariant tensor field $i(X)t$ of order $p - 1$ on M. $i(X)t$ is called the *interior product* of the vector field X and the covariant tensor field t of order $p(\geqq 1)$. If ω is a differential form of order p, then $i(X)\omega$ is a differential form of order $p - 1$.

If ω and θ are differential forms of order p and q, respectively, then, by (14) of §2, we get

$$i(X)(\omega \wedge \theta) = (i(X)\omega) \wedge \theta + (-1)^p\omega \wedge (i(X)\theta). \tag{4}$$

As before, we denote the Lie algebra formed by the set of all vector fields on M by $\mathfrak{X}(M)$, and the associative algebra formed by the set of all C^∞ functions on M by $C^\infty(M)$. As we have said in II, §11, $\mathfrak{X}(M)$ is a $C^\infty(M)$-module. If T is a map from the direct product set $\mathfrak{X}(M)^p = \mathfrak{X}(M) \times \cdots \times \mathfrak{X}(M)$ of p copies of $\mathfrak{X}(M)$ to $C^\infty(M)$ with the property that

$$T(X_1, \ldots, fX_i + gY_i, \ldots, X_p) = fT(X_1, \ldots, X_i, \ldots, X_p)$$
$$+ gT(X_1, \ldots, Y_i, \ldots, X_p)$$
$$\text{for} \quad f, g \in C^\infty(M), \quad i = 1, \ldots, p,$$

then T is called a p-linear function on the $C^\infty(M)$-module $\mathfrak{X}(M)$. As in §2, we define symmetric and alternating p-linear functions on $\mathfrak{X}(M)$.

Let t be a covariant tensor field of order p on M, and let $X_1, \ldots, X_p \in \mathfrak{X}(M)$. At each point a of M, set

$$F(a) = t_a((X_1)_a, \ldots, (X_p)_a).$$

Then F is a C^∞ function on M. This function F is denoted by $t(X_1, \ldots, X_p)$. That is, $t(X_1, \ldots, X_p)$ is a C^∞ function defined by

$$t(X_1, \ldots, X_p)(a) = t_a((X_1)_a, \ldots, (X_p)_a).$$

Problem 3. Verify that $t(X_1, \ldots, X_p)$ is, in fact, of class C^∞.

Since

$$t(X_1, \ldots, fX_i + gY_i, \ldots, X_p)(a)$$
$$= t_a((X_1)_a, \ldots, f(a)(X_i)_a + g(a)(Y_i)_a, \ldots, (X_p)_a)$$
$$= f(a)t(X_1, \ldots, X_i, \ldots, X_p)(a) + g(a)t(X_1, \ldots, Y_i, \ldots, X_p)(a),$$

we have

$$t(X_1, \ldots, fX_i + gY_i, \ldots, X_p)$$
$$= ft(X_1, \ldots, X_i, \ldots, X_p) + gt(X_1, \ldots, Y_i, \ldots, X_p).$$

Hence $(X_1, \ldots, X_p) \to t(X_1, \ldots, X_p)$ is a p-linear function on $\mathfrak{X}(M)$. That is, for a given covariant tensor field t of order p, there is a corresponding p-linear function on $\mathfrak{X}(M)$. It is clear from the correspondence that if t is a symmetric or alternating tensor field, then the corresponding p-linear function on $\mathfrak{X}(M)$ is symmetric or alternating.

Conversely, let T be an arbitrary p-linear function on $\mathfrak{X}(M)$. Let us prove that there is a covariant tensor field t of order p corresponding to T, that is, $T(X_1, \ldots, X_p) = t(X_1, \ldots, X_p)$ holds for arbitrary $X_1, \ldots, X_p \in \mathfrak{X}(M)$.

For this we first prove that, if $X_i = 0$ on an open set U, then the function $T(X_1, \ldots, X_i, \ldots, X_p) \equiv 0$ on U. For simplicity, we let $i = 1$. Let a be a point of U, and choose a C^∞ function on M such that $f(a) = 0$ and $f \equiv 1$ on U^c, the complement of U (Lemma 1 of II, §10). Then we have $fX_1 = X_1$. (In fact, if $b \in U$, then $(fX_1)_b = f(b)(X_1)_b = 0$; if $b \in U^c$, then, since $f(b) = 1$, we have $(fX_1)_b = (X_1)_b$.) Hence $T(X_1, \ldots, X_p) = fT(X_1, \ldots, X_p)$. But $f(a) = 0$, so that the value of the function $T(X_1, \ldots, X_p)$ at a is 0. Since a is an arbitrary point of U, $T(X_1, \ldots, X_p)$ is 0 on U.

Next we shall show that if $(X_i)_a = 0$, then the value of $T(X_1, \ldots, X_p)$ at a is also 0. Let $(x^1, \ldots, x^n)$ be a local coordinate system on a neighborhood of a, and let Y_j ($j = 1, \ldots, n$) be a vector field on M which is equal to $\partial/\partial x^j$ on a sufficiently small neighborhood U of a. Let $X_i = \sum_j \xi^j \partial/\partial x^j$ in a neighborhood of a, and let f^j be a C^∞ function on M which coincides with ξ^j on U. Then $X_i - \sum_j f_j Y_j$ is 0 on U. Since we have assumed that $(X_i)_a = 0$, we have $f_j(a) = 0$ ($j = 1, \ldots, n$). From what we have proved above, we get $T(X_1, \ldots, X_p) = \sum_j f_j T(X_1, \ldots, Y_j, \ldots, X_p)$. But $f_j(a) = 0$ (all j) forces $T(X_1, \ldots, X_p)$ to be 0 at a.

From this we derive the following conclusion. If $(X_1, \ldots, X_p)$ and $(Y_1, \ldots, Y_p)$ are two p-tuples of vector fields such that $(X_i)_a = (Y_i)_a$

$(i = 1, \ldots, p)$, then $T(X_1, \ldots, X_p)$ and $T(Y_1, \ldots, Y_p)$ take the same value at a.

Now, for each point a of M, we define $t_a \in \overset{p}{\otimes} V^*$ as follows. For v_1, $\ldots, v_p \in T_a(M)$, there are vector fields $X_1, \ldots, X_p$ such that $(X_i)_a = v_i$ $(i = 1, \ldots, p)$, and the value of $T(X_1, \ldots, X_p)$ at a depends only on $v_1, \ldots, v_p$ and not on the particular choice of $X_1, \ldots, X_p$. So we let t_a be defined by $t_a(v_1, \ldots, v_p) = T(X_1, \ldots, X_p)(a)$. Then t_a is a p-linear function on $T_a(M)$. If we define a covariant tensor field t of order p by the correspondence $t : a \to t_a$, then t is of class C^∞. In fact, let $(x^1, \ldots, x^n)$ be a local coordinate system on a neighborhood of a, and let Y_j $(j = 1, \ldots, n)$ be vector fields on M coinciding with $\partial/\partial x^j$ on a neighborhood U of a, as before. At each point b of U,

$$t_{i_1 \cdots i_p}(b) = t_b((\partial/\partial x^{i_1})_b, \ldots, (\partial/\partial x^{i_p})_b)$$

is equal to the value of $T(Y_{i_1}, \ldots, Y_{i_p})$ at b. Hence $t_{i_1 \cdots i_p}$ is equal to the C^∞ function $T(Y_{i_1}, \ldots, Y_{i_p})$ on U.

Thus, for T we have found a corresponding covariant tensor field t of order p on M, and, from the definition of t, it is clear that $T(X_1, \ldots, X_p) = t(X_1, \ldots, X_p)$. It is also clear that if T is a symmetric or alternating p-form on $\mathfrak{X}(M)$, then the corresponding t is again symmetric or alternating.

We have proved the following theorem.

THEOREM   *A covariant tensor field of order p on a manifold M can be identified with a p-linear function on the $C^\infty(M)$-module $\mathfrak{X}(M)$. In particular, the differential forms of order p are identified with the alternating p-linear functions on the $C^\infty(M)$-module $\mathfrak{X}(M)$.*

If t and s are covariant tensor fields of order p and q, respectively, then we have

$$(t \otimes s)(X_1, \ldots, X_{p+q}) = t(X_1, \ldots, X_p)s(X_{p+1}, \ldots, X_{p+q}). \qquad (5)$$

If ω and θ are differential forms of order p and q, respectively, then we have

$$(\omega \wedge \theta)(X_1, \ldots, X_{p+q})$$
$$= \frac{1}{p!q!} \sum_{\sigma \in \mathfrak{S}_{p+q}} \varepsilon_\sigma \omega(X_{\sigma(1)}, \ldots, X_{\sigma(p)}) \theta(X_{\sigma(p+1)}, \ldots, X_{\sigma(p+q)}). \qquad (6)$$

Furthermore, if $\omega_1, \ldots, \omega_p$ are differential forms of order 1, then we have

$$(\omega_1 \wedge \cdots \wedge \omega_p)(X_1, \ldots, X_p) = \det(\omega_i(X_j))_{i, j = 1, \ldots, p}. \qquad (7)$$

§4. The Lie Differentiation of Tensor Fields and the Exterior Differentiation of Differential Forms

Let M be an n-dimensional manifold and t a covariant tensor field of order p.

A *The Lie derivative of t.* Let X be a vector field on M. For any p vector fields $X_1, \ldots, X_p$, we set

$$(L_X t)(X_1, \ldots, X_p) = X(t(X_1, \ldots, X_p)) - \sum_{k=1}^{p} t(X_1, \ldots, [X, X_k], \ldots, X_p). \quad (1)$$

Here $X(t(X_1, \ldots, X_p))$ is the derivative of the function $t(X_1, \ldots, X_p)$ by X.

It is easy to see that $L_X t$ is a p-linear function on the $C^\infty(M)$-module $\mathfrak{X}(M)$. Hence $L_X t$ is a covariant tensor field of order p on M. We call $L_X t$ the *Lie derivative* of the covariant tensor field t with respect to X. In what was said above, we have assumed $p > 0$. For $p = 0$, i.e., if t is a scalar field, we define $L_X t$ by

$$L_X t = Xt.$$

That is, using the notation in II, §11, we let $L_X = D_X$ for scalar fields.

The Lie differentiation satisfies the following:

$$L_X(t + s) = L_X t + L_X s \quad (2)$$
$$L_X(t \otimes s) = (L_X t) \otimes s + t \otimes (L_X s). \quad (3)$$

In particular, if f is a scalar field, then

$$L_X(ft) = (L_X f)t + f(L_X t).$$

If ω, θ are differential forms on M, then

$$L_X(\omega \wedge \theta) = (L_X \omega) \wedge \theta + \omega \wedge (L_X \theta). \quad (4)$$

Proof. (2) is clear from the definition of Lie differentiation. Let t be of order p and s be of order q. Then we have

$$(L_X(t \otimes s))(X_1, \ldots, X_{p+q}) = X((t \otimes s)(X_1, \ldots, X_{p+q}))$$
$$- \sum_{k=1}^{p+q} (t \otimes s)(X_1, \ldots, [X, X_k], \ldots, X_{p+q}).$$

Now

$$(t \otimes s)(X_1, \ldots, X_{p+q}) = t(X_1, \ldots, X_p)s(X_{p+1}, \ldots, X_{p+q}).$$

Hence the first term in the right member is equal to

$$X(t(X_1, \ldots, X_p))s(X_{p+1}, \ldots, X_{p+q})$$
$$+ t(X_1, \ldots, X_p)X(s(X_{p+1}, \ldots, X_{p+q})).$$

The second term is equal to

$$-(\sum_{k=1}^{p} t(X_1, \ldots, [X, X_k], \ldots, X_p)s(X_{p+1}, \ldots, X_{p+q})$$

$$- t(X_1, \ldots, X_p)(\sum_{k=p+1}^{p+q} s(X_{p+1}, \ldots, [X, X_k], \ldots, X_{p+q}))$$

From this, we deduce that $L_X(t \otimes s) = (L_X t) \otimes s + t \otimes (L_X s)$. Similarly, using (4) of §2, we can prove (4).

B   *The exterior differentiation of a differential form.* Let ω be a differential form of order $p(\geq 0)$. For any $p + 1$ vector fields $X_1, \ldots,$ X_{p+1}, we let, for $p > 0$,

$$(d\omega)(X_1, \ldots, X_{p+1})$$

$$= \sum_{i=1}^{p+1} (-1)^{i+1} X_i(\omega(X_1, \ldots, \hat{X}_i, \ldots, X_{p+1})$$

$$+ \sum_{i<j} (-1)^{i+j} \omega([X_i, X_j], X_1, \ldots, \hat{X}_i, \ldots, \hat{X}_j, \ldots, X_{p+1}), \quad (5)$$

and for $p = 0$, i.e., when ω is a scalar field, we let

$$(d\omega)(X) = X\omega. \tag{5'}$$

The caret $\wedge$ over a variable in (5) means that the particular variable under the caret should be deleted. (We shall be using this notation often from now on).

It can be proved, by a straightforward, computation, that $(X_1, \ldots,$ $X_{p+1}) \to (d\omega)(X_1, \ldots, X_{p+1})$ is an alternating $(p + 1)$-linear function on the $C^\infty(M)$-module $\mathfrak{X}(M)$. (Since the computation gets too long, we leave it as an exercise for the reader). Hence a differential form $d\omega$ of order $(p + 1)$ is defined by (5) and (5'). We call $d\omega$ the exterior derivative of ω.

If ω and θ are both of order p, then

$$d(\omega + \theta) = d\omega + d\theta$$

is clear from the definition of d. In particular, if ω is of order 1, then we note that

$$(d\omega)(X, Y) = X(\omega(Y)) - Y(\omega(X)) - \omega([X, Y]).$$

C Let us investigate the relation among the operators $i(X)$, L_X, and d that we have defined, and describe the properties of these operators. We shall first prove that

$$i([X, Y]) = L_X i(Y) - i(Y)L_X, \tag{6}$$

i.e., that $i([X, Y])t = L_X(i(Y)t) - i(Y)(L_X t)$ for any covariant tensor t of order $p(> 0)$. Let $X_1, \ldots, X_{p-1} \in \mathfrak{X}(M)$. Then we have

$$
\begin{aligned}
(L_X(i(Y)t))&(X_1, \ldots, X_{p-1}) \\
&= X((i(Y)t)(X_1, \ldots, X_{p-1})) \\
&\quad - \sum_{k=1}^{p-1} (i(Y)t)(X_1, \ldots, [X, X_k], \ldots, X_{p-1}) \\
&= X(t(Y, X_1, \ldots, X_{p-1})) - \sum_{k=1}^{p-1} t(Y, X_1, \ldots, [X, X_k], \ldots, X_{p-1}).
\end{aligned}
$$

The last member of this equation is equal to

$$
\begin{aligned}
(L_X t)&((Y, X_1, \ldots, X_{p-1}) + t([X, Y], X_1, \ldots, X_{p-1}) \\
&= (i(Y)L_X t)(X_1, \ldots, X_{p-1}) + (i([X, Y])t)(X_1, \ldots, X_{p-1}).
\end{aligned}
$$

Hence we have $L_X(i(Y)t) = i(Y)(L_X t) + i([X, Y])t$.

Next we shall prove that

$$L_{[X, Y]} = L_X L_Y - L_Y L_X, \tag{7}$$

i.e., that $L_{[X, Y]}t = L_X(L_Y t) - L_Y(L_X t)$. We prove this by induction on the order of t. If $p = 0$, then because $L_{[X, Y]}t = [X, Y]t = X(Yt) - Y(Xt)$, clearly (7) holds. Thus we assume that (7) holds for covariant tensor fields of order p, and assume that t is of order $p + 1$. For an arbitrary $Z \in \mathfrak{X}(M)$, $i(Z)t$ is of order p, so by the inductive hypothesis, $L_{[X, Y]}i(Z)t = (L_X L_Y i(Z) - L_Y L_X i(Z))t$. By (6), we have

$$L_{[X, Y]}i(Z) = i(Z)L_{[X, Y]} + i([[X, Y], Z]).$$

Again by (6), we have

$$
\begin{aligned}
L_X L_Y i(Z) &= L_X(i(Z)L_Y + i([Y, Z])) \\
&= i(Z)L_X L_Y + i([X, Z])L_Y + i([Y, Z])L_X + i([X, [Y, Z]]).
\end{aligned}
$$

Similarly we have

$$L_Y L_X i(Z) = i(Z)L_Y L_X + i([Y, Z])L_X + i([X, Z])L_Y + i([Y, [X, Z]]).$$

Hence we get

$$L_Y L_X i(Z) - L_X L_Y i(Z)$$
$$= i(Z)(L_X L_Y - L_Y L_X) + i([X, [Y, Z]]) - i([Y, [X, Z]]).$$

But, by the Jacobi identity, we have $[[X, Y], Z] = [X, [Y, Z]] - [Y, [X, Z]]$, and hence $i([X, Y], Z) = i([X, [Y, Z]]) - i([Y, [X, Z]])$. Hence we get

$$i(Z)(L_{[X, Y]}t - L_X L_Y t + L_Y L_X t) = 0$$

for an arbitrary $Z \in \mathfrak{X}(M)$. From this, we conclude that $L_{[X, Y]}t - L_X L_Y t + L_Y L_X t = 0$, and (7) holds for covariant tensor fields of order $p + 1$. Hence, by induction, (7) is proved.

 Problem 1. In the proof above, we have used the following fact: if $i(Z)s = 0$ for all $Z \in \mathfrak{X}(M)$, then $s = 0$. Prove this fact.

Now if ω is a differential form of order p, then

$$L_X \omega = i(X)d\omega + di(X)\omega, \qquad p \geqq 1$$

$$L_X \omega = i(X)d\omega \qquad\qquad\quad p = 0.$$

(8)

These relations are sometimes called the equations of H. Cartan.

 Proof of (8). If $p = 0.$, it is clear. So we suppose $p > 0$. If $X_1, \ldots, X_p \in \mathfrak{X}(M)$, then

$$(i(X)d\omega)(X_1, \ldots, X_p)$$
$$= (d\omega)(X, X_1, \ldots, X_p)$$
$$= X(\omega(X_1, \ldots, X_p)) + \sum_{i=1}^{p} (-1)^{i+2} X_i(\omega(X, X_1, \ldots, \hat{X}_i, \ldots, X_p))$$
$$+ \sum_{j=1}^{p} (-1)^{1+j+1} \omega([X, X_j], \hat{X}, \ldots, \hat{X}_j, \ldots, X_p)$$
$$+ \sum_{i<j} (-1)^{(i+1)+(j+1)} \omega([X_i, X_j], X, \ldots, \hat{X}_i, \ldots, \hat{X}_j, \ldots, X_p).$$

The sum of the first term and the third term in the last member is equal to $(L_X\omega)(X_1, \ldots, X_p)$. The sum of the second term and the fourth term is equal to $-(di(X)\omega)(X_1, \ldots, X_p)$. Hence we get (8).

 Furthermore, we have

$$dL_X = L_X d \qquad (9)$$
$$d^2 = 0. \qquad (10)$$

Proof of (9). If ω is of order 0, then

$$(L_X d\omega)(Y) = X(d\omega(Y)) - (d\omega)([X, Y]) = X(Y\omega) - [X, Y]\omega$$
$$= Y(X\omega) = (dL_X\omega)(Y).$$

Hence (9) holds for ω of order 0. We prove (9) by induction on the order p of ω. Suppose (9) holds for differential forms of order p. Let ω be of order $p + 1$. For an arbitrary vector field Y, $i(Y)\omega$ is of order p, and hence $dL_X(i(Y)\omega) - L_X d(i(Y)\omega) = 0$. By (6), we have $L_X i(Y) = i(Y)L_X + i([X, Y])$. From (8), we have

$$di(Y) = -i(Y)d + L_Y, \qquad di([X, Y]) = -i([X, Y])d + L_{[X, Y]}.$$

Hence

$$dL_X(i(Y)\omega) = -i(Y)(dL_X\omega) + L_Y(L_X\omega) - i([X, Y])(d\omega) + L_{[X, Y]}\omega,$$

and, by (7), we can transform this to $dL_X(i(Y)\omega) = -i(Y)(dL_X\omega) + L_X(L_Y\omega) - i([X, Y])d\omega$. Similarly, we have $L_X d(i(Y)\omega) = -i(Y)(L_X d\omega) + L_X(L_Y\omega) - i([X, Y]) d\omega$. Hence

$$dL_X(i(Y)\omega) - L_X d(i(Y)\omega) = i(Y)(L_X d\omega - dL_X\omega) = 0$$

holds for an arbitrary vector field Y. Thus we have $L_X d\omega = dL_X\omega$. We have proved that (9) holds for differential forms of order $p + 1$, so that, by induction, the proof of (9) is complete.

Proof of (10). If ω is of order 0, then

$$(d^2\omega)(X, Y) = d(d\omega)(X, Y)$$
$$= X((d\omega)(Y)) - Y((d\omega)(X)) - (d\omega)([X, Y])$$
$$= X(Y\omega) - Y(X\omega) - [X, Y]\omega = 0.$$

If ω is of order > 0, then, by (8), $(dL_X - L_X d)\omega = (d^2 i(X) - i(X)d^2)\omega$. Since $(dL_X - L_X d)\omega = 0$ by (9), we obtain $i(X)d^2\omega = d^2 i(X)\omega$. Using this, by induction, we get (10).

Lastly, we shall prove, for differential forms ω and θ of orders p and q, that

$$d(\omega \wedge \theta) = d\omega \wedge \theta + (-1)^p \omega \wedge d\theta. \tag{11}$$

The proof is done by induction on the sum $p + q$ of the orders. If $p = q = 0$, then ω and θ are functions, and it is easy to prove that $d(\omega\theta) = \theta \, d\omega + \omega \, d\theta$. Hence we shall assume that (11) holds when $p + q = r$, and we shall prove that it also holds when the sum of the orders of ω

and θ is $r + 1$. In this case, if ω or θ is of order 0, then the proof is easy, so we assume that p and q are both positive. It suffices to prove that $i(X)(d(\omega \wedge \theta)) = i(X)(d\omega \wedge \theta + (-1)^p \omega \wedge d\theta)$ for an arbitrary vector field X. Since we have $i(X)d = L_X - di(X)$ by (8), it suffices to prove that

$$L_X(\omega \wedge \theta) - d(i(X)(\omega \wedge \theta)) = i(X)(d\omega \wedge \theta + (-1)^p \omega \wedge d\theta). \qquad \text{(i)}$$

From (4) and (8), we get

$$L_X(\omega \wedge \theta)$$
$$= (i(X)d\omega) \wedge \theta + (di(X)\omega) \wedge \theta + \omega \wedge (i(X)d\theta) + \omega \wedge (di(X)\theta).$$

By (4) of §3, we have $\omega \wedge (i(X)d\theta) = (-1)^p i(X)(\omega \wedge d\theta) + (-1)^{p-1}(i(X)\omega) \wedge d\theta$. Substituting this in the right member of the equation above, and using $(di(X)\omega) \wedge \theta + (-1)^{p-1}(i(X)\omega) \wedge d\theta = d((i(X)\omega) \wedge \theta)$, which we get from the inductive hypothesis, we find that

$$L_X(\omega \wedge \theta) = (-1)^p i(X)(\omega \wedge d\theta) + d((i(X)\omega) \wedge \theta)$$
$$+ (i(X)d\omega) \wedge \theta + \omega \wedge (di(X)\theta). \qquad \text{(ii)}$$

On the other hand, from (4) of §3, we have $d(i(X)(\omega \wedge \theta)) = d((i(X)\omega) \wedge \theta) + (-1)^p d(\omega \wedge i(X)\theta)$, and by the inductive hypothesis, $d(\omega \wedge i(X)\theta) = d\omega \wedge i(X)\theta + (-1)^p \omega \wedge (di(X)\theta)$. Hence we have

$$d(i(X)(\omega \wedge \theta))$$
$$= d((i(X)\omega) \wedge \theta) + (-1)^p d\omega \wedge i(X)\theta + \omega \wedge (di(X)\theta). \qquad \text{(iii)}$$

Subtracting (iii) from (ii), the left member of (i) becomes equal to

$$i(X)((-1)^p \omega \wedge d\theta) + (i(X)d\omega) \wedge \theta + (-1)^{p+1} d\omega \wedge i(X)\theta.$$

Using (4) of §3 once more, we see that this is equal to $i(X)(d\omega \wedge \theta + (-1)^p \omega \wedge d\theta)$. Hence (i) is proved, and, by induction, the proof of (11) is complete.

D   Let $(x^1, \ldots, x^n)$ be a local coordinate system on an open set U, and let a differential form ω of order p be written on U as

$$\omega = \sum_{i_1 < \cdots < i_p} \alpha_{i_1 \cdots i_p} dx^{i_1} \wedge \cdots \wedge dx^{i_p}.$$

Then we shall prove that, on U, we have

$$d\omega = \sum_{i_1 < \cdots < i_p} d\alpha_{i_1 \cdots i_p} \wedge dx^{i_1} \wedge \cdots \wedge dx^{i_p}. \qquad \text{(12)}$$

The coefficient of $dx^{j_1} \wedge \cdots \wedge dx^{j_{p+1}}$ ($j_1 < \cdots < j_{p+1}$) in the right member of (12) is equal to

$$\sum_{k=1}^{p+1} (-1)^{k+1} \frac{\partial \alpha_{j_1 \cdots \hat{j}_k \cdots j_{p+1}}}{\partial x^{j_k}}. \tag{*}$$

Hence to prove (12), it suffices to show that the component $(d\omega)_{j_1 \cdots j_{p+1}}$ of $d\omega$ is equal to (*). Let $a \in U$, and let Y_j ($j = 1, \ldots, n$) be a vector field on M which coincides with $\partial/\partial x^j$ on a neighborhood of a. We have $[Y_j, Y_k] = 0$ on a neighborhood of a. The value of $(d\omega)_{j_1 \cdots j_{p+1}}$ at a is equal to $(d\omega)_a((Y_{j_1})_a, \ldots, (Y_{j_{p+1}})_a)$. But, since $[Y_j, Y_k]_a = 0$, from (5) we get

$$(d\omega)_a((Y_{j_1})_a, \ldots, (Y_{j_{p+1}})_a) = \sum_{k=1}^{p+1} (-1)^{k+1}(Y_{j_k})_a \alpha_{j_1 \cdots \hat{j}_k \cdots j_{p+1}}.$$

Since $Y_j = \partial/\partial x^j$ on a neighborhood of a, the value of the right member of this equation is equal to the value of (*) at a. Since a is an arbitrary point of U, we have shown that $(d\omega)_{j_1 \cdots j_{p+1}}$ is equal to (*), and (12) is thus proved.

Problem 2. Use (12) to prove that $d^2 = 0$.

Let us collect the formulas we have proved in §3 and §4.

$$L_{[X, Y]} = [L_X, L_Y] = L_X L_Y - L_Y L_X,\dagger$$
$$i([X, Y]) = [L_X, i(Y)] = L_X i(Y) - i(Y)L_X,$$
$$L_X = di(X) + i(X)d,$$
$$dL_X = L_X d, \qquad d^2 = 0,$$
$$L_X(t \otimes s) = (L_X t) \otimes s + t \otimes (L_X s),$$
$$L_X(\omega \wedge \theta) = (L_X \omega) \wedge \theta + \omega \wedge (L_X \theta),$$
$$i(X)(\omega \wedge \theta) = (i(X)\omega) \wedge \theta + (-1)^p \omega \wedge (i(X)\theta) \quad (\omega \text{ is of order } p),$$
$$d(\omega \wedge \theta) = d\omega \wedge \theta + (-1)^p \omega \wedge d\theta \quad (\omega \text{ is of order } p).$$

§5. Transformations of Covariant Tensor Fields by Maps

Let φ be a C^∞ map from a manifold M to a manifold M', and let t be a covariant tensor field on M' of order $p(> 0)$. For each point a of M, and for any vectors $v_1, \ldots, v_p \in T_a(M)$, set

$$(\varphi^* t)_a(v_1, \ldots, v_p) = t_{\varphi(a)}(\varphi_* v_1, \ldots, \varphi_* v_p).$$

---

†In general, we define the commutator $[A, B]$ of two operators A, B by $[A, B] = AB - BA$.

This defines an element $(\varphi^*t)_a$ of $\overset{p}{\otimes}T_a^*(M)$. We get a covariant tensor field φ^*t of order p on M by the correspondence $a \to (\varphi^*t)_a$. We call φ^*t the *pull back* to M of the covariant tensor field t on M' by the map φ.

If t is of order 0, i.e., if t is a function, then, as before, we define a function φ^*t by

$$(\varphi^*t)(a) = t(\varphi(a)).$$

Problem 1. Show that φ^*t is of class C^∞.

If ω is a differential form of order p on M', then $(\varphi^*\omega)_a$ is alternating. Hence $\varphi^*\omega$ is also a differential form of order p on M. By the definition of φ^*, we have

$$\begin{aligned} \varphi^*(t + s) &= \varphi^*t + \varphi^*s, \\ \varphi^*(t \otimes s) &= (\varphi^*t) \otimes (\varphi^*s). \end{aligned} \tag{1}$$

If ω and θ are differential forms, then it easy to prove that

$$\varphi^*(\omega \wedge \theta) = (\varphi^*\omega) \wedge (\varphi^*\theta). \tag{2}$$

Problem 2. Prove (1) and (2).

If ω is a differential form of order p on M', let us prove that

$$\varphi^*(d\omega) = d(\varphi^*\omega). \tag{3}$$

Let a be an arbitrary point of M, and set $b = \varphi(a)$. Let $(x^1, \ldots, x^m)$ be a local coordinate system of M' on a neighborhood V of b. Let U be a neighborhood of a satisfying $\varphi(U) \subset V$. Regard φ as a map from U to V. Let

$$\omega = \sum_{i_1 < \cdots < i_p} \alpha_{i_1 \ldots i_p} dx^{i_1} \wedge \ldots \wedge dx^{i_p}$$

on V. Then $\varphi^*\omega$ coincides with $\sum_{i_1 < \cdots < i_p} \varphi^*(\alpha_{i_1 \ldots i_p} dx^{i_1} \wedge \ldots \wedge dx^{i_p})$ on U. But, by (2), $\varphi^*(\alpha_{i_1 \ldots i_p} dx^{i_1} \wedge \ldots \wedge dx^{i_p})$ is equal to $\varphi^*(\alpha_{i_1 \ldots i_p}) \varphi^*(dx^{i_1}) \wedge \ldots \wedge \varphi^*(dx^{i_p})$. On the other hand, if f is a function on V, then at any point q of U, we have

$$(\varphi^*df)_q(v) = (df)_{\varphi(q)}(\varphi_*v) = (\varphi_*v)f = v(\varphi^*f) = (d\varphi^*f)_q(v).$$

Hence

$$\varphi^*df = d\varphi^*f. \tag{i}$$

In particular, we have $\varphi^*dx^i = d\varphi^*x^i$, so that on U we have

$$\varphi^*\omega = \sum_{i_1 < \cdots < i_p} \varphi^*\alpha_{i_1 \ldots i_p} d\varphi^*x^{i_1} \wedge \ldots \wedge d\varphi^*x^{i_p}. \tag{ii}$$

Similarly, from (12) of §4, on U we have

$$\varphi^* d\omega = \sum_{i_1 < \cdots < i_p} d\varphi^* \alpha_{i_1 \cdots i_p} \wedge d\varphi^* x^{i_1} \wedge \ldots \wedge d\varphi^* x^{i_p}. \qquad \text{(iii)}$$

On the other hand, by (ii), on U, $d\varphi^* \omega$ is equal to $\sum_{i_1 < \cdots < i_p} d(\varphi^* \alpha_{i_1 \cdots i_p} d\varphi^* x^{i_1} \wedge \ldots \wedge d\varphi^* x^{i_p})$. From (10) and (11) of §4, we have

$$d(\varphi^* \alpha_{i_1 \cdots i_p} d\varphi^* x^{i_1} \wedge \ldots \wedge d\varphi^* x^{i_p})$$
$$= d\varphi^* \alpha_{i_1 \cdots i_p} \wedge d\varphi^* x^{i_1} \wedge \ldots \wedge d\varphi^* x^{i_p}.$$

Hence $d\varphi^* \omega$ is equal to the right member of (iii) on U, so that $\varphi^* d\omega = d\varphi^* \omega$ holds on a neighborhood U of a. Since a is an arbitrary point of M, (3) holds on M.

Let φ be a diffeomorphism of M, s a covariant tensor field on M of order p, and $X_1, \ldots, X_p$ vector fields on M. Set $F = s(\varphi_* X_1, \ldots, \varphi_* X_p)$, $G = (\varphi^* s)(X_1, \ldots, X_p)$. We have $F(a) = s_a((\varphi_* X_1)_a, \ldots, (\varphi_* X_p)_a)$, $(\varphi_* X_i)_a = \varphi_*(X_i)_{\varphi^{-1}(a)}$, and hence we get

$$F(a) = (\varphi^* s)_{\varphi^{-1}(a)}((X_1)_{\varphi^{-1}(a)}, \ldots, (X_p)_{\varphi^{-1}(a)}) = G(\varphi^{-1}(a)).$$

That is, $F = (\varphi^{-1})^* G$. Hence we get

$$(\varphi^* s)(X_1, \ldots, X_p) = \varphi^*(s(\varphi_* X_1, \ldots, \varphi_* X_p)). \qquad (4)$$

Equality (4) can also be written as

$$(\varphi^* s)_a((X_1)_a, \ldots, (X_p)_a)$$
$$= s_{\varphi(a)}((\varphi_* X_1)_{\varphi(a)}, \ldots, (\varphi_* X_p)_{\varphi(a)}) \qquad (a \in M). \qquad (4')$$

Let $\{\varphi_t\}$ be a one-parameter group of transformations on M, and X the infinitesimal transformation of $\{\varphi_t\}$. $(\varphi_t^* s)(X_1, \ldots, X_p)$ is a C^∞ function on $\mathbf{R} \times M$ if $X_1, \ldots, X_p$ are fixed. Let us prove that

$$\lim_{t \to 0} \frac{1}{t}[(\varphi_t^* s)(X_1, \ldots, X_p) - s(X_1, \ldots, X_p)] = (L_X s)(X_1, \ldots, X_p). \qquad (5)$$

To simplify the writing, we let $p = 2$. By (4), the left member of (5) is

$$\lim_{t \to 0} \frac{1}{t}[(\varphi_t^*(s((\varphi_t)_* X_1, (\varphi_t)_* X_2)) - s(X_1, X_2)]$$

$$= \lim_{t \to 0} \{\frac{1}{t}[\varphi_t^*(s((\varphi_t)_* X_1, (\varphi_t)_* X_2)) - \varphi_t^*(s(X_1, X_2))]$$

$$+ \frac{1}{t}[\varphi_t^*(s(X_1, X_2)) - s(X_1, X_2)]\}. \qquad (a)$$

The second term of the right member of (a) is equal to $X(s(X_1, X_2))$.

The value of the first term at a point a of M is equal to

$$\lim_{t \to 0} \frac{1}{t} [s_{\varphi_t(a)}((\varphi_t)_*(X_1)_a, (\varphi_t)_*(X_2)_a) - s_{\varphi_t(a)}((X_1)_{\varphi_t(a)}, (X_2)_{\varphi_t(a)})].$$

This limit is equal to

$$\lim_{t \to 0} s_{\varphi_t(a)}\left(\frac{1}{t}[(\varphi_t)_*(X_1)_a - (X_1)_{\varphi_t(a)}], (\varphi_t)_*(X_2)_a\right)$$

$$+ \lim_{t \to 0} s_{\varphi_t(a)}\left((X_1)_{\varphi_t(a)}, \frac{1}{t}[(\varphi_t)_*(X_2)_a - (X_2)_{\varphi_t(a)}]\right).$$

By (6) of II, §12, this limit is $-s_a([X, X_1]_a, (X_2)_a) - s_a((X_1), [X, X_2]_a)$. Hence the limit in the left member of (5) is equal to

$$X(s(X_1, X_2)) - s([X, X_1], X_2) - s(X_1, [X, X_2]) = (L_X s)(X_1, X_2),$$

so that the equality (5) is proved.

We abbreviate (5) as

$$\lim_{t \to 0} \frac{1}{t}[\varphi_t{}^* s - s] = L_X s. \tag{6}$$

(6) expresses the geometrical meaning of the Lie derivative $L_X s$ of s.

If $\varphi^* s = s$ for a diffeomorphism φ of M, then the covariant tensor field s is said to be invariant by φ. If $\varphi_t{}^* s = s$ for each t, then s is said to be invariant by the one-parameter group of transformations $\{\varphi_t\}$.

THEOREM. *Let $\{\varphi_t\}$ be a one-parameter group of transformations and X the infinitesimal transformation of $\{\varphi_t\}$. A covariant tensor field s on M is invariant by $\{\varphi_t\}$ if and only if $L_X s = 0$.*

Proof. If s is invariant by $\{\varphi_t\}$, from (6), it is clear that $L_X s = 0$. Conversely, if $L_X s = 0$, then it can be proved that s is invariant by $\{\varphi_t\}$, using the uniqueness of solutions of systems of ordinary differential equations, just as we have done in the proof of the theorem of II, §13.

Remark. The condition (1) in II, §13 (the condition for X to be an infinitesimal motion) is nothing but $L_X g = 0$. The theorem there is a special case of the theorem above. See also Theorem 2 of II, §17.

§6. The Cohomology Algebra of a Manifold

Let $D^p(M)$ be the set of all differential forms of order p on an n-dimensional manifold M. The set $D^p(M)$ is a vector space over **R**, and

$D^p(M) = (0)$ if $p > n$. The direct sum of the vector spaces $D^p(M)$ ($p = 0, 1, \ldots, n$) is denoted by $D(M)$. The elements $\omega = \sum \omega^p$ ($\omega^p \in D^p(M)$) are also called differential forms, and ω^p is called the p component of ω. We shall say that ω is of order p if $\omega^q = 0$ ($q \neq p$), i.e., if $\omega = \omega^p$. For an element ω of $D(M)$, we define the exterior derivative $d\omega$ by

$$d\omega = \sum_p d\omega^p.$$

Then d is a linear transformation of $D(M)$, and satisfies

$$dD^p(M) \subset D^{p+1}(M), \qquad d^2 = 0.$$

We define the exterior product of elements of $D(M)$ by

$$\left(\sum_p \omega^p\right) \wedge \left(\sum_q \omega^q\right) = \sum_r \left(\sum_{p+q=r} \omega^p \wedge \omega^q\right).$$

Then $D(M)$ is an associative algebra over **R** with respect to the exterior product, and satisfies

$$D^p(M) \wedge D^q(M) \subset D^{p+q}(M).$$

If we put

$$w\left(\sum_p \omega^p\right) = \sum_p (-1)^p \omega^p,$$

then w is a linear transformation of $D(M)$, and for any two elements ω, θ of $D(M)$,

$$d(\omega \wedge \theta) = (d\omega) \wedge \theta + w\omega \wedge (d\theta) \qquad (1)$$

holds.

An element ω of $D(M)$ satisfying $d\omega = 0$ is called a *closed differential form* or a *closed form*, and the set of all closed forms is denoted by $Z(M)$. The set $Z(M)$ is a subspace of $D(M)$. Since $d(\sum_p \omega^p) = \sum_p d\omega^p = 0$ means that $d\omega^p = 0$ for each p, if we let $Z^p(M)$ be the subspace of $Z(M)$ formed by the closed forms of order p, then we have

$$Z(M) = \sum_p Z^p(M), \qquad Z^p(M) = D^p(M) \cap Z(M). \qquad (2)$$

If, for ω, there is a θ such that $\omega = d\theta$, then ω is called an *exact differential form* or an *exact form*. The set of all exact forms forms a subspace of $D(M)$, denoted by $B(M)$. Since $d(d\theta) = d^2\theta = 0$, we have

$$B(M) \subset Z(M).$$

As in the case of $Z(M)$, we have for $B(M)$

$$B(M) = \sum_p B^p(M), \qquad B^p(M) = D^p(M) \cap B(M). \tag{3}$$

We set

$$H(M) = Z(M)/B(M), \qquad H^p(M) = Z^p(M)/B^p(M),$$

and call the vector spaces $H(M)$ and $H^p(M)$ the *de Rham cohomology group* and the *pth de Rham cohomology group*, respectively. By (2) and (3), $H(M)$ is isomorphic to the direct sum $\sum_{p=0}^n H^p(M)$.

Let the element of $H(M)$ (i.e., the residue classes of $Z(M)$ with respect to the subspace $B(M)$) represented by $\omega \in Z(M)$ be denoted by $[\omega]$. We say that ω and ω' are *cohomologous* if $[\omega] = [\omega']$, i.e., if $\omega - \omega' \in B(M)$, and write $\omega \sim \omega'$. If ω, $\theta \in Z(M)$, and $\theta \sim 0$ (i.e., $\theta \in B(M)$), then $\omega \wedge \theta \sim 0$. In fact, from $\theta \sim 0$, we have $\theta = d\alpha$, so that $\omega \wedge \theta = \omega \wedge d\alpha$. Since $d\omega = 0$, we also have $dw\omega = 0$. Thus using (1), we get $d(w\omega \wedge \alpha) = \omega \wedge d\alpha = \omega \wedge \theta$, and hence $\omega \wedge \theta \sim 0$. Similarly, we also have $\theta \wedge \omega \sim 0$. Using this repeatedly, we can prove easily that if $\omega \sim \omega'$ and $\theta \sim \theta'$, then $\omega \wedge \theta \sim \omega' \wedge \theta'$. Hence the product of two elements $[\omega]$, $[\theta]$ of $H(M)$ can be defined by $[\omega] \wedge [\theta] = [\omega \wedge \theta]$, and the result is independent of the choice of the representatives ω and θ. $H(M)$ becomes an associative algebra with respect to the product $\wedge$, and this associative algebra is called the *de Rham cohomology algebra* of the manifold M. From the definition of the product, it is clear that $H^p(M) \wedge H^q(M) \subset H^{p+q}(M)$, if ω and θ are of orders p and q, respectively, then $[\omega] \wedge [\theta] = (-1)^{pq}[\theta] \wedge [\omega]$.

The definition of the de Rham cohomology algebra of a manifold M, of course, depends on the differentiable structure of M. But if M is paracompact, it really does not depend on the choice of the diffrentiable structure, but only on the topology of M. That is, if M and M' are two paracompact manifolds, and if M and M' are homeomorphic as topological spaces, then the de Rham cohomology algebras of M and M' are isomorphic.

In general, it is not easy to compute the de Rham cohomology groups directly. We shall make a few simple remarks on the de Rham cohomology groups below.

(1) If k is the number of connected components of M, then $H^0(M)$ is a k-dimensional vector space.

In fact, since $B^0(M) = (0)$, we have $H^0(M) = Z^0(M)$. If f is a C^∞ function on M such that $df = 0$, then f is a constant on each connected component M_i $(i = 1, \ldots, k)$ of M. Hence, if f_i is the function that is 1 on M_i and 0 on all the other components, then $f_1, \ldots, f_k$ is a base for $Z^0(M)$.

(2) For the n-dimensional affine space $\mathbf{R}^n$, we have

$$H^p(\mathbf{R}^n) = (0), \qquad p > 0.$$

To prove this, it suffices to prove the following lemma.

LEMMA (*The Poincaré lemma*). Let ω be a differential form of order p (> 0) on $\mathbf{R}^n$, and suppose $d\omega = 0$. Then there is a differential form θ of order $p - 1$ on $\mathbf{R}^n$ satisfying $\omega = d\theta$.

Proof. Let $f(t)$ be a C^∞ function on $\mathbf{R}^1$ satisfying $f \equiv 1$ for $t \geq 1$ and $f \equiv 0$ for $t \leq 0$. Define a map φ from $\mathbf{R}^{n+1}$ to $\mathbf{R}^n$ by $\varphi(t, x^1, \ldots, x^n) = (f(t)x^1, \ldots, f(t)x^n)$.

For $t \geq 1$, we have $\varphi(t, x) = x$, and for $t \leq 0$, we have $\varphi(t, x) = 0$. Set

$$\omega = \sum_{i_1 < \cdots < i_p} \alpha_{i_1 \cdots i_p}(x) \, dx^{i_1} \wedge \cdots \wedge dx^{i_p}.$$

Writing $\varphi^*\omega$ in the form

$$\varphi^*\omega = \sum_{i_1 < \cdots < i_p} \beta_{i_1 \cdots i_p}(t, x) \, dx^{i_1} \wedge \cdots \wedge dx^{i_p}$$
$$+ \sum_{j_1 < \cdots < j_{p-1}} \gamma_{j_1 \cdots j_{p-1}}(t, x) \, dt \wedge dx^{j_1} \wedge \cdots \wedge dx^{j_{p-1}},$$

we have $\beta_{i_1 \cdots i_p}(t, x) = f(t)^p \alpha_{i_1 \cdots i_p}(\varphi(t, x))$. Hence $\beta_{i_1 \cdots i_p}(1, x) = \alpha_{i_1 \cdots i_p}(x)$ and $\beta_{i_1 \cdots i_p}(0, x) = 0$.

Since $d\varphi^*\omega = \varphi^* d\omega = 0$, the coefficient of $dt \wedge dx^{i_1} \wedge \cdots \wedge dx^{i_p}$ in $d\varphi^*\omega$ is 0. Hence we have

$$\frac{\partial \beta_{i_1 \cdots i_p}}{\partial t} = \sum_{k=1}^{p} (-1)^{k-1} \frac{\partial \gamma_{i_1 \cdots i_k \cdots i_p}}{\partial x^{i_k}}.$$

Integrating both members from 0 to 1 with respect to t, we get

$$\alpha_{i_1 \cdots i_p}(x) = \sum_{k=1}^{p} (-1)^{k-1} \int_0^1 \frac{\partial \gamma_{i_1 \cdots i_k \cdots i_p}}{\partial x^{i_k}}(x, t) \, dt. \qquad (*)$$

If we set

$$\theta = \sum_{j_1 < \cdots < j_{p-1}} \left(\int_0^1 \gamma_{j_1 \cdots j_{p-1}}(x, t) \, dt \right) dx^{j_1} \wedge \cdots \wedge dx^{j_{p-1}},$$

then the coefficient of $dx^{i_1} \wedge \cdots \wedge dx^{i_p}$ ($i_1 < \cdots < i_p$) in $d\theta$ is equal to to the right member of (*). Hence, by (*), we have $\omega = d\theta$.

§7. Complex Differential Forms on a Complex Manifold

In the case of complex manifolds, it is important to consider "complex" differential forms. We shall discuss this here.

A   Let V be an n-dimensional vector space over **R**, and let V^C be the complexification of V (II, §16). Let ϕ be a p-linear function on V: ϕ can be extended uniquely to a p-linear function $\tilde{\phi}$ on V^C. Namely, if $\{e_1, \ldots, e_n\}$ is a basis of V, then it is also a basis of V^C, and hence if $\{\phi_{i_1 \cdots i_p}\}$ are the components of ϕ with respect to $\{e_1, \ldots, e_n\}$, then there is a p-form $\tilde{\phi}$ on V^C whose components with respect to $\{e_1, \ldots, e_n\}$ are $\{\phi_{i_1 \cdots i_p}\}$. It is clear from the definition of $\tilde{\phi}$ that if $u_i \in V$, then $\tilde{\phi}(u_1, \ldots, u_p) = \phi(u_1, \ldots, u_p)$, so that $\tilde{\phi}$ is an extension of ϕ. It is clear that the extension of ϕ is unique.

Let ϕ be a p-linear function on V, and ψ a p-linear function on V. It is clear from the definitions that the extension of the $(p + q)$-linear function $\phi \otimes \psi$ on V to V^C is equal to $\tilde{\phi} \otimes \tilde{\psi}$.

If ϕ is symmetric or alternating, then so is $\tilde{\phi}$. Furthermore, if ϕ and ψ are alternating, then the extension of the exterior product of $\phi \wedge \psi$ is equal to $\tilde{\phi} \wedge \tilde{\psi}$.

In particular, if $\{\phi^1, \ldots, \phi^n\}$ is a basis of V^* over **R**, then $\tilde{\phi}^{i_1} \otimes \ldots \otimes \tilde{\phi}^{i_p}$ and $\tilde{\phi}^{i_1} \wedge \ldots \wedge \tilde{\phi}^{i_p}$ are extensions of $\phi^{i_1} \otimes \ldots \otimes \phi^{i_p}$ and $\phi^{i_1} \wedge \ldots \wedge \phi^{i_p}$, respectively. Furthermore, $\{\tilde{\phi}^1, \ldots, \tilde{\phi}^n\}$ is a basis of $(V^C)^*$ over **C**.

Identifying ϕ with its extension $\tilde{\phi}$, we can consider the real vector space $\overset{p}{\otimes} V^*$ formed by the p-linear functions on V as a real subspace of the complex vector space $\overset{p}{\otimes}(V^C)^*$ formed by the p-linear functions on V^C. Similarly, we have $\overset{p}{\wedge} V^* \subset \overset{p}{\wedge}(V^C)^*$. From now on, we shall identify ϕ and $\tilde{\phi}$, and taking off the tilde from $\tilde{\phi}$, we shall write simply ϕ instead of $\tilde{\phi}$. Then the tensor product and exterior product of elements ϕ, ψ of $\sum_p \overset{p}{\otimes} V^*$ are equal to the tensor product and exterior product of ϕ, ψ, considered as elments of $\sum_p \overset{p}{\otimes}(V^C)^*$.

Now let $\theta \in \overset{p}{\otimes}(V^C)^*$. Let $\{\phi^1, \ldots, \phi^n\}$ be a basis of V^*. As this is also a basis of $(V^C)^*$, θ can be written as

$$\theta = \sum \theta_{i_1 \cdots i_p} \phi^{i_1} \otimes \cdots \otimes \phi^{i_p}, \qquad \theta_{i_1 \cdots i_p} \in \mathbf{C}.$$

Let

$$\theta_{i_1 \cdots i_p} = \phi_{i_1 \cdots i_p} + i\psi_{i_1 \cdots i_p} \qquad (\phi_{i_1 \cdots i_p}, \psi_{i_1 \cdots i_p} \in \mathbf{R}),$$

and set

$$\phi = \sum \phi_{i_1 \cdots i_p} \phi^{i_1} \otimes \cdots \otimes \phi^{i_p}, \qquad \psi = \sum \psi_{i_1 \cdots i_p} \phi^{i_1} \otimes \cdots \otimes \phi^{i_p}.$$

Then $\phi, \psi \in \overset{p}{\otimes} V^*$, and θ can be written uniquely as

$$\theta = \phi + i\psi.$$

That is, $\overset{p}{\otimes}(V^C)^*$ is isomorphic to the complexification of $\overset{p}{\otimes} V^*$.

Similarly, $\theta \in \overset{p}{\wedge}(V^C)^*$ is expressed uniquely as $\theta = \phi + i\psi$, where $\phi, \psi \in \overset{p}{\wedge} V^*$, and $\overset{p}{\wedge}(V^C)^*$ is isomorphic to the complexification of $\overset{p}{\wedge} V^*$.

B   Now let W be a $2n$-dimensional real vector space, and I a complex structure on W (II, §17). Define a linear transformation $'I$ on the dual space W^* of W by

$$('I\phi)(u) = \phi(Iu) \quad (\phi \in W^*, u \in W).$$

Then we have $('I)^2 = -1$, and $'I$ is a complex structure of W^*. By II, §17, we have

$$W^C = W^+ + W^-, \qquad W^+ = W^-,$$
$$(W^*)^C = (W^*)^+ + (W^*)^-, \qquad (\overline{W}^*)^+ = (W^*)^-.$$

$(W^*)^C$ and $(W^C)^*$ are identified. Furthermore,

$$\text{if} \quad \phi \in (W^*)^+ \quad \text{and} \quad u \in W^-, \qquad \text{then} \quad \phi(u) = 0.$$

Problem 1. Show that $(W^*)^+$ is the dual space of W^+, and that $(W^*)^-$ is the dual space of W^-.

Now let $\{\theta^1, \ldots, \theta^n\}$ be a basis of $(W^*)^+$. Then $\{\bar{\theta}^1, \ldots, \bar{\theta}^n\}$ is a basis of $(W^*)^-$. Hence $\{\theta^1, \ldots, \theta^n, \bar{\theta}^1, \ldots, \bar{\theta}^n\}$ is a basis of $(W^*)^C = (W^C)^*$. Thus, elements of the form

$$\theta^{i_1} \wedge \cdots \wedge \theta^{i_p} \wedge \bar{\theta}^{j_1} \cdots \wedge \bar{\theta}^{j_q} \qquad (p + q = r, \; i_1 < \cdots < i_p,$$
$$j_1 < \ldots < j_q)$$

form a basis of $\overset{r}{\wedge}(W^C)^*$. That is, an arbitrary alternating r-linear function f on W^C can be written uniquely as

$$f = \sum_{p+q=r} f^{p,q},$$

$$f^{p,q} = \sum_{i_1 < \cdots < i_p} \sum_{j_1 < \cdots < j_q} f_{i_1 \cdots i_p \bar{j}_1 \cdots \bar{j}_q} \theta^{i_1} \wedge \cdots \wedge \theta^{i_p} \wedge \bar{\theta}^{j_1} \wedge \cdots \wedge \bar{\theta}^{j_q}.$$

An element of $\wedge(W^C)^*$ of the form $f^{p,q}$ is said to be of *type* (p, q). Let us denote by $^{p,q}\wedge(W^C)^*$ the subspace of $\wedge(W^C)^*$ formed by elements of type (p, q). Then we have

$$\overset{r}{\wedge}(W^C)^* = \sum_{p+q=r} {}^{p,q}\!\wedge(W^C)^* \qquad \text{(direct sum)}.$$

For $\{\theta^1, \ldots, \theta^n\}$, let $w_1, \ldots, w_n$ be a basis of W^+ such that $\theta^i(w_j) = \delta^i_j (i, j = 1, \ldots, n)$. Then $\{\bar{w}_1, \ldots, \bar{w}_n\}$ is a basis of W^-, and

$$\theta^i(w_j) = \delta^i_j, \qquad \theta^i(\bar{w}_j) = 0,$$
$$\bar{\theta}^i(w_j) = 0, \qquad \bar{\theta}^i(\bar{w}_j) = \delta^i_j. \qquad (i, j = 1, \ldots, n)$$

Hence, by (8), of §2, for $i_1 < \cdots < i_p$, $j_1 < \cdots < j_q$, $k_1 < \cdots < k_s$, $l_1 < \cdots < l_t$, we have

$$(\theta^{i_1} \wedge \cdots \wedge \theta^{i_p} \wedge \bar{\theta}^{j_1} \cdots \wedge \bar{\theta}^{j_q})(w_{k_1}, \ldots, w_{k_s}, \bar{w}_{l_1}, \ldots, \bar{w}_{l_t})$$
$$= 1 \quad \text{if } s = p, t = q \text{ and } k_a = i_a (1 \leq a \leq p), l_b = j_b(1 \leq b \leq q)$$
$$= 0 \quad \text{otherwise.}$$

From this we conclude the following: let f be an alternating r-linear function on W^C, let $u_1, \ldots, u_r \in W^C$, and suppose each u_i belongs either to W^+ or W^-. Let p' and q' be the number of $u_1, \ldots, u_r$ belonging to W^+ and W^-, respectively. f is of type (p, q) if and only if $f(u_1, \ldots, u_r) \neq 0$ implies $p = p'$ and $q = q'$.

C Let M be a manifold. A correspondence ω that assigns to each point a of M an element ω_a of $\wedge(T^C_a(M))^*$ is called a *complex differential form* of degree r. At each point a, ω_a can be written uniquely as $\omega_a = \phi_a + i\psi_a$ ($\phi_a, \psi_a \in \wedge T_a^*(M)$), and $a \to \phi_a$ and $a \to \psi_a$ define (real) differential forms ϕ and ψ. We call ϕ and ψ the real and imaginary parts, respectively, of the complex differential form ω. We can write $\omega = \phi + i\psi$. If $\omega = \phi + i\psi$, set $\bar{\omega} = \phi - i\psi$, and call $\bar{\omega}$ the *conjugate* of ω. The complex vector space formed by the complex differential forms of order r is identified with the complexification of the real vector space formed by the differential forms of order r. ω is said to be of class C^r if its real and imaginary parts are both of class C^r. We shall only treat complex differential forms of class C^∞ hereafter.

The exterior derivative $d\omega$ of a complex differential form $\omega = \phi + i\psi$ is defined by

$$d\omega = d\phi + i \, d\psi.$$

For complex vector fields $X_1, \ldots, X_r$ (II, §17, **C**), we can define a complex-valued function $\omega(X_1, \ldots, X_r)$ on M by setting

$$\omega(X_1, \ldots, X_r)(a) = \omega_a((X_1)_a, \ldots, (X_r)_a)$$

at each point $a \in M$.

The set of all complex vector fields on M is the complexification $\mathfrak{X}^C(M)$ of $\mathfrak{X}(M)$. The set $C^\infty(M)^C$ of all complex-valued functions on M of class C^∞ is the complexification of $C^\infty(M)$, and $\mathfrak{X}^C(M)$ is a $C^\infty(M)^C$-module. Just as in §3, a complex differential form ω of degree r is an alternating r-linear function on the $C^\infty(M)^C$-module $\mathfrak{X}^C(M)$. Conversely, an alternating r-linear function on the $C^\infty(M)^C$-module $\mathfrak{X}^C(M)$ can be regarded as a complex differential form of order r on M. The proof is similar to that of the theorem in §3, and is left to the reader.

D Let M be an almost complex manifold, J the almost complex structure of M, and let the dimension of M be $2n$. A complex differential form ω on M is called a *differential form of type* (p, q) if the value ω_a of ω at each point a of M is an alternating $(p + q)$-linear function of type $(p + q)$ on $T_a^C(M)$. An arbitrary complex differential form ω of order r can be written uniquely as

$$\omega = \sum_{p+q=r} \omega^{p,q}, \qquad \text{where } \omega^{p,q} \text{ is of type } (p, q).$$

That is, if $D^r(M, C)$ denotes the vector space formed by the complex differential forms of order r on M, and if $D^{p,q}(M, C)$ denotes the vector space formed by the differential forms of type (p, q), then we have

$$D^r(M, C) = \sum_{p+q=r} D^{p,q}(M, C) \qquad \text{(direct sum)}.$$

Problem 2. If ω is of type (p, q), show that its conjugate $\bar{\omega}$ is of type (q, p).

From now on, we shall assume that J is integrable (II, §17, **C**). Let us show that, under this assumption, we have

$$dD^{p,q}(M, C) \subset D^{p+1,q}(M, C) + D^{p,q+1}(M, C).$$

That is, if ω is of type (p, q), then $d\omega$ is the sum of differential forms of types $(p + 1, q)$ and $(p, q + 1)$. To show this it suffices to show that for vector fields $X_1, \ldots, X_s$ of holomorphic type and vector fields $Y_1, \ldots, Y_t$ of antiholomorphic type, where $s + t = p + q + 1$, $d\omega(X_1, \ldots, X_s, Y_1, \ldots, Y_t)$ is equal to 0 except for the cases $(s, t) = (p + 1, q)$ and $(s, t) = (p, q + 1)$.

Hence we suppose $(s, t) \neq (p + 1, q)$, $(s, t) \neq (p, q + 1)$. From the definition of $d\omega$, we have

$$(d\omega)(X_1, \ldots, X_s, Y_1, \ldots, Y_t)$$

$$= \sum_{i=1}^{s} (-1)^{i+1} X_i(\omega(X_1, \ldots, \hat{X}_i, \ldots, X_s, Y_1, \ldots, Y_t)$$

$$+ \sum_{j=1}^{t} (-1)^{s+j+1} Y_j(\omega(X_1, \ldots, X_s, Y_1, \ldots, \hat{Y}_j, \ldots, Y_t))$$

$$+ \sum_{i<j} (-1)^{i+j} \omega([X_i, X_j], X_1, \ldots, \hat{X}_i, \ldots, \hat{X}_j, \ldots, X_s,$$
$$Y_1, \ldots, Y_t)$$

$$+ \sum_{i<j} (-1)^{2s+i+j} \omega([Y_i, Y_j], X_1, \ldots, X_s, Y_1, \ldots, \hat{Y}_i, \ldots,$$
$$\hat{Y}_j, \ldots, Y_t)$$

$$+ \sum_{i=1}^{s} \sum_{j=1}^{t} (-1)^{i+s+j} \omega([X_i, Y_j], X_1, \ldots, \hat{X}_i, \ldots, X_s, Y_1, \ldots,$$
$$\hat{Y}_j, \ldots, Y_t).$$

Now, since $(s-1, t) \neq (p, q)$ and $(s, t-1) \neq (p, q)$, and since ω is of type (p, q), we have

$$\omega(X_1, \ldots, \hat{X}_i, \ldots, X_s, Y_1, \ldots, Y_t) = \omega(X_1, \ldots, X_s, Y_1, \ldots,$$
$$\hat{Y}_j, \ldots, Y_t) = 0.$$

Hence the first and second terms of the right member of the above equality are equal to zero. On the other hand, since J is integrable, $[X_i, X_j]$ is holomorphic and $[Y_i, Y_j]$ is antiholomorphic. So, by the same reason as before, the third and fourth terms are zero. But $[X_i, Y_j]$ is the sum of a holomorphic vector field and an antiholomorphic vector field, so by $(s, t-1) \neq (p, q)$ and $(s-1, t) \neq (p, q)$, the last term is also zero.

Let the components of $d\omega$ of types $(p+1, q)$ and $(p, q+1)$ be $d'\omega$ and $d''\omega$, respectively. That is,

$$d\omega = d'\omega + d''\omega, \quad \text{where } d'\omega \text{ is of type } (p+1, q), \quad d''\omega \text{ is of type}$$
$$(p, q+1).$$

In general, for a complex differential form of degree r, let $d'\omega$, $d''\omega$ be defined as follows. Let ω be written as the sum of differential forms of types (p, q): $\omega = \sum_{p+q=r} \omega^{p,q}$. Define $d'\omega$ and $d''\omega$ by $d'\omega = \sum d'\omega^{p,q}$ and $d''\omega = \sum d''\omega^{p,q}$.

Problem 3. Conversely, show that if $dD^{1,0} \subset D^{2,0} + D^{1,1}$, then J is integrable.

Problem 4. Prove $(d')^2 = (d'')^2 = 0$ and $d'd'' + d''d' = 0$.

E Let M be a complex manifold of complex dimension n, and let J be the almost complex structure attached to M. Then J is, of course, integrable. Let $(z^1, \ldots, z^n)$ be a complex local coordinate system on an open set U of M. For each point a of U, $\{dz^1, \ldots, dz^n\}$ is a basis of $(T^*(M))^+$ and $\{d\bar{z}^1, \ldots, d\bar{z}^n\}$ is a basis of $(T^*(M))^-$. Hence, on U, a differential form ω of type (p, q) can be written uniquely as

$$
\omega = \sum_{\alpha_1 < \cdots < \alpha_p} \sum_{\beta_1 < \cdots < \beta_q} \omega_{\alpha_1 \cdots \alpha_p \bar{\beta}_1 \cdots \bar{\beta}_q} \, dz^{\alpha_1} \wedge \cdots \wedge dz^{\alpha_p} \wedge d\bar{z}^{\beta_1} \wedge
$$
$$
\cdots \wedge d\bar{z}^{\beta_q}.
$$

$\{\omega_{\alpha_1 \cdots \alpha_p \bar{\beta}_1 \cdots \bar{\beta}_q}\}$ are called the components of ω with respect to $(z^1, \ldots, z^n)$.

In particular, if f is a C^∞ function on U, then

$$
df = \sum_{\alpha=1}^{n} \frac{\partial f}{\partial z^\alpha} \, dz^\alpha + \sum_{\alpha=1}^{n} \frac{\partial f}{\partial \bar{z}^\alpha} \, d\bar{z}^\alpha,
$$

and hence we have

$$
d'f = \sum_{\alpha=1}^{n} \frac{\partial f}{\partial z^\alpha} \, dz^\alpha, \qquad d''f = \sum_{\alpha=1}^{n} \frac{\partial f}{\partial \bar{z}^\alpha} \, d\bar{z}^\alpha.
$$

But $d\omega$ can be expressed as

$$
d\omega = \sum_{\alpha_1 < \cdots < \alpha_p} \sum_{\beta_1 < \cdots < \beta_q} d\omega_{\alpha_1 \cdots \alpha_p \bar{\beta}_1 \cdots \bar{\beta}_q} \wedge dz^{\alpha_1} \wedge \cdots \wedge dz^{\alpha_p} \wedge
$$
$$
d\bar{z}^{\beta_1} \wedge \cdots \wedge d\bar{z}^{\beta_q}
$$

on U, so that we have

$$
d'\omega = \sum_{1 < \cdots < \alpha_p} \sum_{\beta_1 < \cdots < \beta_q} d'\omega_{\alpha_1 \cdots \alpha_p \bar{\beta}_1 \cdots \bar{\beta}_q} \wedge dz^{\alpha_1} \wedge \cdots \wedge dz^{\alpha_p} \wedge
$$
$$
d\bar{z}^{\beta_1} \wedge \cdots \wedge d\bar{z}^{\beta_q},
$$
$$
d''\omega = \sum_{\alpha_1 < \cdots < \alpha_p} \sum_{\beta_1 < \cdots < \beta_q} d''\omega_{\alpha_1 \cdots \alpha_p \bar{\beta}_1 \cdots \bar{\beta}_q} \wedge dz^{\alpha_1} \wedge \cdots \wedge dz^{\alpha_p} \wedge
$$
$$
d\bar{z}^{\beta_1} \wedge \cdots \wedge d\bar{z}^{\beta_q}
$$

on U.

If ω is of type $(p, 0)$, and the components $\{\omega_{\alpha_1 \cdots \alpha_p}\}$ of ω with respect to complex local coordinate system $(z^1, \ldots, z^n)$ are holomorphic functions, then ω is called a *holomorphic differential form* of degree p on M.

Problem 5. Show that a differential form of type $(p, 0)$ is holomorphic if and only if $d''\omega = 0$.

Now let g be a Hermitian metric on M (II, §16). For arbitrary vector fields X and Y on M, let

$$\omega(X, Y) = g(IX, Y).$$

Then this defines a (real) differential form ω of degree 2 on M (II, §16). If X and Y are both of holomorphic type, then IX is also of holomorphic type, so that, by the property of Hermitian metrics, we have $g(IX, Y) = 0$. Similarly, if X and Y are both of antiholomorphic type, then we have $g(IX, Y) = 0$. Hence ω is a differential form of type $(1, 1)$.

If $g_{\alpha\bar{\beta}}(= g(\partial/\partial z^{\alpha}, \partial/\partial z^{\beta}))$ are the components of g with respect to $(z^1, \ldots, z^n)$, then, by the definition of ω, the components of ω are $\omega_{\alpha\bar{\beta}} = ig_{\alpha\bar{\beta}}$. Hence

$$\omega = \sum_{\alpha, \beta = 1}^{n} ig_{\alpha\bar{\beta}} \, dz^{\alpha} \wedge d\bar{z}^{\beta}.$$

Problem 6. Show that the Hermitian metric g is Kählerian (i.e., $d\omega = 0$) if and only if

$$\frac{\partial g_{\alpha\bar{\beta}}}{\partial z^{\gamma}} - \frac{\partial g_{\gamma\bar{\beta}}}{\partial z^{\alpha}} = 0 \qquad (\alpha, \beta, \gamma = 1, \ldots, n).$$

§8. Differential Systems and Integral Manifolds

Let M be an n-dimensional manifold. Suppose, for each point p of M, an r-dimensional vector subspace $\mathfrak{D}_p$ of the tangent space $T_p(M)$ is assigned to p, satisfying the following condition: there are a neighborhood U of p, and $n - r$ differential forms $\omega_1, \ldots, \omega_{n-r}$ of order 1 on U, such that, at each point q of U, we have

$$\mathfrak{D}_q = \{v \in T_q(M) | (\omega_1)_q(v) = \cdots = (\omega_{n-r})_q(v) = 0\}. \tag{1}$$

The correspondence that assigns $\mathfrak{D}_p$ to p is called an *r-dimensional differential system* on M. In general, if $n - r$ differential forms $\omega_1, \ldots, \omega_{n-r}$ of order 1 on an open set V of M satisfy (1) at each point q of V, then we say that the differential system $\mathfrak{D}$ is defined on V by the *Pfaffian equations*

$$\omega_1 = \cdots = \omega_{n-r} = 0. \tag{2}$$

We call (2) the local equations of $\mathfrak{D}$ on V. We should note that $\omega_1, \ldots, \omega_{n-r}$ are linearly independent at each point of V.[†]

†We mean that $(\omega_1)_q, \ldots, (\omega_{n-r})_q$ are linearly independent at each point q of V. If $(\omega_1)_q, \ldots, (\omega_{n-r})_q$ were not linearly independent, then, by (1), the dimension of $\mathfrak{D}_q$ is greater than r.

Let N be a submanifold of M, and let i be the identity mapping of N into M. If, at each point p of N,

$$i_{*p} T_p(N) \subset \mathfrak{D}_p \tag{3}$$

holds, then N is called an *intergral manifold* of the differential system $\mathfrak{D}$.

If $\omega_1 = \ldots = \omega_{n-r} = 0$ are the local equations of $\mathfrak{D}$ on a neighborhood of a point of N, then condition (3) is equivalent to

$$i^* \omega_i = 0 \qquad (i = 1, \ldots, n - r). \tag{3'}$$

If, at each point p of M, there are a neighborhood U of p, and $n - r$ C^∞ functions $f_1, \ldots, f_{n-r}$ defined on U, such that

$$df_1 = \cdots = df_{n-r} = 0$$

are the local equations of $\mathfrak{D}$ on U, then $\mathfrak{D}$ is said to be *completely integrable*. If this is the case, for each point q of U, set $N = \{x \in U \mid f_i(x) = f_i(q),\ i = 1, \ldots, n - r\}$. Since $(df_1)_q, \ldots, (df_{n-r})_q$ are linearly independent at each point of U, by Theorem 1 of II, §10, N is an r-dimensional submanifold of M containing q. Moreover, by the definition of N, $i^* df_i = di^* f_i = 0$. Hence N is an r-dimensional integral manifold of $\mathfrak{D}$. Hence, if an r-dimensional differential system $\mathfrak{D}$ is completely integrable, then for each point p of M, there is an r-dimensional integral manifold passing through p.

LEMMA 1. *Let $\mathfrak{D}$ be an r-dimensional differential system on M. For each point p of M, we assume that there is a local coordinate system $(x^1, \ldots, x^n)$(with $x^i(p) = 0$, for $i = 1, \ldots, n$) around p, with the following property*: "*For a sufficiently small positive number a, set $Q = \{q \mid |x^i(q)| < a,\ i = 1, \ldots, n\}$. Then for any $(a^{r+1}, \ldots, a^n)$ satisfying $|a^j| < a$, $N_a = \{q \in Q \mid x^{r+1}(q) = a^{r+1}, \ldots, x^n(q) = a^n\}$ is an integral manifold of $\mathfrak{D}$.*" *Then $\mathfrak{D}$ is completely integrable.*

Proof. Let i be the identity map from N_a into M, and let v be a tangent vector to N_a at a point q of N_a. $i_* v$ is a linear combination of $(\partial/\partial x^1)_q$, $\ldots$, $(\partial/\partial x^r)_q$. Hence $i_* T_q(N_a) = \{u \in T_q(M) \mid (dx^j)_q(u) = 0;\ j = r + 1, \ldots, n\}$. Hence $dx^{r+1} = \ldots = dx^n = 0$ are local equations for $\mathfrak{D}$ on Q, and thus $\mathfrak{D}$ is completely integrable.

Problem 1. Conversely, prove that there is a local coordinate system around each point p of M satisfying the condition in Lemma 1, if $\mathfrak{D}$ is completely integrable.

The following theorem of Frobenius, which gives a condition for complete integrability of a differential system $\mathfrak{D}$, has important applications.

THEOREM 1 (*Frobenius*)   *Let $\mathfrak{D}$ be an r-dimensional differential system on an n-dimensional manifold M. Let $\omega_1 = \cdots = \omega_{n-r} = 0$ be local equations of $\mathfrak{D}$ on a sufficiently small neighborhood U of an arbitrary point p of M. For $\mathfrak{D}$ to be completely integrable, it is necessary and sufficient that there are $(n - r)^2$ differential forms $\tilde{\omega}_{\alpha\beta}$ of order 1 on U, satisfying*

$$d\omega_\alpha = \sum_{\beta=1}^{n-r} \tilde{\omega}_{\alpha\beta} \wedge \omega_\beta \qquad (\alpha = 1, \ldots, n - r). \qquad (4)$$

Condition (4) is called the *integrability condition* of $\mathfrak{D}$.

Before going into the proof of the theorem, we shall investigate the relation between two sets of local equations of $\mathfrak{D}$ on an open set V. For $q \in V$, let $\mathfrak{D}_q^\perp = \{\theta \in T_q^*(M) | \theta(v) = 0 \text{ for all } v \in \mathfrak{D}_q\}$. Then $\mathfrak{D}_q^\perp$ is an $(n - r)$-dimensional subspace of $T_q^*(M)$. That $\omega_1 = \cdots = \omega_{n-r} = 0$ are local equations for $\mathfrak{D}$ on V is equivalent to the fact that $(\omega_1)_q, \ldots$ $(\omega_{n-r})_q$ form a base of $\mathfrak{D}_q^\perp$ at each point q of V. Hence, if $\theta_1 = \cdots = \theta_{n-r} = 0$ is another set of local equations for $\mathfrak{D}$ on V, then we have

$$\theta_\alpha = \sum_\beta a_\alpha^\beta \omega_\beta$$

$$\omega_\alpha = \sum_\beta b_\alpha^\beta \theta_\beta.$$

Here, a_α^β, b_α^β are functions on V, and it is easy to see that they are of class C^∞. If (4) holds for ω_α, then we have

$$d\theta_\alpha = \sum_\beta da_\alpha^\beta \wedge \omega_\beta + \sum_{\beta, \lambda} a_\alpha^\beta \tilde{\omega}_{\beta\lambda} \wedge \omega_\lambda$$

$$= \sum_\gamma \left(\sum_\beta b_\beta^\gamma \, da_\alpha^\beta \right) \wedge \theta_\gamma + \sum_\gamma \left(\sum_{\beta, \gamma} a_\alpha^\beta b_\lambda^\gamma \tilde{\omega}_{\beta\lambda} \right) \wedge \theta_\gamma,$$

so that condition (4) holds for θ_α too. That is, condition (4) does not depend on the choice of the local equations for $\mathfrak{D}$.

If $\mathfrak{D}$ is completely integrable, then, on a sufficiently small neighborhood U of p, we have local equations $df_1 = \cdots = df_{n-r} = 0$. If we set $\omega_\alpha = df_\alpha$, then, since $d\omega_\alpha = 0$, we can let $\tilde{\omega}_{\alpha\beta} = 0$, and (4) holds. Hence (4) is a necessary condition for $\mathfrak{D}$ to be completely integrable.

Let us now prove that $\mathfrak{D}$ is completely integrable if (4) holds on a neighborhood of each point p of M. For this, let $(x^1, \ldots, x^n)$ be a local coordinate system on a neighborhood of p, such that $x^i(p) = 0$ ($i = 1, \ldots, n$), and we can suppose that

$$U = \{q | \, |x^i(q)| < \delta\}.$$

Let us identify U with a cube in $\mathbf{R}^n$ of half-width δ and centered at 0. On U, we have

$$\omega_\alpha = \sum_{i=1}^n a_{\alpha i}\, dx^i \qquad (\alpha = 1, \ldots, n - r).$$

Since the ω_α's are linearly independent at each point of U, the $(n - r) \times n$ matrix

$$(a_{\alpha i})_{\substack{\alpha = 1, \ldots, n - r \\ i = 1, \ldots, n}}$$

has rank $n - r$ at each point of U. Taking U sufficiently small, and renumbering the coordinate functions if necessary, we can assume that the square matrix $(a_{\alpha\beta})_{\alpha,\beta = 1, \ldots, n - r}$, formed by the first $n - r$ columns of $(a_{\alpha i})$ is nonsingular at each point of U. Let $(b_{\alpha\beta})$ be the inverse matrix of $(a_{\alpha\beta})$ and set

$$\theta_\alpha = \sum_{\beta = 1}^{n - r} b_{\alpha\beta}\omega_\beta \qquad (\alpha = 1, \ldots, n - r).$$

Then $\theta_1 = \cdots = \theta_{n-r} = 0$ are also local equations of $\mathfrak{D}$ on U. Hence the integrability condition also holds for θ_α. We note that θ_α is of the form

$$\theta_\alpha = dx^\alpha - \sum_{j = n - r + 1}^n \Phi_{\alpha j}\, dx^j \qquad (\alpha = 1, \ldots, n - r).$$

We change notation now. Set $s = n - r$, and write y^α instead of x^α $(\alpha = 1, \ldots, s)$, x^k instead of x^{s+k} $(k = 1, \ldots, r)$, and ω_α instead of θ_α. Then we have

$$\omega_\alpha = dy^\alpha - \sum_{k=1}^r \Phi_{\alpha k}(x, y)\, dx^k \qquad (\alpha = 1, \ldots, s; \ r + s = n).$$

The function $\Phi_{\alpha k}(x, y)$ is C^∞ for $|x^k| < \delta$, $|y^\alpha| < \delta$. Furthermore (4) holds for ω_α.

We write $\Phi_{\alpha k}(t, x, y)$ for the function we obtain from $\Phi_{\alpha k}(x, y)$ by replacing x^k by $x^k t$. This function is C^∞ for $|t| < 2$, $|x^k| < \delta/2$, $|y^\alpha| < \delta$.

Let us consider a system of ordinary differential equations of the unknown functions $(y^1, \ldots, y^s)$ given by

$$\frac{dy^\alpha}{dt} = \sum_{j=1}^r \Phi_{\alpha j}(t, x, y)x^j \qquad (\alpha = 1, \ldots, s). \tag{5}$$

We consider $(x^1, \ldots, x^r)$ in the right member of (5) as parameters. Choose positive numbers δ_1, δ_2 and ε sufficiently small, and set

$$U_1(\delta_1) = \{(x^1, \ldots, x^r) \mid |x^k| < \delta_1\}, \quad U_2(\delta_2) = \{(y^1, \ldots, y^s) \mid |y^\alpha| < \delta_2\},$$
$$I_\varepsilon = (-\varepsilon, +\varepsilon).$$

By the existence theorem of ordinary differential equations (I, §3, E), there are s C^∞ functions $\varphi^\alpha(t, x, a)$ defined on $I_\varepsilon \times U_1(\delta_1) \times U_2(\delta_2)$, which form a unique solution of (5), such that

$$\varphi^\alpha(0, x, a) = a^\alpha \qquad (\alpha = 1, \ldots, s) \qquad (6)$$

holds. From (6) we get

$$\frac{\partial \varphi^\alpha}{\partial a^\beta}(0, x, a) = \delta^\alpha_\beta.$$

Hence, choosing ε sufficiently small, we can assume

$$\det \frac{\partial \varphi^\alpha}{\partial a^\beta}(t, x, a) \neq 0 \qquad (7)$$

for $(t, x, a) \in I_\varepsilon \times U_1(\delta_1) \times U_2(\delta_2)$. Now fix an s such that $|s| \leq \varepsilon/2$, and let $|t| < 2$. Then, since $|st| < \varepsilon$, $\varphi^\alpha(st, x, a)$ is defined for $|t| < 2$, and $(x, a) \in U_1(\delta_1) \times U_2(\delta_2)$. If we write $y^\alpha(t)$ for $\varphi^\alpha(st, x, a)$ considered as a function of t, then we have

$$\frac{dy^\alpha(t)}{dt} = \sum_{k=1}^{r} \Phi_{\alpha k}(st, x, y(t)) x^k s$$

$$= \sum_{k=1}^{r} \Phi_{\alpha k}(t, sx, y(t)) x^k s \qquad (sx = (sx^1, \ldots, sx^r)),$$

so that $\{y^\alpha(t)\}$ is a solution of (5), coinciding with $\{\varphi^\alpha(t, sx, a)\}$ for $|t| < \varepsilon$. Hence, for $(t, x, a) \in I_2 \times U_1(\delta_1') \times U_2(\delta_2)$ $(\delta_1' = \varepsilon\delta_1/2)$, if we set

$$\psi^\alpha(t, x, a) = \varphi^\alpha\left(\frac{\varepsilon}{2} t, \frac{2}{\varepsilon} x, a\right), \qquad (8)$$

then $\psi^\alpha(t, x, a)$ is a solution of (5). From (7) and (8), if $|t| < 2$, $|x^k| < \delta_1'$ and $|a^\alpha| < \delta_2$, then

$$\det\left(\frac{\partial \psi^\alpha(t, x, a)}{\partial a^\beta}\right) \neq 0. \qquad (9)$$

Now define the map

$$\Psi_a : I_2 \times U_1(\delta_1') \to U$$

by

$$\Psi_a(t, x) = (tx^1, \ldots, tx^r, \psi^1(t, x, a), \ldots, \psi^s(t, x, a)).$$

That is, Ψ_a is given by

$$x^i \circ \Psi_a = tx^i,$$
$$y^\alpha \circ \Psi_a = \psi^\alpha(t, x, a).$$

Here we consider a as a parameter. Then

$$\Psi_a^*(\omega_\alpha) = d\psi^\alpha - \sum_{k=1}^{r} \Phi_{\alpha k}(tx, \psi(t, x, a))(x^k \, dt + t \, dx^k)$$

$$= \frac{d\psi^\alpha}{dt}(t, x, a) \, dt + \sum_{k=1}^{r} \frac{\partial \psi^\alpha}{\partial x^k}(t, x, a) \, dx^k$$

$$- \sum_{k=1}^{r} \Phi_{\alpha k}(tx, \psi(t, x, a))(x^k \, dt + t \, dx^k).$$

Since $\psi^\alpha(t, x, a)$, as a function of t, is a solution of (5), the coefficient of dt is 0, and we have

$$\Psi_a^*(\omega_\alpha) = \sum_{k=1}^{r} A_{\alpha k}(t, x, a) \, dx^k,$$

$$A_{\alpha k} = \frac{\partial \psi^\alpha}{\partial x^k}(t, x, a) - t\Phi_{\alpha k}(tx, \psi(t, x, a))t. \tag{10}$$

Since $\psi^\alpha(0, x, a) = a^\alpha$, we have

$$A_{\alpha k}(0, x, a) = 0. \tag{11}$$

Using the integrability condition (4), we shall prove that $A_{\alpha k} \equiv 0$. From (10), we have

$$\Psi_a^*(d\omega_\alpha) = d\Psi_a^*(\omega_\alpha)$$

$$= \sum_{k=1}^{r} \frac{\partial A_{\alpha k}}{\partial t} \, dt \wedge dx^k + \sum_{j,k=1}^{r} \frac{\partial A_{\alpha k}}{\partial x^j} \, dx^j \wedge dx^k, \tag{12}$$

while, from (4), we have

$$\Psi_a^*(d\omega_\alpha) = \sum_{\beta=1}^{s} \Psi_a^*(\tilde{\omega}_{\alpha\beta}) \wedge \Psi_a^*(\omega_\beta). \tag{13}$$

Hence, setting $\Psi_a^*(\tilde{\omega}_{\alpha\beta}) = P_{\alpha\beta} \, dt + \sum_{k=1}^{r} Q_{\alpha\beta k} \, dx^k$ and comparing the coefficients of $dt \wedge dx^k$ in (12) and (13), we get

$$\frac{\partial A_{\alpha k}}{\partial t} = \sum_{\beta=1}^{s} P_{\alpha\beta} A_{\beta k} \qquad (k = 1, \ldots, r; \alpha = 1, \ldots, s). \tag{14}$$

That is, $\{A_{\alpha k}\}$, as functions of t, is a solution of the system of differential equations (14) and satisfies the initial condition (11). However, it is clear, from the form of (14), that the set of functions that are constantly zero is also a solution of (14) satisfying the initial condition (11). Hence, by the uniqueness of the solution, $A_{\alpha k} \equiv 0$ on $I_2 \times U(\delta_1')$. Thus we have

$\Psi_a^*(\omega_\alpha) = 0$ $(\alpha = 1, \ldots, s)$. Set

$$f^\alpha(x, a) = \psi^\alpha(1, x, a) \qquad (\alpha = 1, \ldots, s),$$

and let N_a be the set of points in U satisfying

$$y^\alpha - f^\alpha(x, a) = 0 \qquad (\alpha = 1, \ldots, s) \quad (a \text{ is a parameter}).$$

By Theorem 1 of II, §10, N_a is a submanifold of U. It is an integral manifold of $\mathfrak{D}$, because $i^*\omega_\alpha$ is obtained by setting $t = 1$ in (10), and since $\Psi_a^*(\omega_\alpha) = 0$, we get $i^*\omega_\alpha = 0$.

Now consider $(y^1, \ldots, y^s, x^1, \ldots, x^r, a^1, \ldots, a^s)$ as $2s + r$ variables, and let $F^\alpha(x, y, a) = y^\alpha - f^\alpha(x, a)$ $(\alpha = 1, \ldots, s)$. By (9), we see that $dF^\alpha, dy^\beta, dx^\gamma$ $(\alpha, \beta = 1, \ldots, s; \gamma = 1, \ldots, r)$ are linearly independent on a neighborhood of the origin in $\mathbf{R}^{2s + r}$. Hence $(y^1, \ldots, y^s, x^1, \ldots, x^r, F^1, \ldots, F^s)$ is a local coordinate system of $\mathbf{R}^{2s + r}$ around the origin. Hence $F^\alpha(x, y, a) = 0$ $(\alpha = 1, \ldots, s)$ defines an $n(= r + s)$-dimensional submanifold $\tilde{N}$ of $\mathbf{R}^{2s + r}$, and $(x^1, \ldots, x^r, y^1, \ldots, y^s)$ is a local coordinate system of $\tilde{N}$ around 0. Hence there are s C^∞ functions $g^\alpha(x, y)$ of (x, y), and $a^\alpha = g^\alpha(x, y)$ holds on $\tilde{N}$. That is,

$$y^\alpha - f^\alpha(x, g(x, y)) = 0, \qquad g^\alpha(0, 0) = 0 \quad (\alpha = 1, \ldots, s)$$

holds identically. Hence we also have

$$dy^\alpha - \sum_\gamma \frac{\partial f^\alpha}{\partial x^\gamma}(x, g(x, y))\, dx^\gamma - \sum_\gamma \sum_\beta \frac{\partial f^\alpha}{\partial a^\beta}(x, g(x, y)) \frac{\partial g^\beta}{\partial x^\gamma}(x, y)\, dx^\gamma$$

$$- \sum_\lambda \sum_\beta \frac{\partial f^\alpha}{\partial a^\beta}(x, g(x, y)) \frac{\partial g^\beta}{\partial y^\lambda}(x, y)\, dy^\lambda = 0,$$

so that we get

$$\sum_\beta \frac{\partial f^\alpha}{\partial a^\beta}(x, g(x, y)) \frac{\partial g^\beta}{\partial y^\lambda}(x, y) = \delta_\lambda^\alpha.$$

Since we have (9), this leads to

$$\det\left(\frac{\partial g^\beta}{\partial y^\lambda}(x, y)\right) \neq 0.$$

Hence we have

$$D(x^1, \ldots, x^r, g^1, \ldots, g^s)/D(x^1, \ldots, x^r, y^1, \ldots, y^s)_0 \neq 0,$$

so that $(x^1, \ldots, x^r, g^1, \ldots, g^s)$ is a local coordinate system of $\mathbf{R}^n$

around 0, and N_a is given by the equations

$$g^\alpha = a^\alpha \qquad (\alpha = 1, \ldots, s).$$

Hence, by Lemma 1, $\mathfrak{D}$ is completely integrable. Thus Frobenius' theorem has been proved.

In what was said above, all of the ω_α's were real differential forms, but if we consider holomorphic differential forms on a neighborhood of the origin in $\mathbf{C}^n$, by arguments similar to those in the real case, we can prove the following theorem:

THEOREM 1'. *Let ω_α $(\alpha = 1, \ldots, n - r)$ be holomorphic differential forms of order 1, defined on a neighborhood U of the origin 0 in $\mathbf{C}^n$. That is, ω_α is of the form $\omega_\alpha = \sum_{j=1}^n f_{\alpha j}(z)\, dz^j$, where $f_{\alpha j}$ is a holomorphic function on $(z^1, \ldots, z^n)$. Furthermore, we assume that the ω_α's are linearly independent at each point of U, and there are $(n - r)^2$ holomorphic differential forms $\tilde{\omega}_{\alpha\beta}$ on U such that*

$$d\omega_\alpha = \sum_{\beta=1}^{n-r} \tilde{\omega}_{\alpha\beta} \wedge \omega_\beta \qquad (\alpha = 1, \ldots, n - r)$$

holds. Then there are a neighborhood V of 0 contained in U, $n - r$ holomorphic functions g_α defined on V, and an $(n - r) \times (n - r)$ matrix $(a_{\alpha\beta}(z))$ whose entries are $(n - r)^2$ holomorphic functions defined on V, such that $\det(a_{\alpha\beta}(z)) \neq 0$ at each point of V, and $\omega_\alpha = \sum_{\beta=1}^{n-r} a_{\alpha\beta}\, dg_\beta$ holds on V.

The proof of Theorem 1' is completely analagous to that for $\mathbf{R}^n$, except that we should replace the existence theorem of solutions of differential equations of real variables and the inverse function theorem by the following existence theorem of solutions of holomorphic ordinary differential equations and the inverse function theorem for holomorphic functions.

a   Let $f^i(z, w^1, \ldots, w^k; t^1, \ldots, t^j)$ $(i = 1, \ldots, k)$ be complex-valued functions of $k + j + 1$ complex variables $(z, w^1, \ldots, w^k; t^1, \ldots, t^j)$, and suppose that they are holomorphic on a neighborhood of the origin of $\mathbf{C}^{k+j+1}$. If $(a^1, \ldots, a^k)$ is an arbitrary point in $\mathbf{C}^k$ which is sufficiently close to the origin 0, then the system of differential equations

$$\frac{dw^i}{dz} = f^i(z, w, t)$$

has a unique holomorphic solution such that $w^i = a^i$ for $z = 0$. This solution depends on the initial value (a^i) and the parameter (t^i), so that if we write

$$w^i = \varphi^i(z, a^1, \ldots, a^k; t^1, \ldots, t^j),$$

then the φ^i's, as functions of $k + j + 1$ variables, are holomorphic on a neighborhood of the origin of $\mathbf{C}^{k+j+1}$.†

b Let ψ be a holomorphic map from a neighborhood of the origin of $\mathbf{C}^n$ to a neighborhood of the origin of $\mathbf{C}^n$, and suppose that $\psi(0) = 0$. Let $\psi(z) = (\psi^1(z), \ldots, \psi^n(z))$, and assume that $\det (\partial\psi^i(0)/\partial z^j) \neq 0$. Then ψ is a homeomorphism from a sufficiently small neighborhood U of the origin in $\mathbf{C}^n$ to $\psi(U)$, and ψ^{-1} is also a holomorphic map.

The statement **b** can be proved easily by using the inverse function theorem for real variables (I, §4) (cf. H. Cartan, *loc, cit.*, p. 136).

Returning to the case of real variables, we shall now use vector fields instead of differential forms to express a differential system and its integrability condition.

Let $X_1, \ldots, X_n$ be n vector fields on an open set V of M, and $\omega_1, \ldots, \omega_n$ be n differential forms of order 1 on V. Suppose that these vector fields and also these differential forms are both linearly independent at each point of V, and suppose that $\omega_i(X_j) = \delta_{ij}$ $(i, j = 1, \ldots, n)$ holds at each point of V. Then we say that $\{X_1, \ldots, X_n\}$ and $\{\omega_1, \ldots, \omega_n\}$ are dual systems.

Suppose we are given n differential forms $\omega_1, \ldots, \omega_n$ which are linearly independent at each point of V. We define vector fields $X_i : q \to (X_i)_q$ $(i = 1, \ldots, n)$ on V by the condition that $\{(X_1)_q, \ldots, (X_n)_q\}$ and $\{(\omega_1)_q, \ldots, (\omega_n)_q\}$ are bases dual to each other. Then X_i is of class C^∞, and $\{X_1, \ldots, X_n\}$ and $\{\omega_1, \ldots, \omega_n\}$ are dual systems. Similarly, we can start from n vector fields $\{X_1, \ldots, X_n\}$ on V which are linearly independent at each point of V, and obtain a dual system $\{\omega_1, \ldots, \omega_n\}$.

LEMMA 2. *Let $\omega_{r+1}, \ldots, \omega_n$ (rep. $X_1, \ldots, X_r$) be differential forms of order 1 (resp. vector fields) which are linearly independent at each point of a neighborhood U of p. If U is taken sufficiently small, then we can choose r differential forms $\omega_1, \ldots, \omega_r$ of order 1 on U (resp. $n - r$ vector fields $X_{r+1}, \ldots, X_n$ on U) such that $\omega_1, \ldots, \omega_n$ (resp. $X_1, \ldots, X_n$) are linearly independent at each point of U.*

The proof of this lemma is easy and is left to the reader.

LEMMA 3. *Let $\mathfrak{D}$ be a correspondence that assigns an r-dimensional subspace $\mathfrak{D}_p$ of $T_p(M)$ to each point p of M. For $\mathfrak{D}$ to be a differential*

†For the proof, cf. H. Cartan, *loc. cit.*, Chapter VII.

system, it is necessary and sufficient that for each point p of M, there are a neighborhood U of p, and vector fields $X_1, \ldots, X_r$ on U, such that, $\{(X_1)_q, \ldots, (X_r)_q\}$ is a basis of $\mathfrak{D}_q$ at each point q of U.

Proof. If there are such vector fields $X_1, \ldots, X_r$, then $X_1, \ldots, X_r$ are linearly independent at each point of U. By Lemma 2, we can pick $n - r$ vector fields $X_{r+1}, \ldots, X_n$ on a neighborhood $U'(\subset U)$ of p such that $X_1, \ldots, X_n$ are linearly independent at each point of U. Furthermore, let $\omega_1, \ldots, \omega_n$ be n differential forms of order 1 on U that are dual to $X_1, \ldots, X_n$. At each point q of U, $\mathfrak{D}_q$ is spanned by $\{(X_1)_q, \ldots, (X_r)_q\}$, and $(\omega_i)_q((X_j)_q) = \delta_{ij}$ holds. Hence we have

$$\mathfrak{D}_q = \{v \in T_q(M) | (\omega_{r+1})_q(v) = \cdots = (\omega_n)_q(v) = 0\}.$$

Hence $\mathfrak{D}$ is a differential system defined by $\omega_{r+1} = \cdots = \omega_n = 0$ on U. Conversely, if $\mathfrak{D}$ is a differential system, then we can prove, in a quite similar manner, that there are $X_1, \ldots, X_r$ satisfying the condition in the lemma.

In general, if vector fields $X_1, \ldots, X_r$ on an open set V of M satisfy the condition: "$\{(X_1)_q, \ldots, (X_r)_q\}$ is a basis of $\mathfrak{D}_q$ at each point q of V", then $\{X_1, \ldots, X_r\}$ is called a *local basis* of $\mathfrak{D}$ on V.

We can rephrase the complete integrability of $\mathfrak{D}$ as follows: $\mathfrak{D}$ is said to be completely integrable if, at each point $p \in M$, there is a local coordinate system $\{x^1, \ldots, x^n\}$, defined in a neighborhood U of p, such that $\partial/\partial x^1, \ldots, \partial/\partial x^r$ form a local basis of $\mathfrak{D}$ on U.

Using local basis, we can also rephrase Frobenius' theorem as follows:

THEOREM 2. *Let $\mathfrak{D}$ be an r-dimensional differential system on an n-dimensional manifold M. Then $\mathfrak{D}$ is completely integrable if and only if for every local basis $\{X_1, \ldots, X_r\}$ of $\mathfrak{D}$ on any open set V of M, there are C^∞ functions c_{ij}^k on V such that*

$$[X_i, X_j] = \sum_{k=1}^{r} c_{ij}^k X_k \qquad (i, j = 1, \ldots, r) \tag{15}$$

holds.

Proof. Sufficiency: Let $\omega_{r+1} = \cdots = \omega_n = 0$ be local equations of $\mathfrak{D}$ on a sufficiently small neighborhood U of p. Pick r differential forms $\omega_1, \ldots, \omega_r$ of order 1 such that $\{\omega_1, \ldots, \omega_n\}$ is linearly independent at each point of U. If $X_1, \ldots, X_n$ are n vector fields on U dual to $\omega_1, \ldots, \omega_n$, then $\{X_1, \ldots, X_r\}$ is a local basis of $\mathfrak{D}$ on U. Then, by the hypothesis of Theorem 2, we have

$$[X_i, X_j] = \sum_{k=1}^{r} c_{ij}^k X_k.$$

Now for $\alpha = r + 1, \ldots, n$, $d\omega_\alpha$ can be written uniquely as

$$d\omega_\alpha = \sum_{1 \leq i < j \leq r} a_{ij}^\alpha \, \omega_i \wedge \omega_j + \sum_{i=1}^{r} \sum_{\beta = \gamma + 1}^{n} b_{i\beta}^\alpha \, \omega_i \wedge \omega_\beta$$
$$+ \sum_{r+1 \leq \beta < \gamma \leq n} d_{\beta\gamma}^\alpha \, \omega_\beta \wedge \omega_\gamma.$$

We want to show that $a_{ij}^\alpha = 0$, i.e., $(d\omega_\alpha)(X_i, X_j) = 0$ $(i, j = 1, \ldots, r)$, because then $d\omega_\alpha = \sum_{\beta = r+1}^{n} \tilde{\omega}_{\alpha\beta} \wedge \omega_\beta$, and by Frobenius' theorem, $\mathfrak{D}$ is completely integrable. So we observe that

$$(d\omega_\alpha)(X_i, X_j) = X_i(\omega_\alpha(X_j)) - X_j(\omega_\alpha(X_i)) - \omega_\alpha([X_i, X_j])$$
$$= X_i(\omega_\alpha(X_j)) - X_j(\omega_\alpha(X_i)) - \sum_{k=1}^{r} c_{ij}^k \omega_\alpha(X_k).$$

However, $\omega_\alpha(X_i) = \delta_{\alpha i} = 0$ $(\alpha = r+1, \ldots, n; i = 1, \ldots, r)$. Hence $(d\omega_\alpha)(X_i, X_j) = 0$. *Necessesity*: Let $\mathfrak{D}$ be completely integrable, $X_1, \ldots, X_r$ a local base of $\mathfrak{D}$ on V. We shall prove that (15) holds. For that, it suffices to prove (15) holds for a sufficiently small neighborhood U of each point p of V. As before, take dual systems $\{X_1, \ldots, X_n\}$, $\{\omega_1, \ldots, \omega_n\}$ on U. Then $\omega_{r+1} = \cdots = \omega_n = 0$ are local equations of $\mathfrak{D}$. From the integrability condition of $\mathfrak{D}$, we know that $d\omega_\alpha$ $(\alpha = r+1, \ldots, n)$ is of the form

$$d\omega_\alpha = \sum_{i=1}^{r} \sum_{\beta=r+1}^{n} b_{i\beta}^\alpha \, \omega_i \wedge \omega_\beta + \sum_{r+1 \leq \beta < \gamma \leq n} d_{\beta\gamma}^\alpha \, \omega_\alpha \wedge \omega_\gamma.$$

Hence we have $(d\omega_\alpha)(X_i, X_j) = 0$ $(i, j = 1, \ldots, r)$. Following the previous proof backwards, we get

$$\omega_\alpha([X_i, X_j]) = 0 \qquad (\alpha = r+1, \ldots, n; i, j = 1, \ldots, r). \qquad (16)$$

We can write $[X_i, X_j]$ on U as $[X_i, X_j] = \sum_{k=1}^{r} c_{ij}^k X_k + \sum_{\alpha=r+1}^{n} e_{ij}^\alpha X_\alpha$, but, from (16), we get $e_{ij}^\alpha = 0$ $(\alpha = r+1, \ldots, n)$, and thus we have $[X_i, X_j] = \sum_k c_{ij}^k X_k$. That is, (15) holds.

Problem 2. Show that a 1-dimensional differential system is always completely integrable.

Problem 3. Let X be a vector field on M which is never 0. By defining $\mathfrak{D}_p = \{X_p\}$, we get a 1-dimensional differential system $\mathfrak{D}$ on M. Study the relation between the integral curves of the vector field X and the integral manifolds of $\mathfrak{D}$.

§9. An Application to Integrable Almost Complex Structures

Let M be a $2n$-dimensional *analytic* manifold, and J an almost complex structure on M (II, §17). Let $(x^1, \ldots, x^{2n})$ be a C^ω-local coordinate system on M. J is said to be of class C^ω if all components J_k^i ($i, k = 1,$ $\ldots, 2n$) of J with respect to $(x^1, \ldots, x^{2n})$ are of class C^ω. As an application of Theorem 1 of §8, we shall prove the following theorem.

THEOREM. *Let M be a 2n-dimensional analytic manifold, and let J be a C^ω almost complex structure on M. If J is integrable, then M is a complex manifold, and J is the almost complex structure attached to M.*

In proving this theorem, we shall use freely the notation in II, §17.

LEMMA. *For each point p of an almost complex manifold M of dimension $2n$, suppose that we have a neighborhood U of p, and n complex-valued functions f^α ($\alpha = 1, \ldots, n$) defined on U, satisfying (1) $\{df^\alpha\}$ are linearly independent at each point of U, (2) for each point q of U, and each $a \in T_q^-(M)$, $(df^\alpha)_q(a) = 0$ holds.*

Then M is a complex manifold, and J is the almost complex structure attached to M.

Proof. Let $\bar{f}^\alpha$ be the conjugate of f^α. For an arbitrary $b \in T_q^C(M)$, we have $\overline{(d\bar{f}^\alpha)_q(b)} = (df^\alpha)_q(\bar{b})$. Hence, by condition (2), if $b \in T_q^+(M)$, then $(d\bar{f}^\alpha)_q(b) = 0$ holds at each point q of U. From this and condition (1), we see that $df^1, \ldots, df^n, d\bar{f}^1, \ldots, d\bar{f}^n$ are linearly independent at each point of U. Hence, if we let $f^\alpha = u^\alpha + iv^\alpha$ ($\alpha = 1, \ldots, n$), then $du^1, dv^1, \ldots, du^n, dv^n$ are linearly independent at each point of U, and hence $(u^1, v^1, \ldots, u^n, v^n)$ is a local coordinate system of M on U. But

$$(df^\alpha)(\partial/\partial u^\beta + i\partial/\partial v^\beta) = \partial f^\alpha/\partial u^\beta + i\,\partial f^\alpha/\partial v^\beta = \delta_\beta^\alpha + i^2\delta_\beta^\alpha = 0.$$

Similarly we have $(d\bar{f}^\alpha)(\partial/\partial u^\beta - i\,\partial/\partial v^\beta) = 0$. Hence, by conditions (1) and (2), we have $(\partial/\partial u^\beta + i\,\partial/\partial v^\beta)_q \in T_q^-(M)$, $(\partial/\partial u^\beta - i\,\partial/\partial v^\beta)_q \in T_q^+(M)$. That is,

$$J_q(\partial/\partial u^\beta + i\,\partial/\partial v^\beta)_q = -i(\partial/\partial u^\beta + i\,\partial/\partial v^\beta)_q,$$
$$J_q(\partial/\partial u^\beta - i\,\partial/\partial v^\beta)_q = i(\partial/\partial u^\beta - i\,\partial/\partial v^\beta)_q.$$

From this we get $J_q(\partial/\partial u^\beta)_q = (\partial/\partial v^\beta)_q$, $J_q(\partial/\partial v^\beta)_q = -(\partial/\partial u^\beta)_q$ ($\beta = 1, \ldots, n$). Hence, by the Lemma of II, §17, M is a complex manifold and J is the almost complex structure attached to M.

Now let us show that, under the hypothesis of the theorem, we can choose n functions in a neighborhood of each point p of M satisfying the conditions in the lemma above. As our problem is a local one, it suffices to take p as the origin of $\mathbf{R}^{2n}$, and argue on a neighborhood U of the origin in $\mathbf{R}^{2n}$. Thus we let $(x^1, \ldots, x^{2n})$ be the coordinate system on $\mathbf{R}^{2n}$, and express J as a matrix formed by its components, $J = (J_k^i(x))$, and let all $J_k^i(x)$'s be of class C^ω. If U is sufficiently small, we can find

n linearly independent C^ω vector fields† X_α ($\alpha = 1, \ldots, n$) of holomorphic type. In fact, $(\partial/\partial x^i)^+ = \frac{1}{2}((\partial/\partial x^i) - iJ(\partial/\partial x^i))$ is of class C^ω and is of holomorphic type. Now, at the origin, there are n of the $(\partial/\partial x_i)^+$ ($i = 1, \ldots, 2n$) which are linearly independent. Let them be X_α ($\alpha = 1, \ldots, n$). If one takes a sufficiently small neighborhood U of the origin, the X_α's are linearly independent at each point of U. Let the conjugate $\bar{X}_\alpha$ of X_α be denoted by $X_{\bar{\alpha}}$ ($\alpha = 1, \ldots, n$). Then $X_1, \ldots, X_n, X_{\bar{1}}, \ldots, X_{\bar{n}}$ are linearly independent at each point of U. Moreover, at each point q of U, $(X_1)_q, \ldots, (X_n)_q$ is a base of $T_q^+(M)$, and $(X_{\bar{1}})_q, \ldots, (X_{\bar{n}})_q$ is a base of $T_q^-(M)$. Now define $2n$ complex differential forms $\omega_\alpha, \omega_{\bar{\alpha}}$ ($\alpha = 1, \ldots, n$) of degree 1 on U by

$$\omega_\alpha(X_\beta) = \delta_{\alpha\beta}, \qquad \omega_\alpha(X_{\bar{\beta}}) = 0,$$
$$\omega_{\bar{\alpha}}(X_\beta) = 0 \ , \qquad \omega_{\bar{\alpha}}(X_{\bar{\beta}}) = \delta_{\alpha\beta}. \tag{1}$$

Then $\omega_{\bar{\alpha}} = \bar{\omega}_\alpha$. Moreover $\omega_\alpha, \omega_{\bar{\alpha}}$ are of class C^ω, i.e., if

$$\omega_\alpha = \sum_{i=1}^{2n} f_{\alpha i}(x) \, dx^i,$$

then the complex-valued functions $f_{\alpha i}$ are of class C^ω. This is because the $2n \times 2n$ matrix formed by the components of $\{\omega_\alpha, \omega_{\bar{\alpha}}\}$ is the inverse matrix of the $2n \times 2n$ matrix formed by the components of $\{X_\alpha, X_{\bar{\alpha}}\}$, and because the components of $X_\alpha, X_{\bar{\alpha}}$ are C^ω functions.

Now, since J is integrable by hypothesis, $[X_\alpha, X_\beta]$ is of holomorphic type, and $[X_{\bar{\alpha}}, X_{\bar{\beta}}]$ is of antiholomorphic type. That is, on U, we have

$$[X_\alpha, X_\beta] = \sum_\gamma c_{\alpha\beta}^\gamma X_\gamma, \quad [X_{\bar{\alpha}}, X_{\bar{\beta}}] = \sum_\gamma \bar{c}_{\alpha\beta}^\gamma X_{\bar{\gamma}} \qquad (\alpha, \beta = 1, \ldots, n). \tag{2}$$

The $2n$ forms $\{\omega_\alpha, \omega_{\bar{\alpha}}\}$ are linearly independent at each point of U, so that we can write, on U, uniquely

$$d\omega_\alpha = \sum_{\beta < \gamma} a_{\beta\gamma}^\alpha \, \omega_\alpha \wedge \omega_\gamma + \sum_{\beta, \gamma} b_{\beta\bar{\gamma}}^\alpha \, \omega_\beta \wedge \omega_{\bar{\gamma}} + \sum_{\beta < \gamma} d_{\bar{\beta}\bar{\gamma}}^\alpha \, \omega_{\bar{\beta}} \wedge \omega_{\bar{\gamma}}. \tag{3}$$

By (1) and (3), we have $d_{\bar{\beta}\bar{\gamma}}^\alpha = (d\omega_\alpha)(X_{\bar{\beta}}, X_{\bar{\gamma}})$. On the other hand, from (1) and (2), we get

$$(d\omega_\alpha)(X_{\bar{\beta}}, X_{\bar{\gamma}}) = X_{\bar{\beta}}\omega_\alpha(X_{\bar{\gamma}}) - X_{\bar{\gamma}}\omega_\alpha(X_{\bar{\beta}}) - \sum_\lambda \bar{c}_{\beta\gamma}^\lambda \, \omega_\alpha(X_{\bar{\lambda}}) = 0.$$

Hence we have $d_{\bar{\beta}\bar{\gamma}}^\alpha = 0$,

$$d\omega_\alpha = \sum_{\beta < \gamma} a_{\beta\gamma}^\alpha \, \omega_\beta \wedge \omega_\gamma + \sum_{\beta, \gamma} b_{\beta\bar{\gamma}}^\alpha \, \omega_\beta \wedge \omega_{\bar{\gamma}} \tag{3'}$$

---

†In general, a complex vector field $X + iY$ is said to be of class C^ω if both X and Y are of class C^ω.

and its conjugate

$$d\omega_{\bar{\alpha}} = \sum_{\beta < \gamma} \bar{a}^{\alpha}_{\bar{\beta}\bar{\gamma}}\, \omega_{\bar{\beta}} \wedge \omega_{\bar{\gamma}} + \sum_{\beta,\gamma} \bar{b}^{\alpha}_{\bar{\beta}\gamma}\, \omega_{\bar{\beta}} \wedge \omega_{\gamma}. \tag{3''}$$

Now imbed $\mathbf{R}^{2n}$ in $\mathbf{C}^{2n}$ in the natural way. The coefficients $f_{\alpha i}(x)$ of ω_α and their conjugates $\bar{f}_{\alpha i}(x) = f_{\bar{\alpha} i}(x)$ are C^ω functions of the real variables $(x^1, \ldots, x^{2n})$ on a neighborhood U of 0 in $\mathbf{R}^{2n}$. Taking U sufficiently small, and considering the Taylor expansions of these functions around 0, we see the following. There are a neighborhood $\tilde{U}$ of 0 in $\mathbf{C}^{2n}$, and functions $\tilde{f}_{\alpha i}, \tilde{f}_{\bar{\alpha} i}$ holomorphic on $\tilde{U}$, such that $\tilde{U} \cap \mathbf{R}^{2n} = U$, and the restrictions of $\tilde{f}_{\alpha i}$ and $\tilde{f}_{\bar{\alpha} i}$ to U are equal to $f_{\alpha i}$ and $f_{\bar{\alpha} i}$, respectively. Now let

$$\tilde{\omega}_\alpha = \sum_{i=1}^{2n} \tilde{f}_{\alpha i}(z)\, dz^i, \qquad \tilde{\omega}_{\bar{\alpha}} = \sum_{i=1}^{2n} \tilde{f}_{\bar{\alpha} i}(z)\, dz^i.$$

Then $\tilde{\omega}_\alpha$ and $\tilde{\omega}_{\bar{\alpha}}$ are holomorphic differential forms on U, and we have

$$i^*\tilde{\omega}_\alpha = \omega_\alpha, \qquad i^*\tilde{\omega}_{\bar{\alpha}} = \omega_{\bar{\alpha}}. \tag{4}$$

Here i is the injection from U into $\tilde{U}$. Now the $2n \times 2n$ matrix formed by the coefficients of $\{\omega_\alpha, \omega_{\bar{\alpha}}\}$ has a nonzero determinant at the origin. Hence, by taking $\tilde{U}$ sufficiently small, we can assume that $\{\tilde{\omega}_\alpha, \tilde{\omega}_{\bar{\alpha}}\}$ are linearly independent at each point of $\tilde{U}$. Since $d\tilde{\omega}_\alpha, d\tilde{\omega}_{\bar{\alpha}}$ are holomorphic, they can be written as linear combinations of $dz^\beta \wedge dz^{\gamma}$'s, and, conversely, since $\{\tilde{\omega}_\alpha, \tilde{\omega}_{\bar{\alpha}}\}$ are linearly independent at each point of $\tilde{U}$, dz^β ($\beta = 1, \ldots, 2n$) can be written as a linear combination of the $2n$ forms $\{\tilde{\omega}_\alpha, \tilde{\omega}_{\bar{\alpha}}\}$. Hence $d\tilde{\omega}_\alpha$ can be written as

$$d\tilde{\omega}_\alpha = \sum_{\beta < \gamma} \tilde{a}^{\alpha}_{\beta\gamma}\, \tilde{\omega}_\beta \wedge \tilde{\omega}_\gamma + \sum_{\beta,\gamma} \tilde{b}^{\alpha}_{\beta\bar{\gamma}}\, \tilde{\omega}_\beta \wedge \tilde{\omega}_{\bar{\gamma}} + \sum_{\beta < \gamma} \tilde{c}^{\alpha}_{\bar{\beta}\bar{\gamma}}\, \tilde{\omega}_{\bar{\beta}} \wedge \tilde{\omega}_{\bar{\gamma}},$$

and the coefficients $\tilde{a}^{\alpha}_{\beta\gamma}$, $\tilde{b}^{\alpha}_{\beta\bar{\gamma}}$, $\tilde{c}^{\alpha}_{\bar{\beta}\bar{\gamma}}$ are holomorphic functions. Now if we apply i^* on both sides of this equation, we obtain $i^*\tilde{c}^{\alpha}_{\bar{\beta}\bar{\gamma}} = 0$, by (3) and (4). That is, for the holomorphic functions $\tilde{c}^{\alpha}_{\bar{\beta}\bar{\gamma}}$, if we restrict the values of the variables $z^1, \ldots, z^{2n}$ to real numbers, then $\tilde{c}^{\alpha}_{\bar{\beta}\bar{\gamma}}$ becomes 0. Hence we conclude that $\tilde{c}^{\alpha}_{\bar{\beta}\bar{\gamma}} \equiv 0$. Hence we have

$$d\tilde{\omega}_\alpha = \sum_{\beta < \gamma} \tilde{a}^{\alpha}_{\beta\gamma}\, \tilde{\omega}_\beta \wedge \tilde{\omega}_\gamma + \sum_{\beta,\gamma} \tilde{b}^{\alpha}_{\beta\bar{\gamma}}\, \tilde{\omega}_\beta \wedge \tilde{\omega}_{\bar{\gamma}} \qquad (\alpha = 1, \ldots, n). \tag{5}$$

Hence, by Theorem 1 of §8, there exist n holomorphic functions $\tilde{g}_\alpha$ on $\tilde{U}$, and an $n \times n$ matrix $(\tilde{A}_{\alpha\beta}(z))$ whose entries $\tilde{A}_{\alpha\beta}$ are holomorphic functions on $\tilde{U}$, such that

$$d\tilde{g}_\alpha = \sum \tilde{A}_{\alpha\beta}\tilde{\omega}_\beta, \quad \det(\tilde{A}_{\alpha\beta}(z)) \neq 0 \qquad (z \in \tilde{U}) \tag{6}$$

hold. Let $A_{\alpha\beta}$, g_α be the functions on U obtained by restricting $\tilde{A}_{\alpha\beta}$, $\tilde{g}_\alpha$, respectively, to U. Then, by (4) and (6), we have

$$dg_\alpha = \sum A_{\alpha\beta}\omega_\beta, \quad \det(A_{\alpha\beta}(x)) \neq 0 \qquad (x \in U).$$

Hence dg_α $(\alpha = 1, \ldots, n)$ are linearly independent at each point of U, and if $a \in T_q^-(M)$, then, since $\omega_\beta(a) = 0$, we have $(dg_\alpha)_q(a) = \sum_\beta A_{\alpha\beta}(q) \omega_\beta(a) = 0$, i.e., $g_1, \ldots, g_n$ satisfy the conditions of the lemma. Thus the theorem is proved.

§10. Maximal Connected Integral Manifolds

Let $\mathfrak{D}$ be a completely integrable r-dimensional differential system on an n-dimensional manifold M. An r-dimensional connected integral manifold N of $\mathfrak{D}$ is said to be *maximal* if the only r-dimensional connected integral manifold containing N as a subset is N itself. For maximal connected integral manifolds, we have the following theorem of Chevalley and Ehresmann.

THEOREM 1.   *Let $\mathfrak{D}$ be a completely integrable r-dimensional differential system on an n-dimensional manifold M. For each point p of M, there is one and only one r-dimensional maximal connected integral manifold F_p of $\mathfrak{D}$ passing through p. Every r-dimensional connected integral manifold passing through p is an open submanifold of F_p. An r-dimensional maximal connected integral manifold of $\mathfrak{D}$ is called a leaf of $\mathfrak{D}$.*

To prove this theorem, we let $\{N_\alpha\}_{\alpha \in A}$ be the set of all r-dimensional connected integral manifolds passing through p, and set

$$F_p = \bigcup_{\alpha \in A} N_\alpha.$$

Suppose we can put some topology and differentiable structure on F_p so that (1) F_p is a submanifold of M, and (2) each N_α is an open submanifold of F_p. Then F_p is the desired leaf. In fact, from (2), we see that each N_α is a connected open set of F_p, and since $\cap_{\alpha \in A} N_\alpha \neq \varnothing$, F_p, which is the union of the N_α's, is also connected. Now let $q \in F_p$. Then $q \in N_\alpha$ for some $\alpha \in A$. Since N_α is an open set of F_p, the tangent spaces at q of F_p and N_α coincide. But N_α is an integral manifold, so that $T_q(F_p) = \mathfrak{D}_q$, and hence F_p is an r-dimensional connected integral manifold. It is clear from the definition of F_p and condition (2) that F_p satisfies the condition of the theorem.

In order to define a topology and a differentiable structure on F_p, we introduce the following notation and terminology. At a point q of

M, choose a neighborhood Q of q, and a local coordinate system $(x^1, \ldots, x^n)$ on Q, satisfying the condition of §8, Lemma 1, and $x^i(q) = 0$, $i = 1, \ldots, n$. Then

$$W_q = \{w \in Q | x^{r+1}(w) = \cdots = x^n(w) = 0\}$$

is an r-dimensional connected integral manifold passing through q. W_q is called a *slice* of Q passing through q. W_q is a regular submanifold of M.

LEMMA 1.   *Let N be an r-dimensional integral manifold passing through q. If we take a sufficiently small neighborhood Q of q in M, then the slice W_q of Q is an open submanifold of N.*

Proof.   Since the identity map from N into M is continuous, $Q \cap N$ is an open set of the manifold N. Hence the connected component V_q of $Q \cap N$ containing q is a neighborhood of q in N. Hence V_q is also an r-dimensional integral manifold passing through q. By the choice of the local coordinate system $(x^1, \ldots, x^n)$ on Q, at each point w of Q, $(\partial/\partial x^1)_w, \ldots, (\partial/\partial x^r)_w$ is a basis of $\mathfrak{D}_w$. Hence, writing $y^1, \ldots, y^n$ for the restrictions of $x^1, \ldots, x^n$ to V_q, we have $\partial/\partial y^1, \ldots, \partial/\partial y^r$ spanning the tangent space of V_q at each point of V_q. Thus $(y^1, \ldots, y^r)$ is a local coordinate system around each point of V_q. On the other hand, since $\partial y^j/\partial y^i = 0$ $(i = 1, \ldots, r; j = r + 1, \ldots, n)$, $y^{r+1}, \ldots, y^n$ are constants on V_q, and since $y^j(q) = 0$, we conclude that $y^{r+1} = \ldots = y^n = 0$. Hence V_q is contained in the slice W_q of Q. Taking $\varepsilon > 0$ sufficiently small, and setting $V_q' = \{w \in V_q | \, |y^i(w)| < \varepsilon, \, i = 1, \ldots, r\}$, $Q' = \{w \in Q | \, |x^i(w)| < \varepsilon; \, i = 1, \ldots, n\}$, V_q' becomes the connected component of $Q' \cap N$ containing q, and by the definition of V_q', V_q' coincides with the slice W_q' of Q'.

Let us define a topology and a differentiable structure on F_p. We define the open sets O of F_p as follows: O is either the empty set, or O is not empty and for each point q of O, there is an r-dimensional integral manifold N passing through q and contained in O. If we define the open sets of F_p in this way, then (1) F_p itself and the empty set are open sets, (2) if O_i $(i \in I)$ are open sets, then $\cup_{i \in I} O_i$ is also an open set, and (3) if O_1, O_2 are open sets, then $O_1 \cap O_2$ is an open set. (1), (2) are clear, so let us prove (3). Let $q \in O_1 \cap O_2$. By the definition of open sets, there are r-dimensional integral manifolds N_1, N_2 containing q such that $N_1 \subset O_1$, $N_2 \subset O_2$. By Lemma 1, we can choose a sufficiently small neighborhood Q of q in M so that the slice W_q of Q is contained in N_1 and N_2. Then we have $W_q \subset N_1 \cap N_2 \subset O_1 \cap O_2$, and since W_q is an

r-dimensional integral manifold passing through q, $O_1 \cap O_2$ is an open set. Since conditions (1), (2), and (3) hold for the open sets, they define a topology on F_p, and by the definition of open sets, each N_α is an open set of F_p.

Furthermore, F_p is a Hausdorff space. In fact, let q, $q' \in F_p$, $q \neq q'$, $q \in N_\alpha$, $q' \in N_\beta$. Let Q, Q' be neighborhoods of q, q' in M, taken sufficiently small so that $Q \cap Q' = \varnothing$. Let W_q, $W_{q'}$ be slices of Q, Q' passing through q, q', respectively. Then W_q, $W_{q'}$ are neighborhoods of q, q' in F_p, and clearly $W_q \cap W_{q'} = \varnothing$, so that F_p is a Hausdorff space.

We give F_p a differentiable structure as follows. For a point q in F_p, we choose a slice $W_q(W_q \subset F_p)$. The slice W_q is an r-dimensional differentiable manifold and is diffeomorphic to a cube in $\mathbf{R}^r$ by the map $w(\in W_q) \to \psi_q(w) = (x^1(w), \ldots, x^r(w))$. Hence $\{W_q, \psi_q\}_{q \in F_p}$ is a coordinate neighborhood system of F_p, and one can check easily that the coordinate transformations are of class C^∞. Hence F_p is an r-dimensional differentiable manifold. It is clear from the definition of the differentiable structure of F_p and Lemma 1 that each N_α is an open submanifold of F_p and that F_p is a submanifold of M.

Problem. If F and F' are maximal connected integral manifolds, then show that $F \cap F' = \varnothing$ or $F = F'$.

Example. Let φ be a differentiable map from an n-dimensional manifold M onto an m-dimensional manifold M' $(n > m)$, and let $\mathrm{rank}_p(\varphi) = m$ at each point p of M. There is a completely integrable $(n - m)$-dimensional differential system on M, whose maximal connected integral manifolds are the connected components of the inverse images $\varphi^{-1}(q')$, where $q' \in M'$.

An outline of a proof. Since $\mathrm{rank}_p(\varphi) = m$ for each point p of M, the differential $(\varphi_*)_p$ of the map φ at p is a linear map from $T_p(M)$ onto $T_{\varphi(p)}(M')$. Hence, if we set

$$\mathfrak{D}_p = \{v \in T_p(M) | (\varphi_*)_p(v) = 0\},$$

then $\mathfrak{D}_p$ is an $(n - m)$-dimensional subspace of $T_p(M)$. One can prove that $\mathfrak{D} : p \to \mathfrak{D}_p$ is a completely integrable differential system, and that the connected components of $\varphi^{-1}(q')$ are the maximal connected integral manifolds of $\mathfrak{D}$, using Problem 5 of II, §7.

LEMMA 2. *Let M be n-dimensional manifold with a countable base, let $\mathfrak{D}$ be a completely integrable r-dimensional differential system on M, and let F be a leaf of $\mathfrak{D}$. Let $q \in F$, let Q be a coordinate neighborhood of* *q in M, taken as in Lemma 1, and let W_q be the slice of Q passing through q. Then W_q is the connected component of $Q \cap F$ containing q with respect to the topology of M.*

Proof. The local coordinate system $(x^1, \ldots, x^n)$ is taken so that $x^i(q) = 0$ ($i = 1, \ldots, n$) and $Q = \{q| |x^i(q)| < a, i = 1, \ldots, n\}$ are satisfied. Every slice $\{w \in Q|x^{r+1}(w) = a^{r+1}, \ldots, x^n(w) = a^n\}$, where $|a^i| < a$, is an integral manifold, and $W_q = \{w \in Q|x^{r+1}(w) = 0, \ldots, x^n(w) = 0\}$. Since M has a countable base, by Theorems 1 and 2 of II, §15, the connected submanifold F of M is σ-compact. Since F is the maximal connected integral manifold passing through each point of F, $Q \cap F$ is the union of slices of Q, and these slices are mutually disjoint open subsets of F. Hence a compact subset of F will intersect only a finite number of these slices. But F is σ-compact, so we conclude that $Q \cap F$ is the union of at most a countable number of slices $\{W_k\}_{k=1, 2, \ldots}$. Now let $\tilde{W}_q$ be the connected component of $Q \cap F$, with respect to the topology of M, containing W_q. Define a C^∞ map ψ from Q to $\mathbf{R}^{n-r}$ by $\psi(q) = (x^{r+1}(q), \ldots, x^n(q))$. Then the image of each slice by ψ is a point. Hence $\psi(Q \cap F)$ is a countable set in $\mathbf{R}^{n-r}$, and since $\psi(\tilde{W}_q)$ is connected, $\psi(\tilde{W}_q)$ is a point. Hence $\tilde{W}_q$ consists of only one slice, and we have $\tilde{W}_q = W_q$.

THEOREM 2. *Let $\mathfrak{D}$ be a completely integrable differential system on a manifold M with a countable base, and let F be a leaf of $\mathfrak{D}$. Let φ be a differentiable map from a manifold V into M such that $\varphi(V) \subset F$. If φ is considered as a map from V into F, φ is still differentiable.*

Proof. As one can see from the proof of the Proposition of II, §10, it suffices to show that φ, from V into F, is a continuous map. Let $p \in V$, and pick a neighborhood Q of $q = \varphi(p)$ in M as in Lemma 2. Given an arbitrary neighborhood W of q in F, we can choose Q sufficiently small so that the slice W_q of Q is contained in W. Since φ is a continuous map from V into M, there is a neighborhood U of p such that $\varphi(U) \subset Q$. Here we can assume that U is connected. Since $\varphi(U) \subset Q \cap F$, $\varphi(U)$ is contained in the connected component of $Q \cap F$ (with respect to the topology of M) containing q. Hence, from Lemma 2, $\varphi(U) \subset W_q \subset W$, which shows that φ is continuous.

Lie Groups
and Homogeneous Spaces

§1. Topological Groups

If a set G has the following properties, G is called a *topological group*:

(1) The set G is a group and a topological space simultaneously.

(2) The map $(x, y) \to xy$ from the direct product space $G \times G$ to G is continuous. Here xy denotes the product of two elements x and y of the group G.

(3) The map $x \to x^{-1}$ from G to G is continuous. Here x^{-1} is the inverse element of x in G.

Let G be a topological group, and $g \in G$. Define the maps L_g and R_g from G to G by:

$$L_g(x) = gx, \quad R_g(x) = xg \quad (x \in G).$$

Then L_g and R_g are homeomorphisms of G. In fact, by the definition of a topological group, L_g and R_g are both $1:1$ continuous maps from G onto G, and their inverse maps are $L_{g^{-1}}$ and $R_{g^{-1}}$ respectively. The inverse maps are also continuous, so L_g and R_g are homeomorphisms of G. The maps L_g and R_g are called the *left translation* and the *right translation* of G, respectively, by the element g of G.

Problem 1. Show that the map $x \to x^{-1}$ from G to G is a homeomorphism.

Problem 2. The conditions (2) and (3) together in the definition of a topological group G are equivalent to the following single condition: the map $(x, y) \to xy^{-1}$ from $G \times G$ to G is continuous. Prove this.

Examples of Topological Groups

(1) $\mathbf{R}^n$ is a topological group with respect to addition.

(2) Let $GL(n, \mathbf{R})$ (resp. $GL(n, \mathbf{C})$) be the set of nonsingular $n \times n$ matrices with real (resp. complex) entries. $GL(n, \mathbf{R})$ (resp. $GL(n, \mathbf{C})$) is a group with respect to multiplication of matrices. The $n \times n$ unit matrix is the identity element, and the inverse of a is the inverse matrix of a. On the other hand, we can identify the set of all $n \times n$ real matrices with $\mathbf{R}^{n^2}$. The determinant det a of a matrix a is a continuous function of $a \in \mathbf{R}^{n^2}$. We have $GL(n, \mathbf{R}) = \{a \in \mathbf{R}^{n^2} | \det a \ne 0\}$. Hence $GL(n, \mathbf{R})$ is an open set of $\mathbf{R}^{n^2}$, and hence can be considered to be a topological space. The group $GL(n, \mathbf{R})$ is a topological group with respect to this topology. Similarly, $GL(n, \mathbf{C})$ is also a topological group. We call $GL(n, \mathbf{R})$ the (*real*) *general linear group* of degree n, and $GL(n, \mathbf{C})$ the (*complex*) *general linear group* of degree n.

(3) Let G be an arbirary group. If we define any subset of G to be an open set, then G becomes a topological space. (This topology is called the *discrete topology* of G.) G is a topological group with respect to the discrete topology.

For subsets A, B of a topological group G, set $A^{-1} = \{a^{-1} | a \in A\}$, $AB = \{ab | a \in A, b \in B\}$, $gAg^{-1} = \{gag^{-1} | a \in A\}$ $(g \in G)$. Since $AB = \cup_{b \in B} R_b(A) = \cup_{a \in A} L_a(B)$, and since R_b, L_a are homeomorphisms of G, if A(resp. B) is an open set of G, then $R_b(A)$ (resp. $L_a(B)$) is an open set. Since AB is the union of these sets, it follows that if A or B is an open set, then AB is also an open set. Similarly, if A is an open set, A^{-1} and gAg^{-1} are also open sets.

Now let $\mathfrak{U}$ be the set of all neighborhoods of the identity element e of G. For $g \in G$, $g\mathfrak{U} = \{gU | U \in \mathfrak{U}\}$. Then $g\mathfrak{U}$ is the set of all neighborhoods of g. In fact, let V be any neighborhood of g, and set $U = g^{-1}V$. U is an open set containing $e = g^{-1}g$, so $U \in \mathfrak{U}$, and moreover $V = gU$. Conversely, if $U \in \mathfrak{U}$, then gU is an open set containing $g = ge$, and hence a neighborhood of g.

$\mathfrak{U}$ has the following properties:

(1) $\mathfrak{U}$ is not an empty set. If $U \in \mathfrak{U}$, then $e \in U$.

(2) For U_1, $U_2 \in \mathfrak{U}$, there is a $U_3 \in \mathfrak{U}$ such that $U_3 \subset U_1 \cap U_2$.

(3) For any $U \in \mathfrak{U}$, there is a $V \in \mathfrak{U}$ such that $VV^{-1} \subset U$.

(4) For any $U \in \mathfrak{U}$, and any element a of U, there is a $V \in \mathfrak{U}$ such that $aV \subset U$.

(5) For any $U \in \mathfrak{U}$, and any element g of G, there is a $V \in \mathfrak{U}$ such that $gVg^{-1} \subset U$.

Proof. (1) and (2) are clear. The map φ from $G \times G$ to G, given by $(x, y) \to xy^{-1}$, is continuous and satisfies $\varphi(e, e) = e$, so that for a neighborhood U of e, there is a neighborhood V of e, such that $\varphi(V, V) \subset U$. Since $\varphi(V, V) = VV^{-1}$, we have $VV^{-1} \subset U$, and hence (3) is proved. L_a is a continuous map from G to G, and $L_a(e) = a$ holds. U is a neighborhood for an element a of U, so there is a neighborhood V of e such that $L_a(V) \subset U$, i.e., $aV \subset U$ holds. Thus (4) is proved. $L_g R_{g^{-1}}$ is a continuous map from G to G, and we have $(L_g R_{g^{-1}})(e) = geg^{-1} = e$. So for a neighborhood U of e, there is a neighborhood V of e, such that $(L_g R_{g^{-1}})(V) = gVg^{-1} \subset U$ holds, and this proves (5).

Problem 3. For a group G, suppose there is a family $\mathfrak{U}$ of subsets of G satisfying the conditions (1) through (5) above. For each $g \in G$, set $\mathfrak{U}(g) = \{gU \mid U \in \mathfrak{U}\}$. There is a topology on G which has $\mathfrak{U}(g)$ as the fundamental neighborhood system of g for each $g \in G$. The group G is a topological group with respect to this topology. Prove this.

If a topological group G is a Hausdorff space as a topological space, then G is called a *Hausdorff group*. A topological group is Hausdorff if and only if

$$\bigcap_{U \in \mathfrak{U}} U = \{e\}. \tag{6}$$

Here $\mathfrak{U}$ denotes the set of all neighborhoods of the identity element e of G.

Proof. Suppose G is a Hausdorff space. For an element g of G distinct from e, there is a $U \in \mathfrak{U}$ such that $g \notin U$. Hence (6) holds. Conversely, if (6) holds, then let $g, h \in G$, $g \neq h$. Since $h^{-1}g \neq e$, by (6) there is a $U \in \mathfrak{U}$ such that $h^{-1}g \notin U$. Let V be a neighborhood of e such that $VV^{-1} \subset U$. Then $h^{-1}g \notin VV^{-1}$, so we have $gV \cap hV = \varnothing$. Hence any G satisfying (6) is a Hausdorff space.

§2. Subgroups and Quotient Spaces of Topological Groups

Let G be a topological group and H a subgroup of the group G. If we give H the topology as a subspace of G, then H is also a topological group. The topological group H is called a *topological subgroup* of G, or simply a *subgroup* of G. In particular, if H is a closed subset of G, then H is called a *closed subgroup* of G.

PROPOSITION.   *Let G be a topological group and H a subgroup of G. Then*

(1) *The closure $\bar{H}$ of H is a closed subgoup of G.*

(2) *If H is a normal subgoup of G, then $\bar{H}$ is also a normal subgroup of G.*

(3) *If H is an open set of G, then H is also a closed set of G.*

Proof. (1) It suffices to show that $gh^{-1} \in \bar{H}$ for g, $h \in \bar{H}$. For this, we only have to show that $gh^{-1}U \cap H \neq \emptyset$ for any neighborhood U of e in G. From (1), (3) and (5), for a given neighborhood U of e, there is a neighborhood V of e such that $hVV^{-1}h^{-1} \subset U$. Then we have $VV^{-1}h^{-1} \subset h^{-1}U$. Since g, $h \in \bar{H}$, we have $gV \cap H \neq \emptyset$, $hV \cap H \neq \emptyset$, so that there are u, $v \in V$ satisfying $gu \in H$, $hv \in H$. Since H is a subgroup, we have $(gu)(hv)^{-1} = guv^{-1}h^{-1} \in H$. On the other hand, $guv^{-1}h^{-1} \in gVV^{-1}h^{-1} \subset g(h^{-1}U)$, and hence $gh^{-1}U \cap H \neq \emptyset$.

(2) For any $g \in G$, the map $x \to gxg^{-1}$ is a homeomorphism of G. Hence for any subset A of G, the closure of gAg^{-1} is equal to $g\bar{A}g^{-1}$. In particular, if we set $A = H$, then, since H is a normal subgroup of G, we have $gHg^{-1} = H$. Hence $\bar{H} = g\bar{H}g^{-1}$. This shows that $\bar{H}$ is also a normal subgroup of G.

(3) Let the left coset decomposition of G be written as the disjoint union $G = H \cup (\cup_{\alpha \in A} a_\alpha H)$. Since H is an open set by hypothesis, $a_\alpha H$ is also an open set, and hence $H' = \cup_{\alpha \in A} a_\alpha H$ is also an open set of G. Since H is the complementary set of the open set H', H is a closed set of G.

Problem 1. (a) Let H be a closed subgroup of G. Set $N(H) = \{x \in G \,|\, xHx^{-1} = H\}$. Show that $N(H)$ is a closed subgroup of G. (b) The set of elements of a topological group G that commute with all the elements in G is called the *center* of G. Show that the center of G is a closed subgroup of G.

For a topological group G and a subgroup H of G, we denote the set of all left cosets of G with respect to H by G/H, and the left coset containing the element x of G by $\pi(x)$. That is, we let $\pi(x) = xH$. So π is a map from G to G/H. We define a subset O of G/H to be open if the inverse image $\pi^{-1}(O) = \{x \in G \,|\, \pi(x) \in O\}$ of O by π is an open set of G. This defines a topology on G/H, and G/H becomes a topological space. This topological space G/H is called the *quotient space* of the topological group G by the subgroup H. The map π is clearly a continuous map from G to G/H. Furthermore, π is an *open map** from G to G/H. In fact, if U is an open set of G, then $\pi(U)$ is the subset of G/H formed by the left cosets of the form xH $(x \in U)$. Hence $\pi^{-1}(\pi(U)) = UH$,

---

*A map f from a topological space X to a topological space Y is called an open map if, for any open set U of X, $f(U)$ is an open set in Y.

and since U is an open set of G, UH is also an open set of G, and by the definition of open sets in G/H, $\pi(U)$ is an open set of G/H. We call π the *natural* map from G to G/H.

Problem 2. Show that the open sets of G/H are of the form $\pi(U)$, where U is an open set of G.

Problem 3. We can define a map φ from the product space $G \times (G/H)$ to G/H by

$$\varphi(g, \pi(h)) = \pi(gh).$$

Show that φ is a continuous map. For $g \in G$, $a \in G/H$, set $T_g a = \varphi(g, a)$. Show that T_g is a homomorphism of G/H, and that $T_g T_h = T_{gh}$ holds.

Now if H is a normal subgroup of G, then G/H is a group. Let us show that the group G/H becomes a topological group by the topology of the quotient space. For this, it suffices to show that the map ψ from $(G/H) \times (G/H)$ to G/H, given by $\psi : (xH, yH) \to xy^{-1}H$, is continuous. If O is an arbitrary neighborhood of $xy^{-1}H$, then O is of the form $O = \pi(xy^{-1}U)$, where U is a neighborhood of e in G. For U, there is a neighborhood V of e such that $(xV)(yV)^{-1} \subset xy^{-1}U$. Then $\pi(xV)$, $\pi(yV)$ are neighborhoods of xH, yH, respectively, and we have $\psi(\pi(xV), \pi(yV)) \subset O$. Hence ψ is continuous, and G/H is a topological group. The topological group G/H is called the *quotient group* of the topological group G by the normal subgroup H.

Let us find the condition for the quotient space G/H to be a Hausdorff space. If G/H is Hausdorff, then the one point set $\pi(e)$ is a closed set. But $H = \pi^{-1}(\pi(e))$, and since π is continuous, H is a closed subgroup. Conversely, if H is a closed subgroup of G, we want to show that G/H is a Hausdorff space. Let aH and bH be two distinct points of G/H. Then $a^{-1}b \notin H$. Since H is a closed set, there is a neighborhood U of e satisfying $a^{-1}bU \cap H = \varnothing$. The map $(x, y) \to x^{-1}y$ from $G \times G$ to G is continuous, and since $(a, b) \to a^{-1}b$, for the neighborhood $a^{-1}bU$ of $a^{-1}b$ there is a neighborhood V of e such that $(aV)^{-1}bV \subset a^{-1}bU$. Then the intersection of $(aV)^{-1}bV$ and H is empty, so that the intersection of $\pi(aV)$ and $\pi(bV)$ is also empty. Since $\pi(aV)$ and $\pi(bV)$ are neighborhoods of aH and bH, respectively, G/H is a Hausdorff space.

Hence if G is a *topological group, for G/H to be a Hausdorff space, it is necessary and sufficient that H be a closed subgroup of G.*

§3. Isomorphisms and Homomorphisms of Toplogical Groups

Let G_1 and G_2 be topological groups, and let φ be a continuous map from G_1 to G_2. If φ satisfies

$$\varphi(xy) = \varphi(x)\varphi(y)$$

for any x, $y \in G_1$, then φ is called a *homomorphism* from the topological group G_1 to the topological group G_2.

If φ is a homomorphism from G_1 to G_2, and if φ is a homeomorphism from G_1 onto G_2, then φ is called an *isomorphism* from G_1 onto G_2. If there is an isomorphism from G_1 onto G_2, then the topological groups G_1 and G_2 are said to be isomorphic, and we write $G_1 \cong G_2$.

Problem 1. Show that the relation of one topological group being isomorphic to another is an equivalence relation.

If φ is a homomorphism from G_1 to G_2, and if φ is an open map, then φ is called an *open homomorphism*. If H is a normal subgroup of the topological group G, then $\pi : G \to G/H$ is an open homomorphism from G onto G/H.

If $\varphi : G_1 \to G_2$ is a homomorphism, then

$$N = \{x \in G_1 | \varphi(x) = e, \; e \text{ the identity element of } G_2\}$$

is called the *kernel* of the homomorphism. The kernel is a normal subgroup of G_1. If G_2 is Hausdorff, then $\{e\}$ is a closed set of G_2, and since φ is continuous and $N = \varphi^{-1}(e)$, N is a closed normal subgroup of G_1.

THE HOMOMORPHISM THEOREM. *Let G_1, G_2 be topological groups, and φ an open homomorphism from G_1 onto G_2. Let N be the kernel of φ. Then we have*

$$G_1/N \cong G_2.$$

The proof of this theorem is left to the reader as an exercise.

§4. The Connected Component of a Toplogical Group

Let G be a topological group. If G is connected as a topological space, then we call G a *connected topological group*.

THEOREM 1. *Let G be a connected topological group, and let U be a neighborhood of the identity element e of G satisfying $U = U^{-1}$.* Then an arbitrary element g of G can be written in the form*

$$g = g_1 \cdots g_k, \qquad g_i \in U \quad (i = 1, \ldots, k).$$

---

*If V is any neighborhood of e, then if we set $U = V \cap V^{-1}$, U is a neighborhood of e satisfying $U = U^{-1}$.

Proof. If we let $U^k = \{g_1 \ldots g_k | g_i \in U\ (i = 1, \ldots, k)\}$, then U^k $(k = 1, 2, \ldots)$ is an open set of G. In fact, for $k = 1$, the assertion is clear, and for $k > 1$, since $U^k = UU^{k-1}$ and U is an open set, so is U^k. Hence if we set $G' = \cup_{k=1}^{\infty} U^k$, then G' is also an open set. But G' is a subgroup of G. In fact, if $a, b \in G'$, then there are positive integers k, j such that $a \in U^k$, $b \in U^j$ hold. We have $b^{-1} \in (U^j)^{-1}$, but since $U = U^{-1}$ by hypothesis, we have $(U^j)^{-1} = U^j$, and hence $ab^{-1} \in U^k U^j = U^{k+j}$, so that $ab^{-1} \in G'$ and G' is a subgroup of G. Since G' is an open set, by the proposition of §2, G' is also a closed set of G. By the definition of G', G' is clearly nonempty, and since G is connected, we have $G = G'$. Hence for an arbitrary element g of G, there is some positive integer k such that $g \in U^k$. That is, g is of the form $g = g_1 \ldots g_k$, $g_i \in U$.

Problem 1. If G is compact in Theorem 1, show that there is a positive integer n such that every $g \in G$ can be expressed as a product $g = g_1 \cdots g_n$ $(g_i \in U)$ of n elements of U.

THEOREM 2.   *Let G be a topological group, and H a subgroup of G. If H and G/H are both connected, then G is connected.*

Proof.   Let O_1, O_2 be open sets of G, satisfying $G = O_1 \cup O_2$ and $O_1 \cap O_2 = \varnothing$. It suffices to show that O_1 or O_2 is empty. The natural map π from G to G/H is an open map and $\pi(O_1) \cup \pi(O_2) = G/H$ holds. Moreover, we have $\pi(O_1) \cap \pi(O_2) = \varnothing$. In fact, if $\pi(g) \in \pi(O_1) \cap \pi(O_2)$, then we can write $gH = g_1 H = g_2 H$ for some $g_i \in O_i$. However, H is connected, so is $g_1 H$, and since $g_1 H \cap O_1 \neq \varnothing$, we have $g_1 H \cap O_2 = \varnothing$, i.e., $g_1 H \subset O_1$. Similarly, we have $g_2 H \subset O_2$. Hence we get $g_1 H = g_2 H \subset O_1 \cap O_2$, contradicting $O_1 \cap O_2 = \varnothing$. Hence $\pi(O_1) \cap \pi(O_2) = \varnothing$. But we are assuming that G/H is connected, so that $\pi(O_1)$ or $\pi(O_2)$ is empty, and hence O_1 or O_2 is empty.

THEOREM 3.   *Let G be a topological group, and let G_0 be the connected component of G containing the identity element e of G. Then*

(1) *G_0 is a closed normal subgroup of G.*
(2) *The connected component $(G/G_0)_0$ of G/G_0 containing the identity e of G/G_0 consists of the identity element only.**

---

*In general, if the connected component containing the identity e is $\{e\}$, then G is called a totally disconnected topological group.

(3) *If G is locally connected, then G/G_0 is a discrete topological group.**

Proof. (1) For any $g \in G_0$, $x \to xg^{-1}$ is a homeomorphism of G, so that $G_0 g^{-1}$ is the connected component of G containing $gg^{-1} = e$. Hence $G_0 g^{-1} = G_0$, so that if h, $g \in G_0$, then $hg^{-1} \in G_0$, and G_0 is a subgroup of G. For any $g \in G$, $x \to gxg^{-1}$ is a homeomorphism of G, so that $gG_0 g^{-1}$ is the connected component of G containing $geg^{-1} = e$. Hence $gG_0 g^{-1} = G_0$, and G_0 is a normal subgroup of G. Since the closure $\bar{G}_0$ of G_0 is also a connected set containing e, we have $G_0 = \bar{G}_0$, and hence G_0 is a closed set.

(2) Let π be the natural map from G to G/G_0, and set $\pi^{-1}((G/G_0)_0) = H$. Then H is a subgroup of G which contains G_0. Moreover, $H/G_0 = (G/G_0)_0$ holds. Hence, by Theorem 2, H is connected. But H contains G_0, and G_0 is the connected component containing e, so that we have $H = G_0$. Hence $(G/G_0)_0$ consists of the identity element only.

(3) If G is locally connected, then G_0 is an open set of G. Hence for any $g \in G$, gG_0 is an open set of G. For an arbitrary subset A of G/G_0, $\pi^{-1}(A)$ is the union of sets gG_0, where $\pi(g) \in A$, and hence $\pi^{-1}(A)$ is an open set of G, so that A is an open set of G/G_0. Hence the topology of G/G_0 is the discrete topology.

Problem 2. Show that the connected component of G containing the element g of G is gG_0.

PROPOSITION. *Let $GL^+(n, \mathbf{R})$ denote the set of all $n \times n$ real matrices with positive determinants. $GL^+(n, \mathbf{R})$ is the connected component of the general linear group $GL(n, \mathbf{R})$ containing the identity 1.*

Proof. Let G_0 be the connected component of $G = GL(n, \mathbf{R})$ containing 1. Let R^* be the multiplicative group formed by the set of all nonzero real numbers, that is, let $GL(1, \mathbf{R}) = \mathbf{R}^*$. The connected component of $\mathbf{R}^*$ containing the identity element 1 coincides with the set $\mathbf{R}^+$ of all positive real numbers. If we let φ be the map from $GL(n, \mathbf{R})$ to $\mathbf{R}^*$ defined by $\varphi(a) = \det a$, then φ is a homomorphism (of topological groups) from $GL(n, \mathbf{R})$ to $\mathbf{R}^*$. Since $\varphi(G_0)$ is a connected subset of $\mathbf{R}^*$ containing 1, we have $\varphi(G_0) \subset \mathbf{R}^+$. That is, the matrices belonging to G_0 have positive determinant. Hence we have $G_0 \subset GL^+(n, \mathbf{R})$. To show $GL^+(n, \mathbf{R}) = G_0$, it suffices to show that $GL^+(n, \mathbf{R})$ is connected. We shall do this by induction on n. For $n = 1$, $GL^+(1, \mathbf{R})$ is equal to $\mathbf{R}^+$, and is connected. Let $n > 1$, and suppose that we have proved that $GL^+(n - 1, \mathbf{R})$ is connected. Let H be the subgroup of $GL^+(n, \mathbf{R})$ formed

*This means that the topology of G/G_0 is the discrete topology (§1, Example 3).

by the matrices of the form

$$\begin{bmatrix} 1 & * & \cdots & * \\ 0 & & & \\ \cdot & & C & \\ \cdot & & & \\ \cdot & & & \\ 0 & & & \end{bmatrix}, \qquad C \in GL^+(n-1, \mathbf{R}).$$

As a topological space, H is homeomorphic to $\mathbf{R}^{n-1} \times GL^+(n-1, \mathbf{R})$. Hence, by the inductive assumption, H is connected. Now consider the quotient space $GL^+(n, \mathbf{R})/H$. Let π be the natural map from $GL^+(n, \mathbf{R})$ to $GL^+(n, \mathbf{R})/H$. For two matrices A, B in $GL^+(n, \mathbf{R})$, $\pi(A) = \pi(B)$ if and only if the first columns in A and B coincide. Hence the points of $GL^+(n, \mathbf{R})/H$ and the points of $\mathbf{R}^n - \{0\}$ correspond in a one-to-one manner. Furthermore, this $1:1$ correspondence is a homeomorphism from $GL^+(n, \mathbf{R})/H$ onto $\mathbf{R}^n - \{0\}$. But, since $n > 1$, $\mathbf{R}^n - \{0\}$ is connected, and hence $GL^+(n, \mathbf{R})/H$ is connected. Since H is connected, by Theorem 2 we conclude that $GL^+(n, \mathbf{R})$ is also connected.

Problem 3. Show that $GL(n, \mathbf{R})$ has exactly two connected components.

Problem 4. Show that $GL(n, \mathbf{C})$ is connected.

§5. Homogeneous Spaces of Topological Groups, Locally Compact Groups

DEFINITION. Let G be a topological group and X a topological space. If G and X satisfy the following condition, then G is called a *topological transformation group* on X. A continuous map φ from $G \times X$ to X is defined, for which, writing $\varphi(g, x) = g \cdot x$, the following holds:

(1) $(g \cdot h) \cdot x = g \cdot (h \cdot x)$ for g, $h \in G$, $x \in X$,
(2) For the identity element e of G and all $x \in X$, $e \cdot x = x$.

If this is the case, $x \to g \cdot x$ is a continuous map from X to X, and by (1), (2) we have $g^{-1} \cdot (g \cdot x) = g \cdot (g^{-1} \cdot x) = e \cdot x = x$. From this, we see that $x \to g \cdot x$ is a homeomorphism of X.

If the only element g of G satisfying $g \cdot x = x$ for all $x \in X$ is the identity element e, then G is said to act *effectively* on X.

If for any two points x, y of X, there is an element g of G satisfying $g \cdot x = y$, then G is said to act *transitively* on X. If G acts transitively on X, then X is called a *homogeneous space* of the topological group G.

If G is a topological transformation group on X, for a point x of X, set $H = \{g \in G \,|\, g \cdot x = x\}$. Then H is a subgroup of G. In fact, if $g \in H$, then from $g \cdot x = x$, we get $g^{-1} \cdot (g \cdot x) = g^{-1} \cdot x$, and because $g^{-1} \cdot (g \cdot x) = (g^{-1} \cdot g) \cdot x = e \cdot x = x$, we get $g^{-1} \cdot x = x$, and hence $g^{-1} \in H$; also, if $g, h \in H$, then $(g \cdot h) \cdot x = g \cdot (h \cdot x) = g \cdot x = x$, so that $g \cdot h \in H$. Hence H is a subgroup of G. The subgroup H is called the *isotropy subgroup* of G at x. If X is a Hausdorff space, the isotropy subgroup H of G at a point x of X is a closed subgruoup of G. In fact, for a fixed x, $\psi : g \to g \cdot x$ is a continuous map from G to X, and we have $H = \psi^{-1}(x)$. But X is a Hausdorff space, so that a subset of X consisting of a single point x is a closed set of X, and hence H is a closed set of G.

Example 1. Let G be a topological group and H a subgroup of G. G is a topological transformation group on G/H, and G acts transitively on G/H.

Problem 1. The set N of all elements g of G satisfying $g \cdot x = x$ for all $x \in X$ is a normal subgroup of G. Show this.

Problem 2. If G acts transitively on X, show that isotropy subgroups at any two points of X are conjugate to each other.*

DEFINITION. Let G be a Hausdorff topological group.

(1) If G is locally compact as a topological space, then G is called a *locally compact group*.

(2) If G is compact as a topological space, then G is called a *compact group*.

THEOREM 1. *Let X be a locally compact Hausdorff space which is a homogeneous space of a locally compact group G with a countable base. Let H be the isotropy subgroup of G at a point x of X. Then the map α from G/H to X given by*

$$\alpha : gH \to g \cdot x$$

is a homeomorphism from G/H onto X, and $\alpha(h \cdot \xi) = h \cdot \alpha(\xi)$ holds for all $\xi \in G/H$, $g \in G$.

To prove this theorem, let us prove the following lemma first.

LEMMA.† *Let X be a locally compact Hausdorff space and suppose $X = \cup_{n=1}^{\infty} X_n$, where each X_n is a closed set of X. Then at least one X_n contains an open set of X.*

*Two subgroups H, H' of a group G are said to be conjugate to each other if there is an element g of G satisfying $gHg^{-1} = H'$.

†This is called the Baire category theorem.

Proof. Let us suppose that no X_n contains an open set of X. Since X is locally compact, there is an open set U_0 such that $\bar{U}_0$ is compact. Since X_1 does not contain an open set, $U_0 - X_1$ is not empty. Hence we can choose a neighborhood U_1 of a point y_1 of $U_0 - X_1$ such that $\bar{U}_1 \subset U_0 - X_1$. The set $\bar{U}_1$ is compact, and we have $\bar{U}_1 \cap X_1 = \varnothing$, $\bar{U}_1 \subset U_0$. Using U_1, X_2 instead of U_0, X_1, and repeating the argument, we find a nonempty open set U_2 of X, such that $\bar{U}_2 \cap X_2 = \varnothing$, $\bar{U}_2 \subset U_1$. Repeating this process, we can find a sequence $\{U_n\}$ of nonempty open sets of X, such that $\bar{U}_n \cap X_n = \varnothing$, $\bar{U}_n \subset U_{n-1}$ ($n = 1, 2, \ldots$). Since $\bar{U}_0$ is compact and $\bar{U}_0 \supset \bar{U}_1 \supset \cdots \supset \bar{U}_n \supset \cdots$, $\cap_{n=1}^{\infty} \bar{U}_n$ is not empty. On the other hand, since $\bar{U}_n \cap X_n = \varnothing$, the intersection of $\cap_{n=1}^{\infty} \bar{U}_n$ and $\cup_{n=1}^{\infty} X_n$ is empty. However, since $X = \cup_{n=1}^{\infty} X_n$, we have that $\cap_{n=1}^{\infty} \bar{U}_n$ is empty, and this is a contradiction. Since this contradiction arose from the supposition that every X_n contained no open set, at least one X_n has to contain an open set.

Proof of Theorem 1.   Fix an $x \in X$, and define a continuous map ψ from G to X by $\psi(g) = g \cdot x$. Since G acts transitively on X, ψ is a map from G onto X. In Fig. 4.1, for an arbitrary $g \in G$, we have $\alpha(\pi(g)) = \psi(g)$.

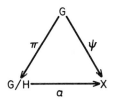

Fig. 4.1

That is, we have $\psi = \alpha \cdot \pi$. Hence if O is any open set in X, then $\pi^{-1}(\alpha^{-1}(O)) = \psi^{-1}(O)$. Since ψ is continuous, $\psi^{-1}(O)$ is an open set of G, so that $\pi^{-1}(\alpha^{-1}(O))$ is an open set of G. By the definition of open sets in G/H, $\alpha^{-1}(O)$ is an open set of G/H. This shows that α is a continuous map. It is easy to see that α is a $1:1$ map from G/H onto X. Hence if we can show that α is an open map, then we have shown that α is a homeomorphism. Let O be an arbitrary open set of G/H. We can choose an open set U of G such that $\pi(U) = 0$. Then we have $\alpha(O) = \alpha(\pi(U)) = \psi(U)$.

Thus it suffices to show that ψ is an open map. Let V be an open set of G and g_0 an arbitrary point of V. We can choose a neighborhood U of the identity element e of G such that $\bar{U}$ is compact and $g_0 \cdot \bar{U}^{-1} \cdot \bar{U} \subset V$. $\{gU \,|\, g \in G\}$ is an open covering of G and, since G has a countable base, we can choose from $\{gU \,|\, g \in G\}$ a countable subset $\{g_n U \,|\, n = 1, 2, \ldots\}$ to cover G (cf. II, §15). Then we certainly have $\cup_{n=1}^{\infty} \psi(g_n \overline{U}) = X$, and each $\psi(g_n \overline{U})$ is a closed set of X, because $g_n \overline{U}$ is compact. Hence, by the lemma, there is an n such that $\psi(g_n \overline{U})$ contains an open set of X. Since $\psi(g_n \overline{U}) = g_n \psi(\overline{U})$, we have $g_n^{-1} \psi(g_n \overline{U}) = \psi(\overline{U})$, and since $y \to g_n^{-1} y$ is a homeomorphism of X, $\psi(\overline{U})$ also contains an open set of X. Hence for some $h \in \overline{U}$, $\psi(h) = h \cdot x$ is an interior point of $\psi(\overline{U})$. Hence $g_0 \cdot x = g_0(h^{-1}\psi(h))$ is an interior point of $g_0 h^{-1}\psi(\overline{U}) = \psi(g_0 h^{-1}\overline{U})$. But since $g_0 h^{-1}\overline{U} \subset g_0 \overline{U}^{-1}\overline{U} \subset V$, $g_0 \cdot x$ is an interior point of $\psi(V)$. Since $g_0 \cdot x$ is an arbitrary point of $\psi(V)$, $\psi(V)$ is an open set of X. Hence ψ is an open map.

It is clear from the definition of α that $\alpha(h \cdot \xi) = h\alpha(\xi)$ holds for all $\xi \in G/H$, $h \in G$.

THEOREM 2. *Let G, G' be locally compact groups, and let G have a countable base. A homomorphism from G onto G' is always an open homomorphism.*

Proof Let φ be a homomorphism from G onto G'. For $g \in G$, $x \in G'$, define $g \cdot x$ by $g \cdot x = \varphi(g) \cdot x$. The map $(g, x) \to g \cdot x$ is a continuous map from $G \times G'$ to G', by which G becomes a topological transformation group on G'. Since φ is onto, G is transitive on G'. Hence, by the proof of Theorem 1, $g \to g \cdot e'$ (where e' is the identity element of G') is an open map. But $g \cdot e' = \varphi(g)$, so that φ is an open map.

COROLLARY 1. *Let G, G' be locally compact groups, and let G have a countable base. A $1:1$ homomorphism from G onto G' is an isomorphism from G onto G'.*

COROLLARY 2 *Let G, G' be as above, and let φ be a homomorphism from G onto G'. Let N be the kernel of φ. Then we have*

$$G/N \cong G'.$$

THEOREM 3. *A connected locally compact group G is σ-compact.*

Proof. Let U be a neighborhood of the identity element such that $U = U^{-1}$ and $\overline{U}$ is compact. By Theorem 1 of §4, we have $G = \cup_{k=1}^{\infty}(\overline{U})^k$. Define a map φ_k from the direct product $G \times \cdots \times G$ of k copies of G to G by $\varphi_k(g_1, \ldots, g_k) = g_1 g_2 \cdots g_k$. Then φ_k is continuous and $(\overline{U})^k = \varphi_k(\overline{U} \times \cdots \times \overline{U})$. Since $\overline{U} \times \cdots \times \overline{U}$ is a compact subset of $G \times \cdots \times G$, its image $(\overline{U})^k$ by φ_k is a compact subset of G. Hence G is σ-compact.

Example 2. Let $O(n)$ be the set of all $n \times n$ orthogonal matrices, and let $SO(n)$ be the set of all $n \times n$ orthogonal matrices with determinate 1. The sets $O(n)$ and $SO(n)$ are subgroups of $GL(n, \mathbf{R})$, and they are both *compact*. The groups $O(n)$ and $SO(n)$ are called the *orthogonal group* and the *special orthogonal group*, respectively. Let A be a matrix belonging to $O(n)$ and let x be an element of $\mathbf{R}^n$. Then we have $\|Ax\| = \|x\|$.* Hence the elements of $O(n)$ map the unit sphere S^{n-1} in $\mathbf{R}^n$ to itself, and $O(n)$ is a topological transformation group on S^{n-1}. The isotropy subgroup of $O(n)$ at the point $e_1 = (1, 0, \ldots, 0)$ on S^{n-1} consists of the orthogonal matrices of the form

$$\begin{bmatrix} 1 & 0 & \ldots & 0 \\ 0 & & & \\ \cdot & & B & \\ \cdot & & & \\ \cdot & & & \\ 0 & & & \end{bmatrix}, \qquad B \in O(n - 1),$$

and can be identified with $O(n - 1)$. By Theorem 1, $O(n)/O(n - 1)$ and S^{n-1} are homeomorphic.

Similarly, $SO(n)$ is also a transitive topological transformation group on S^{n-1}, and the isotropy subgroup of $SO(n)$ at e_1 can be identified with $SO(n - 1)$. The sphere S^{n-1} is homeomorphic to $SO(n)/SO(n - 1)$.

Problem 3. Following the proof of the proposition of §4, show that $SO(n)$ is the connected component of $O(n)$ containing the identity element.

Example 3. The affine transformation group $A(n, \mathbf{R})$ of the n-dimensional affine space $\mathbf{A}^n$ is locally compact and is a transitive transformation group on $\mathbf{A}^n$. The isotropy subgroup at the origin O is $GL(n, \mathbf{R})$. Hence, by Theorem 1, $A(n, \mathbf{R})/GL(n, \mathbf{R})$ is homeomorphic to $\mathbf{A}^n$.

§6. Lie Groups and Lie Algebras

DEFINITION 1.   Let G be a C^∞ manifold with a countable base. If G is a group, and if the map $(x, y) \to x \cdot y$ from the product manifold $G \times G$ to G and the map $x \to x^{-1}$ from G to G are both differentiable, then G is called a *Lie group*.

A Lie group is clearly a locally compact group with a countable base. Let G_0 be the connected component of G containing the identity

*$\|x\|$ denotes the length $(\sum_{i=1}^{n} (x^i)^2)^{1/2}$ of an n-dimensional column vector.

element e. By §4, G_0 is a closed normal subgroup of G. Moreover, since G_0 is locally connected, G_0 is an open submanifold of G. For x, $y \in G_0$, the map $(x, y) \to x \cdot y$ from $G_0 \times G_0$ to G_0 and the map $x \to x^{-1}$ from G_0 to G_0 are differentiable maps. Hence G_0 is a Lie group.

Remark 1. Let G be a connected C^∞ manifold which is a group, and suppose the maps $(x, y) \to x \cdot y$ and $x \to x^{-1}$ are both differentiable. Since G is a connected locally compact group, by Theorem 3 of §5, and Theorem 1 of II, §15, G has a countable base, and G is a Lie group.

Remark 2. In Definition 1, the assumption that G has a countable base is not necessary for establishing the elementary properties of Lie groups. However, to discuss the rather essential properties of Lie groups, we must almost always assume that G has a countable base. For this reason we added this assumption in the definition of Lie groups.

Let L_g and R_g denote the left and right translation, respectively, by an element g of Lie group G. Then L_g and R_g are diffeomorphisms of G. From the definition of L_g and R_g, it is easy to see that the following holds:

$$L_g \cdot L_h = L_{gh}, \qquad R_g \cdot R_h = R_{hg},$$
$$L_{g^{-1}} = L_g^{-1}, \qquad R_{g^{-1}} = R_g^{-1}, \qquad L_g \cdot R_h = R_h \cdot L_g. \tag{1}$$

For $g \in G$, set

$$A_g = L_g \cdot R_{g^{-1}}. \tag{2}$$

The transformation A_g is a diffeomorphism of G and, by definition, we have $A_g(x) = gxg^{-1}$. Hence for two elements x, y of G, we have $A_g(xy) = A_g(x)A_g(y)$. We call A_g the *inner automorphism of G* by the element g of G.

Problem 1. Let φ be a $1:1$ map from G onto G. If $\varphi \cdot L_g = L_g \cdot \varphi$ is satisfied for all $g \in G$, then there is an $h \in G$ such that $\varphi = R_h$. Similarly, if $\varphi \cdot R_g = R_g \cdot \varphi$ holds for all $g \in G$, then there is an $h \in G$ such that $\varphi = L_h$. Prove this.

If a vector field X on a Lie group G satisfies

$$(L_g)_* X = X$$

for all $g \in G$, then X is called a *left invariant vector field*. If, instead, X satisfies

$$(R_g)_* X = X$$

for all $g \in G$, then X is called a *right invariant vector field*.

Let $\mathfrak{g}$ be the set of all left invariant vector fields on G. If X, $Y \in \mathfrak{g}$ and λ, $\mu \in R$, then $\lambda X + \mu Y$, $[X, Y]$ also belong to $\mathfrak{g}$. In fact, from II, §11, we have

$$(L_g)_*(\lambda X + \mu Y) = \lambda (L_g)_* X + \mu (L_g)_* Y = \lambda X + \mu Y,$$
$$(L_g)_*[X, Y] = [(L_g)_* X, (L_g)_* Y] = [X, Y].$$

Hence, by II, §11, $\mathfrak{g}$ becomes a Lie algebra of vector fields with respect to the commutator product $[X, Y]$.

DEFINITION 2.   The Lie algebra $\mathfrak{g}$ formed by the set of all left invariant vector fields on G is called the *Lie algebra of the Lie group G*.

THEOREM 1.   *If a Lie group G has dimension n, then the Lie algebra of G also has dimension n.*

Proof. To $X \in \mathfrak{g}$, we assign X_e, the value of X at the identity element e, and obtain a linear map from $\mathfrak{g}$ to the tangent space $T_e(G)$ of G at e. Denote this linear map by α. If we can show that α is a $1 : 1$ map from $\mathfrak{g}$ onto $T_e(G)$, then, since the tangent space $T_e(G)$ has dimension n, $\mathfrak{g}$ will also have dimension n. First let us show that α is $1 : 1$. Let X, $Y \in \mathfrak{g}$, and suppose that $X_e = Y_e$. Since X, Y are left invariant, by II, §11 we have $X_g = ((L_g)_* X)_g = (L_g)_* X_e$ and similarly $Y_g = (L_g)_* Y_e$. Since $X_e = Y_e$ by assumption, we get $X_g = Y_g$ for all $g \in G$. That is, we get $X = Y$. Hence α is $1 : 1$. Next we shall see that α is a map from $\mathfrak{g}$ onto $T_e(G)$. For this it suffices to show that for any given $v \in T_e(G)$, there is an $X \in \mathfrak{g}$ such that $X_e = v$. If we define a vector field $X : g \to X_g$ by

$$X_g = (L_g)_* v,$$

then clearly $X_e = v$. We must show that this vector field X is of class C^∞ and left invariant. Let W be an open set of G, $(y^1, \ldots, y^n)$ a local coordinate system of G on W, and $g_0 \in W$. The map $(g, h) \to \varphi(g, h) = g \cdot h$ from $G \times G$ to G is differentiable and, since we can write $g_0 = g_0 \cdot e$, there is a sufficiently small neighborhood W_0 of g_0 $(W_0 \subset W)$ and a sufficiently small neighborhood U of e such that $W_0 \cdot U \subset W$. Take a local coordinate system $(x^1, \ldots, x^n)$ on U, where $x^i(e) = 0$ $(i = 1, \ldots, n)$. We have

$$y^i(\varphi(g, s)) = \varphi^i(y^1(g), \ldots, y^n(g); x^1(s), \ldots, x^n(s)) \qquad (g \in W_0, s \in U).$$

Here $\varphi^i(u^1, \ldots, u^n; v^1, \ldots, v^n)$ is a C^∞ function of $2n$ variables. Now let $v = \sum_{i=1}^n a^i (\partial/\partial x^i)_e$ and $X = \sum_{i=1}^n \xi^i \, \partial/\partial y^i$ on W. Since $\varphi(g, s) =$

$L_g(s)$, we have

$$\xi^i(g) = \sum_{j=1}^{n} \frac{\partial \varphi^i}{\partial v^j} (y^1(g), \ldots, y^n(g); 0, \ldots, 0) \, a^j.$$

Hence ξ^i is of class C^∞, and thus X is of class C^∞. On the other hand, for arbitrary $s, g \in G$, we have

$$((L_s)_* X)_g = (L_s)_* X_{s^{-1}g} = (L_s)_* (L_{s^{-1}g})_* v.$$

But $L_s L_{s^{-1}g} = L_g$, so that $(L_s)_*(L_{s^{-1}g})_* v = (L_g)_* v$. Hence, from the equality above and the definition of X_g, we have $((L_s)_* X)_g = X_g$. Hence $(L_s)_* X = X$ holds for all $s \in G$, and X is left invariant. This proves Theorem 1.

Problem 2. Prove that for any given $v \in T_g(G)$, there is one and only one element X of $\mathfrak{g}$ such that $X_g = v$.

Now take a basis $\{X_1, \ldots, X_n\}$ of the Lie algebra $\mathfrak{g}$ of G. Then $[X_i, X_j]$ is uniquely expressed as

$$[X_i, X_j] = \sum_{k=1}^{n} c_{ij}^k X_k \qquad (i, j = 1, \ldots, n), \quad c_{ij}^k \in \mathbf{R}.$$

The n^3 constants c_{ij}^k are called the *structure constants* of the Lie group G with respect to the basis $\{X_1, \ldots, X_n\}$ of $\mathfrak{g}$.

Because of the properties

$$[X_i, X_j] = -[X_j, X_i],$$
$$[X_i, [X_j, X_k]] + [X_j, [X_k, X_i]] + [X_k, [X_i, X_j]] = 0,$$

of the commutator product, the following relations hold between the structure constants c_{ij}^k:

$$c_{ij}^k = -c_{ji}^k \qquad (i, j, k = 1, \ldots, n)$$
$$\sum_{s=1}^{n} (c_{is}^t c_{jk}^s + c_{js}^t c_{ki}^s + c_{ks}^t c_{ij}^s) = 0 \qquad (i, j, k, t = 1, \ldots, n)$$

DEFINITION 3. Let $\mathfrak{g}_1$, $\mathfrak{g}_2$ be Lie algebras over a field K. If a linear map α from $\mathfrak{g}_1$ into $\mathfrak{g}_2$ satisfies

$$\alpha([X, Y]) = [\alpha(X), \alpha(Y)],$$

for arbitrary $X, Y \in \mathfrak{g}$, then α is called a *homomorphism* from $\mathfrak{g}_1$ to $\mathfrak{g}_2$. If $\alpha(\mathfrak{g}_1) = \mathfrak{g}_2$, then α is called a homomorphism from $\mathfrak{g}_1$ *onto* $\mathfrak{g}_2$. If the homomorphism α is a 1:1 map, then α is called an *isomorphism* from

$\mathfrak{g}_1$ into $\mathfrak{g}_2$; if furthermore $\alpha(\mathfrak{g}_1) = \mathfrak{g}_2$, then α is called an isomorphism from $\mathfrak{g}_1$ onto $\mathfrak{g}_2$. If there is an isomorphism from $\mathfrak{g}_1$ onto $\mathfrak{g}_2$, then $\mathfrak{g}_1$ and $\mathfrak{g}_2$ are said to be isomorphic, and we write $\mathfrak{g}_1 \cong \mathfrak{g}_2$. An isomorphism from a Lie algebra $\mathfrak{g}$ onto itself is called an *automorphism* of $\mathfrak{g}$.

For $u, v \in T_e(G)$, there are unique elements $X, Y \in \mathfrak{g}$ such that $u = X_e$, $v = Y_e$. Setting $w = [X, Y]_e$, and defining the product $[u, v]$ of u and v by $[u, v] = w$, the tangent space $T_e(G)$ becomes a Lie algebra, which is isomorphic to $\mathfrak{g}$ by the map $X \to X_e$. The Lie algebra thus obtained, by defining a product in the tangent space $T_e(G)$ of G at the identity element e, is often identified with the Lie algebra $\mathfrak{g}$.

The same argument as above shows that the set $\mathfrak{g}'$ of all right invariant vector fields on G also forms an n-dimensional Lie algebra. Moreover $\mathfrak{g}$ and $\mathfrak{g}'$ are isomorphic.

Problem 3. Let ψ be the diffeomorphism of G defined by $\psi(x) = x^{-1}$. Show that $\psi \cdot L_g = R_{g^{-1}} \cdot \psi$, $\psi \cdot R_g = L_{g^{-1}} \cdot \psi$ hold. Use this to prove that $X \to \psi_* X$ is an isomorphism from $\mathfrak{g}$ onto $\mathfrak{g}'$.

If G_0 is the connected component of G containing the identity element e, then, as was said before, G_0 is also a Lie group. The Lie algebra $\mathfrak{g}_0$ formed by the left invariant vector fields on G_0, and the Lie algebra $\mathfrak{g}$ formed by the left invariant vector fields on G, are isomorphic for the following reason. Since G_0 is an open submanifold of G, it follows that if we restrict a vector field X of G to G_0, then we obtain a vector field rX on G_0. If $X \in \mathfrak{g}$, then $rX \in \mathfrak{g}_0$, and clearly r is a linear map from $\mathfrak{g}$ to $\mathfrak{g}_0$, and $r[X, Y] = [rX, rY]$ holds. Moreover, if $rX = 0$, then in particular $X_e = 0$, and since X is left invariant, we get $X = 0$. Hence r is a $1:1$ map from $\mathfrak{g}$ to $\mathfrak{g}_0$. However, $\mathfrak{g}$ and $\mathfrak{g}_0$ have the same dimension and, since r is a linear map, $\mathfrak{g}$ and $\mathfrak{g}_0$ are isomorphic. From now on, the Lie algebras of G and G_0 are identified by the isomorphism r.

Remark. It is known that if G is a Lie group, then the C^∞ structure of G is subordinate to a C^ω structure, and that the maps $(x, y) \to x \cdot y$ and $x \to x^{-1}$ are of class C^ω with respect to this C^ω structure. However, we shall not prove this fact in this book.

§7. Invariant Differential Forms on Lie Groups

Let ω be a (differential) p-form on a Lie group G. If, for all $g \in G$, we have

$$L_g^*\omega = \omega,$$

then we call ω a *left invariant p-form*, while if, for all $g \in G$, we have

$$R_g^*\omega = \omega,$$

then we call ω a *right invariant p-form*. If ω is simultaneously left and right invariant, then ω is said to be *two-sided invariant*. If ω is two-sided invariant, then, for all $g \in G$, we have

$$A_g^*\omega = \omega.$$

LEMMA.   *A p-form on G is left invariant if and only if for any choice of p left invariant vector fields $X_1, \ldots, X_p$, the function $\omega(X_1, \ldots, X_p)$ on G is a constant function.*

Proof.   Let $X_1, \ldots, X_p$ be left invariant vector fields on G. Then we have

$$(L_g^*\omega)_s((X_1)_s, \ldots, (X_p)_s) = \omega_{gs}((L_g)_*(X_1)_s, \ldots, (L_g)_*(X_p)_s)$$
$$= \omega_{gs}((X_1)_{gs}, \ldots, (X_p)_{gs}). \qquad (*)$$

Hence if ω is left invariant, then in (*) let $s = e$, and we have

$$\omega_e((X_1)_e, \ldots, (X_p)_e) = \omega_g((X_1)_g, \ldots, (X_p)_g).$$

That is, $\omega(X_1, \ldots, X_p)$ is a constant. Conversely if, for any choice of $X_1, \ldots, X_p \in \mathfrak{g}$, the function $\omega(X_1, \ldots, X_p)$ on G is a constant function, then

$$\omega_{gs}((X_1)_{gs}, \ldots, (X_p)_{gs}) = \omega_s((X_1)_s, \ldots, (X_p)_s).$$

Hence, by (*), we have

$$(L_g^*\omega)_s((X_1)_s, \ldots, (X_p)_s) = \omega_s((X_1)_s, \ldots, (X_p)_s).$$

But for any p-tuple of vectors $v_1, \ldots, v_p$ of $T_s(G)$, there are left invariant vector fields $X_1, \ldots, X_p$ such that $(X_i)_s = v_i$ $(i = 1, \ldots, p)$. Hence $(L_g^*\omega)_s = \omega_s$ holds for arbitrary g, s. Thus $L_g^*\omega = \omega$ holds for arbitrary g, and ω is left invariant.

If ω is left invariant, then by the lemma, for $X_i \in \mathfrak{g}$ $(i = 1, \ldots, p)$, we have $\omega(X_1, \ldots, X_p) \in \mathbf{R}$. Hence we can define an alternating p-linear function f_ω on the vector space $\mathfrak{g}$ by

$$f_\omega(X_1, \ldots, X_p) = \omega(X_1, \ldots, X_p).$$

If we let $\mathfrak{A}^p$ denote the vector space over $\mathbf{R}$ formed by the left invariant p-forms on G, then $f : \omega \to f_\omega$ is an isomorphism from the vector space $\mathfrak{A}^p$ onto the vector space $\overset{p}{\wedge} \mathfrak{g}^*$ formed by the alternating p-linear functions on $\mathfrak{g}$. Let us prove this.

If $\{Z_1, \ldots, Z_n\}$ is a base of $\mathfrak{g}$, then $(Z_1)_g, \ldots, (Z_n)_g$ are linearly independent for each $g \in G$. Hence any vector field X on G can be written uniquely as $X = \xi^1 Z_1 + \cdots + \xi^n Z_n$, where $\xi^1, \ldots, \xi^n$ are C^∞ functions on G. Now let $Y_1, \ldots, Y_p$ be any p-tuple of vector fields on G, and let $Y_i = \sum_{j=1}^n \eta_i^j Z_j$. If $\omega \in \mathfrak{A}^p$, then we have

$$\omega(Y_1, \ldots, Y_p) = \sum_{j_1, \ldots, j_p = 1}^n \eta_1^{j_1} \cdots \eta_p^{j_p} f_\omega(Z_{j_1}, \ldots, Z_{j_p}).$$

Hence if $f_\omega = 0$, then $\omega(Y_1, \ldots, Y_p) = 0$, and thus $\omega = 0$. That is, f is 1:1. Conversely, for $h \in \overset{p}{\wedge} \mathfrak{g}^*$, let

$$\omega(Y_1, \ldots, Y_p) = \sum_{j_1, \ldots, j_p = 1}^n \eta_1^{j_1} \cdots \eta_p^{j_p} h(Z_{j_1}, \ldots, Z_{j_p}).$$

Then ω is an alternating p-linear function on the $C^\infty(M)$-module $\mathfrak{X}(G)$. Hence ω is a p-form on G (the theorem of III, §3). Moreover, if $Y_i \in \mathfrak{g}$ $(i = 1, \ldots, n)$, then η_i^j are constants, so that $\omega(Y_1, \ldots, Y_p)$ is a constant on G, and hence by the lemma, ω is left invariant. By the definition of ω, we have $f_\omega = h$, so that f is a map from $\mathfrak{A}^p$ onto $\overset{p}{\wedge} \mathfrak{g}^*$. Thus we have proved the following theorem.

THEOREM. *The vector space $\mathfrak{A}^p$ formed by the left invariant p-forms on G, and the vector space $\overset{p}{\wedge} \mathfrak{g}^*$ formed by the alternating p-linear functions on $\mathfrak{g}$, are isomorphic by the correspondence $\omega \to f_\omega$. In particular, the vector space formed by the left invariant 1-forms on G and the dual space $\mathfrak{g}^*$ of the vector space $\mathfrak{g}$ are isomorphic.*

If $\{X_1, \ldots, X_n\}$ is a base of $\mathfrak{g}$, then there are n linearly independent left invariant 1-forms $\{\omega^1, \ldots, \omega^n\}$ such that

$$\omega^i(X_j) = \delta_j^i \qquad (i, j = 1, \ldots, n).$$

Then we have

$$(d\omega^i)(X_j, X_k) = X_j(\omega^i(X_k)) - X_k(\omega^i(X_j)) - \omega^i([X_j, X_k])$$

$$= -\sum_{l=1}^n c_{jk}^l \omega^i(X_l) = -c_{jk}^i.$$

Hence we obtain the equations

$$d\omega^i = -\frac{1}{2} \sum_{j, k = 1}^{n} c^i_{jk} \, \omega^j \wedge \omega^k \qquad (i = 1, \ldots, n),$$

which are called the *Maurer-Cartan equations*.

If ω is a left invariant p-form, and $Y_1, \ldots, Y_{p+1} \in \mathfrak{g}$, then, by (5) of III, §4, we have

$$(d\omega)(Y_1, \ldots, Y_{p+1}) = \sum_{1 \leq 1 < j \leq p+1} (-1)^{i+j} \, \omega([Y_i, Y_j], Y_1, \ldots,$$
$$\hat{Y}_i, \ldots, \hat{Y}_j, \ldots, Y_{p+1}).$$

Problem 1. If ω amd θ are left invariant, then show that $\omega \wedge \theta$ and $d\omega$ are also left invariant.

Problem 2. Show that the set of all left invariant differential forms on G forms an algebra (with respect to the exterior product), which is isomorphic to $\wedge \mathfrak{g}^*$.

§8. One-Parameter Subgroups and the Exponential Map

Let G be a Lie group, and $a : t \to a(t)$ a differentiable curve of G defined on $(-\infty, +\infty)$. If, for any $s, t \in \mathbf{R}$, we have

$$a(s) \cdot a(t) = a(s + t), \qquad (1)$$

then $\{a(t) | t \in \mathbf{R}\}$ is called a *one-parameter subgorup* of G. By (1), we have $a(0)a(t) = a(t)$, so that multiplying by the inverse element of $a(t)$ on the right, we have

$$a(0) = e.$$

Also, since $a(t)a(-t) = a(-t)a(t) = a(0) = e$, we have

$$a(t)^{-1} = a(-t).$$

Furthermore, since $a(s)a(t) = a(s + t) = a(t)a(s)$, $a(s)$ and $a(t)$ commute. Hence a one-parameter subgroup of G is a commutative subgroup of G.

$\{L_{a(t)}; t \in \mathbf{R}\}$, $\{R_{a(t)}; t \in \mathbf{R}\}$ are both one-parameter groups of transformations of G, and the orbits of the identity element e by these transformation groups coinside with $a(t)$. For the infinitesimal transformations of these one-parameter groups of transformations, the following lemma holds.

LEMMA 1.   *Let X be the infinitiesimal transformation of $R_{a(t)}$, and let Y be the infinitesimal transformation of $L_{a(t)}$. Then X is left invariant and*

Y is right invariant, and $X_e = Y_e = a'(0)$ *holds. Here* $a'(t)$ *denotes the tangent vector to the curve* a *at* $a(t)$.

Proof. If f is a C^∞ function on a neighborhood of a point h of G, then

$$(L_{g*}X)_h f = X_{g^{-1}h} (f \circ L_g).$$

On the other hand, by the definition of X and by the commutativity of L_g and $R_{a(t)}$, we have

$$X_{g^{-1}h} (f \circ L_g) = \lim_{t \to 0} \frac{1}{t} [(f \circ L_g)(R_{a(t)}g^{-1}h) - (f \circ L_g)(g^{-1}h)]$$

$$= \lim_{t \to 0} \frac{1}{t} [f(R_{a(t)}h) - f(h)] = X_h f.$$

Hence $(L_{g*}X)_h = X_h$ holds at each point h of G, and X is left invariant. Similarly, we can show that Y is right invariant. Since $R_{a(t)}(e) = a(t)$, we have that $a(t)$ is an integral curve of X, and hence $X_{a(t)} = a'(t)$ holds. In particular, we have $X_e = a'(0)$. Similarly, we have $Y_e = a'(0)$.

LEMMA 2.   *Let* $\{\varphi_t ; t \in R\}$ *be a one-parameter group of transformations of* G, *and set* $\varphi_t(e) = a(t)$. *If* $\varphi_t \cdot L_g = L_g \cdot \varphi_t$ *holds for all* $g \in G$ *and for all* $t \in \mathbf{R}$, *then* $a(t)$ *is a one-parameter subgroup of* G, *and* $\varphi_t = R_{a(t)}$ *holds for all* $t \in \mathbf{R}$. *If* $\varphi_t \cdot R_g = R_g \cdot \varphi_t$ *holds for all* $g \in G$ *and all* $t \in \mathbf{R}$, *then* $a(t)$ *is a one-parameter subgroup of* G, *and* $\varphi_t = L_{a(t)}$ *holds for all* $t \in \mathbf{R}$.

Proof. The map $t \to a(t)$ is differentiable and, moreover, $a(s + t) = \varphi_{s+t}(e) = \varphi_t(\varphi_s(e)) = \varphi_t(L_{a(s)}(e)) = L_{a(s)}(\varphi_t(e)) = a(s)a(t)$. Hence $a(t)$ is a one-parmeter subgroup of G. On the other hand, for any $g \in G$, we have $\varphi_t(g) = \varphi_t(L_g(e)) = L_g(\varphi_t(e)) = g \cdot a(t) = R_{a(t)}(g)$. Hence $\varphi_t = R_{a(t)}$. We can argue similarly for the case $\varphi_t \cdot R_g = R_g \cdot \varphi_t$. Conversely, we have the following lemma.

LEMMA 3.   *If X is a left invariant vector field on G, then X is complete. The one-parameter group of transformations Exp tX of G, generated by X, satisfies $(Exp\ tX) \cdot L_g = L_g \cdot (Exp\ tX)$ for all $t \in \mathbf{R}$, $g \in G$. The corresponding facts hold for a right invariant vector field.*

Proof. By II, §12, for X, there is a neighborhood U_e of e and a local transformation $\varphi_t^{(e)}(|t| < \varepsilon)$ of G, and these satisfy the following three conditions: (1) The domain of definition of $\varphi_t^{(e)}$ contains U_e, $\varphi_0^{(e)}$ is the identity transformation on U_e, and $(t, g) \to \varphi_t^{(e)}(g)$ is a differentiable map from $(-\varepsilon, \varepsilon) \times U_e$ to G. (2) If $|s|$, $|t|$, $|s + t| < \varepsilon$ and $g \in U_e$, then $\varphi_s^{(e)}(\varphi_t^{(e)}(g)) = \varphi_{s+t}^{(e)}(g)$. (3) $X_g f = (df(\varphi_t^{(e)}(g))/dt)_{t=0}$ holds at each point g of U_e.

Now, for any $h \in G$, set $U_h = L_h(U_e)$ and $\varphi_t^{(h)} = L_h \cdot \varphi_t^{(e)} \cdot L_h^{-1}$ ($|t| < \varepsilon$). Then, using the left invariance of X, we can show easily that $\{U_h, \varepsilon, \varphi_t^{(h)}\}_{h \in G}$ is a local one-parameter group of local transformations generated by X. Since ε is a constant independent of h, by Theorem 2 of II, §12, X is complete. By the corollary to Theorem 3 of II, §12, we have $L_g \cdot (\text{Exp } tX) \cdot L_g^{-1} = \text{Exp } t((L_g)_* X)$, but, since X is left invariant, $(L_g)_* X = X$, and thus $L_g \cdot (\text{Exp } tX) = (\text{Exp } tX) \cdot L_g$ holds for all g, t.

By Lemmas 1, 2, and 3, we see that there is a $1:1$ correspondence between the one-parameter subgroups of G and the left invariant vector fields on G as follows:

If $a(t)$ is one-parameter subgroup of G, then there is an $X \in \mathfrak{g}$ such that $a(t) = (\text{Exp } tX)(e)$; conversely, for an arbitrary $X \in \mathfrak{g}$, $(\text{Exp } tX)(e) = a(t)$ is a one-parameter subgroup of G and $\text{Exp } tX = R_{a(t)}$.

DEFINITION.   For $X \in \mathfrak{g}$, set

$$\exp tX = (\text{Exp } tX)(e).$$

The map $X \to \exp X$ is a map from $\mathfrak{g}$ to G, and is called the *exponential map*.

By definition, $\exp tX$ is a one-parameter subgroup of G, and we have

$$\exp(t + s)X = \exp(tX) \exp(sX)$$
$$R_{\exp tX} = \text{Exp } tX.$$

By Theorem 4 of II, §12, for X, $Y \in \mathfrak{g}$, we have

$$[X, Y]_g = \lim_{t \to 0} \frac{1}{t}\{Y_g - ((R_{\exp tX})_* Y)_g\}. \tag{2}$$

Since, for an arbitrary element g of G, we have $A_g = R_{g^{-1}} \cdot L_g$, it follows that for $Y \in \mathfrak{g}$, we have

$$A_{g*} Y = R_{g^{-1}*} Y.$$

However, since L_h and $R_{g^{-1}}$ commute, we have $L_{h*}(R_{g^{-1}*} Y) = R_{g^{-1}*}(L_{h*} Y) = R_{g^{-1}*} Y$, so that $R_{g^{-1}*} Y \in \mathfrak{g}$. Hence, for $g \in \mathfrak{g}$, we have $A_{g*} Y \in \mathfrak{g}$. The map $Y \to A_{g*} Y$ is a linear transformation of the vector space $\mathfrak{g}$, and we denote this linear transformation by $\text{Ad}(g)$. That is,

$$\text{Ad}(g) Y = A_{g*} Y = R_{g^{-1}*} Y \qquad (g \in G, \ Y \in \mathfrak{g}).$$

Furthermore, since $A_{gh} = A_g A_h$, we have

$$\text{Ad}(gh) = \text{Ad}(g)\text{Ad}(h)$$

for any two elements g, h of G. In particular, it is clear from the definition that $\text{Ad}(e)$ is the identity transformation 1 of the vector space $\mathfrak{g}$. Hence we have $\text{Ad}(g^{-1}) \cdot \text{Ad}(g) = 1$. Hence $\text{Ad}(g)$ is a nonsingular linear transformation of $\mathfrak{g}$, and

$$\text{Ad}(g)^{-1} = \text{Ad}(g^{-1})$$

holds. The map $g \rightarrow \text{Ad}(g)$ is called the *adjoint representation* of the Lie group G.*

Since $A_{g*}[X, Y] = [A_{g*}X, A_{g*}Y]$, we have

$$\text{Ad}(g)[X, Y] = [\text{Ad}(g)X, \text{Ad}(g)Y] \qquad (X, Y \in \mathfrak{g}).$$

That is, $\text{Ad}(g)$ is an automorphism of the Lie algebra $\mathfrak{g}$.

If we let

$$A_X(t) = \text{Ad}(\exp tX) \qquad (X \in \mathfrak{g}),$$

then we have $A_X(t + s) = A_X(t)A_X(s)$. That is, $A_X(t)$ is a one-parameter group of linear transformations of the vector space $\mathfrak{g}$. If we set

$$C_X = \left[\frac{d}{dt} A_X(t)\right]_{t=0},\dagger$$

then we have

$$A_X(t) = \exp tC_X.$$

In fact from $A_X(t + s) = A_X(t)A_X(s)$, we obtain

$$\frac{d}{dt} A_X(t) = C_X A_X(t)$$

$$A_X(0) = 1$$

which shows that $A_X(t)$ is a solution to a system of differential equations and satisfies a given initial condition. However, clearly $\exp tC_X$ satisfies the same system of differential equations and the same initial condition, hence by uniqueness of solutions, we conclude that $A_X(t) = \exp tC_X$.‡

---

*We shall give the definitions of representations of Lie groups and Lie algebras in §15.

†We consider $A_X(t)$ to be a matrix, and differentiate it (i.e., differentiate its entries).

‡Note that the definitions of C_X and $\exp tC_X$ really do not depend on the choice of basis of $\mathfrak{g}$.

On the other hand, from (2), in $\mathfrak{g}$ we have

$$[X, Y] = \lim_{t \to 0} \frac{1}{t} \{Y - \mathrm{Ad}(\exp(-tX)) \cdot Y\}$$

$$= -\left[\frac{d}{dt} A_X(-t)\right]_{t=0} \cdot Y$$

$$= C_X \cdot Y.$$

Hence C_X is equal to the linear transformation $\mathrm{ad}(X)$ of $\mathfrak{g}$ defined by $Y \to [X, Y]$. That is, if we set

$$\mathrm{ad}(X) Y = [X, Y] \qquad (X, Y \in \mathfrak{g}),$$

then $A_X(t) = \exp t \, \mathrm{ad}(X)$. Hence

$$\mathrm{Ad}(\exp tX) = \exp t \, \mathrm{ad}(X) \tag{3}$$

holds.

The map $X \to \mathrm{ad}(X)$ is called the *adjoint representation* of the Lie algebra $\mathfrak{g}$. From the Jacobi identity for Lie algebras, we have

$$\mathrm{ad}(X)[Y, Z] = [\mathrm{ad}(X)Y, Z] + [Y, \mathrm{ad}(X)Z].$$

That is, using the terminology in II, §11, $\mathrm{ad}(X)$ is a derivation of the Lie algebra $\mathfrak{g}$. The formula (3), proved above, gives the relation between the automorphism $\mathrm{Ad}(\exp tX)$ of the Lie algebra $\mathfrak{g}$ and the derivation $\mathrm{ad}(X)$ of $\mathfrak{g}$.

By the corollary to Theorem 3 of II, §12, for an arbitrary diffeomorphism θ of G, we have

$$\theta \cdot (\mathrm{Exp}\ tX) \cdot \theta^{-1} = \mathrm{Exp}\ t(\theta_* X).$$

If we let $\theta = A_g$, then $\mathrm{Exp}\ t(\mathrm{Ad}(g)X) = A_g \cdot (\mathrm{Exp}\ tX) \cdot A_{g^{-1}}$. Hence $\exp t(\mathrm{Ad}(g)X) = A_g(\mathrm{Exp}\ tX)(A_{g^{-1}}e)$. Since $A_{g^{-1}}(e) = g^{-1}eg = e$, we get

$$\exp t(\mathrm{Ad}(g)X) = g(\exp tX)g^{-1}. \tag{4}$$

Using this formula, we can see that if $[X, Y] = 0$, then $\exp tX$ and $\exp sY$ commute for arbitrary $t, s \in \mathbf{R}$.

In fact, if $\mathrm{ad}(Y) \cdot X = 0$, then, from (3), we have

$$\mathrm{Ad}(\exp sY) \cdot X = (\exp s \, \mathrm{ad}(Y)) \cdot X = \sum_{n=0}^{\infty} \frac{s^n}{n!} (\mathrm{ad}\ Y)^n \cdot X = X.$$

Hence, in (4), if we set $g = \exp sY$, then the left member if $\exp tX$, while the right member is $(\exp sY)(\exp tY)(\exp sY)^{-1}$. Hence $\exp tX$ and $\exp sY$ commute.

Problem 1. Conversely, show that if $\exp tX$ and $\exp sY$ commute for arbitrary $s, t \in \mathbf{R}$, then $[X, Y] = 0$.

Problem 2. Prove that if X is left invariant and Y is right invariant, then $[X, Y] = 0$.

Problem 3. If Y is a vector field on G, G is connected, and $[X, Y] = 0$ for all $X \in \mathfrak{g}$, then show that Y is right invariant.

Problem 4. If $X, Y \in \mathfrak{g}$ are such that $[X, Y] = 0$, then show that $t \to \exp tX \cdot \exp tY$ is a one-parameter subgroup of G, and that its tangent vector at $t = 0$ is $X_e + Y_e$. Show that, hence, $(\exp tX) \cdot (\exp tY) = \exp t(X + Y)$.

§9. Examples of Lie Groups

(1) For two elements $x = (x^1, \ldots, x^n)$, $y = (y^1, \ldots, y^n)$ of $\mathbf{R}^n$, define the sum $x + y = (x^1 + y^1, \ldots, x^n + y^n)$. Then $\mathbf{R}^n$ is a commutative group. With this group structure, and with its differentiable structure as an affine space, $\mathbf{R}^n$ becomes a Lie group. Since $\mathbf{R}^n$ is commutative, $L_g = R_g$, and L_g is nothing but the parallel translation $x \to x + g$. The left invariant vector fields on $\mathbf{R}^n$ are of the form

$$\sum_{i=1}^{n} a^i \frac{\partial}{\partial x^i} \quad (a^i \in \mathbf{R}).$$

(2) The set of all complex numbers of absolute value 1, form a commutative group T^1, with respect to multiplication. There is a 1:1 correspondence between this group and the circle S^1 of radius 1, so that transporting the differentiable structure of S^1 onto T^1, T^1 becomes a Lie group.

(3) Let G, G' be two groups. If we define the product of two elements (a, b), (a', b') of the direct product set $G \times G'$ by $(a, b)(a', b') = (a\,a', b\,b')$, then $G \times G'$ becomes a group. If G, G' are Lie groups, then $G \times G'$ is a manifold as a direct product manifold of G and G', and $G \times G'$ is a Lie group with respect to the group structure mentioned above and this differentiable structure. This Lie group is called the *direct product* of the Lie group G and the Lie group G'. Similarly, we can define the direct product $G_1 \times \cdots \times G_r$ of r Lie groups $G_1, \ldots, G_r$.

(4) The direct product $T^r = T^1 \times \cdots \times T^1$ of r copies of the Lie group T^1, which was given in (2), is a Lie group, and is called the r-

dimensional torus. The torus T^r is a compact, connected, and commutative Lie group.

(5) The general linear group $GL(n, \mathbf{R})$ can be considered to be an open set of $\mathbf{R}^{n^2}$, so $GL(n, \mathbf{R})$ has a structure of a manifold. If $a = (a_j^i)_{i,j=1,\ldots,n}$ is a matrix in $GL(n, \mathbf{R})$, set

$$x_j^i(a) = a_j^i \qquad (i, j = 1, \ldots, n).$$

Then (x_j^i) is a coordinate system on $GL(n, \mathbf{R})$. The map $\varphi:(a, b) \to a \cdot b$ from $GL(n, \mathbf{R}) \times GL(n, \mathbf{R})$ to $GL(n, \mathbf{R})$ and the map $\varphi:a \to a^{-1}$ from $GL(n, \mathbf{R})$ to $GL(n, \mathbf{R})$ are both differentiable. In fact, if we set $x_j^i \circ \varphi = \varphi_j^i$, then $\varphi_j^i(a, b)$ is the (i, j)th entry of matrix $a \cdot b$. Hence we have

$$\varphi_j^i(a, b) = \sum_{k=1}^n x_k^i(a) \cdot x_j^k(b),$$

which shows that $\varphi_j^i(a, b)$ is a polynomial in the coordinates of a and b. Hence φ_j^i is of class C^∞, and the map φ is of class C^∞. Similarly, if we set $x_j^i \circ \psi = \psi_j^i$, then $\psi_j^i(a)$ is the (i, j)th entry of the inverse matrix a^{-1} of a. By Cramer's rule, we find the (i, j)th entry of a^{-1} to be of the form

$$\psi_j^i(a) = f_j^i(a)/\det a,$$

where the denominator and the numerator are both polynomials in the coordinates $x_j^i(a)$ of a, and the denominator is not 0 at any point of $GL(n, \mathbf{R})$. Hence ψ_j^i is of class C^∞, and hence ψ is also of class C^∞. Thus we have shown that $GL(n, \mathbf{R})$ is an n^2-dimensional Lie group.

Now let us consider the complex general linear group $GL(n, \mathbf{C})$. As in the case of $GL(n, \mathbf{R})$, we can consider $GL(n, \mathbf{C})$ as an open set of $\mathbf{C}^{n^2}$. Hence, as an open submanifold of $\mathbf{C}^{n^2}$, $GL(n, \mathbf{C})$ is a complex n^2-dimensional complex manifold. We can show, as in the case of $GL(n, \mathbf{R})$, that the maps $(a, b) \to a \cdot b$ and $a \to a^{-1}$ are holomorphic maps. Hence $GL(n, \mathbf{C})$ is a $2n^2$-dimensional Lie group.*

Let us investigate the Lie algebras of $GL(n, \mathbf{R})$ and $GL(n, \mathbf{C})$.

Let $A(t)$ be a one-parameter subgroup of $GL(n, \mathbf{R})$. As we have seen in §8, we can write

$$A(t) = \exp tC,$$

---

*The group $GL(n, \mathbf{C})$ is a complex Lie group in the sense that will be defined in §11.

where C is determined uniquely by

$$C = \left[\frac{dA(t)}{dt}\right]_{t=0}.$$

Conversely, if C is an arbitrary $n \times n$ real matrix, then the exponential function $\exp tC$ is a one-parameter subgroup of $GL(n, \mathbf{R})$.

Now let $\mathfrak{g}$ be the Lie algebra of $GL(n, \mathbf{R})$. For $X \in \mathfrak{g}$, consider the one-parameter subgroup $\exp tX$ of $GL(n, \mathbf{R})$. Then there is an $n \times n$ matrix $C(X)$ such that

$$\exp tX = \exp tC(X).$$

Applying

$$X_a f = \lim_{t \to 0} \frac{1}{t} \left[f(\mathrm{Exp}\ tX(a)) - f(a)\right]$$

to $f = x_j^i$, and using $\mathrm{Exp}\ tX = R_{\exp tX} = R_{\exp tC(X)}$, we get for the matrix $(X_a x_j^i)$:

$$(X_a x_j^i) = \lim_{t \to 0} \frac{1}{t}\left[a \exp tC(X) - a\right]$$
$$= a \cdot C(X).$$

That is, if we set

$$C(X) = (c_j^i(X)), \qquad a = (a_j^i),$$

then we have

$$X_a x_j^i = \sum_{k=1}^{n} a_k^i c_j^k(X).$$

Hence the vector field X is expressed as

$$X = \sum_{i,j=1}^{n} \left(\sum_{k=1}^{n} (x_k^i c_j^k(X))\right)\frac{\partial}{\partial x_j^i} \tag{1}$$

with respect to the coordinate system (x_j^i). Again, if we compute $[X, Y]x_j^i$, we see that it is equal to

$$\sum_{k=1}^{n} x_k^i \left(\sum_{t=1}^{n} (c_t^k(X)c_j^t(Y) - c_t^k(Y)c_j^t(X))\right).$$

Hence we have

$$[X, Y] = \sum_{i,j=1}^{n}\left(\sum_{k=1}^{n} x_k^i \left(\sum_{t=1}^{n} (c_t^k(X)\ c^t{}_j(Y) - c_t^k(Y)c_j^t(X))\right)\right)\frac{\partial}{\partial x_j^i},$$

and we obtain

$$c_j^i([X, Y]) = \sum_{i=1}^{n} (c_t^i(X)c_j^t(Y) - c_t^i(Y)c_j^t(X)). \tag{2}$$

Now, if we define the commutator product $[A, B]$ in the associative algebra of all $n \times n$ real matrices to be

$$[A, B] = AB - BA,$$

then we obtain a Lie algebra, which will be denoted by $\mathfrak{gl}(n, R)$.

The formula (2) then becomes

$$C([X, Y]) = [C(X), C(Y)].$$

From (1), it is clear that the correspondence $X \to C(X)$ is $1:1$ and onto, and that $C(\lambda X) = \lambda C(X)$ for $\lambda \in R$, and $C(X + Y) = C(X) + C(Y)$. Hence, the map $X \to C(X)$ is an isomorphism from the Lie algebra $\mathfrak{g}$ of $GL(n, R)$ onto the Lie algebra $\mathfrak{gl}(n, R)$. We shall identify $\mathfrak{g}$ and $\mathfrak{gl}(n, R)$ by this isomorphism from now on. Then, the exponential map

$$\exp : \mathfrak{gl}(n, R) \to GL(n, R)$$

is nothing but the exponential function, which assigns to each matrix X, belonging to $\mathfrak{gl}(n, R)$, the value $\exp X$.

From (4) of §8, we have

$$\exp t(\mathrm{Ad}(a)X) = a(\exp tX)a^{-1} \qquad (X \in \mathfrak{gl}(n, R), a \in GL(n, R)).$$

Differentiating both sides with respect to t and setting $t = 0$, we obtain

$$\mathrm{Ad}(a)X = a \cdot X \cdot a^{-1}. \tag{4}$$

That is, if we consider $\mathfrak{gl}(n, R)$ to be the Lie algebra of $GL(n, R)$, then the adjoint representation of $GL(n, R)$ is given by (4).

Similarly, the set $\mathfrak{gl}(n, C)$ of all $n \times n$ complex matrices is a Lie algebra with respect to the commutator product $[A, B] = AB - BA$. As in the case of $GL(n, R)$, we can prove that $\mathfrak{gl}(n, C)$ is isomorphic to the Lie algebra of $GL(n, C)$.

§10. The Canonical Coordinate Systems of Lie Groups

Let us study the property of the exponential map $\exp$ from $\mathfrak{g}$ to G, which we have defined in §8. Since $\mathfrak{g}$ is an n-dimensional real vector space, we can consider it as an n-dimensional affine space. Let us show that exp is a differentiable map from $\mathfrak{g}$ to G, and that $\mathrm{rank}_0(\exp) = n$, where 0 denotes the zero vector in $\mathfrak{g}$. Let $(x^1, \ldots, x^n)$ be a local coordinate system of G around the identity element e such that $x^i(e) = 0$, $i = 1, \ldots, n$, and set $U = \{q \mid |x^i(q)| < c, \; i = 1, \ldots, n\}$. Take a basis $\{X_1, \ldots, X_n\}$ of $\mathfrak{g}$, and set

$$X_k = \sum_{i=1}^{n} \xi_k^i(x) \frac{\partial}{\partial x^i} \qquad (k = 1, \ldots, n)$$

on U. Consider the solution of the system of differential equations

$$\frac{dx^i}{dt} = \sum_{k=1}^{n} y^k \xi_k^i(x) \qquad (i = 1, \ldots, n) \tag{1}$$

containing parameters $(y^1, \ldots, y^n)$, and satisfying the initial conditions

$$x^i(0) = 0 \qquad (i = 1, \ldots, n). \tag{2}$$

Take positive numbers ε, δ sufficiently small, and set $I_\varepsilon = (-\varepsilon, \varepsilon)$ and $U(\delta) = \{(y^1, \ldots, y^n) \mid |y^k| < \delta, \; k = 1, \ldots, n\}$. By the existence theorem for solutions of differential equations, there is a unique solution $x^i(t; y)$ $(i = 1, \ldots, n)$ of (1) satisfying the initial conditions (2), which is of class C^∞ on $I_\varepsilon \times U(\delta)$, and which satisfies $|x^i(t; y)| < c$. Fix an s such that $|s| \leq \varepsilon/2$. Then, since $|st| < \varepsilon$ for $|t| < 2$, the functions $x^i(st; y)$ are defined for $|t| < 2$ and $y \in U(\delta)$. But we have

$$\frac{dx^i(st; y)}{dt} = \sum_{k=1}^{n} sy^k \xi_k^i(x(st; y)).$$

So, by the uniqueness of the solution, $x^i(st; y)$ coincides with $x^i(t; sy)$ for $|t| < \varepsilon$, provided that $|sy^i| < \delta$ for $i = 1, \ldots, n$. Hence, for $|t| < 2$ and $|y^i| < \varepsilon\delta/2$, set

$$\varphi^i(t; y) = x^i\left(\frac{\varepsilon}{2} t; \frac{2}{\varepsilon} y\right).$$

Then $\{\varphi^i(t; y)\}$ is a solution of (1) satisfying the initial conditions (2), and we have $|\varphi^i(t; y)| < c$ $(i = 1, \ldots, n)$. Now, if we denote the point of U whose coordinates are $(\varphi^1(t; y), \ldots, \varphi^n(t; y))$ by $\varphi(t; y)$, then the fact that $\{\varphi^i(t; y)\}$ is a solution of (1) means that the curve $t \to \varphi(t; y)$ is an integral curve of the vector field

$$X = \sum_{k=1}^{n} y^k X_k,$$

and the fact that $\{\varphi^i(t;y)\}$ satisfies the initial condition (2) means that $\varphi(0;y) = e$, and hence $\varphi(t;y) = (\text{Exp } tX)\,(e) = \exp tX$. If we set $t = 1$ in this relation, we get $\varphi(1;y) = \exp X$, and we have

$$x^i(\exp X) = \varphi^i(1;y^1, \ldots, y^n).$$

The functions $\varphi^i(1;y^1, \ldots, y^n)$ are C^∞ functions with respect to the coordinate system $(y^1, \ldots, y^n)$ on a neighborhood of 0 in $\mathfrak{g}$. Hence the exponential map exp is a differentiable map on a neighborhood of 0. Since $\varphi^i(1;0, \ldots, y^k, \ldots, 0) = x^i(\exp y^k X_k)$, we have

$$\frac{\partial \varphi^i}{\partial y^k}(1;0) = \xi_k^i(0).$$

But $(X_1)_e, \ldots, (X_n)_e$ are linearly independent, so that the matrix $(\xi_k^i(0))$ is of rank n, and hence $\text{rank}_0(\exp)$ is equal to n.

Let Y be an arbitrary element of $\mathfrak{g}$. Let V be a sufficiently small neighborhood of 0 in $\mathfrak{g}$. Then there are a positive number m and a neighborhood W of Y such that if $X \in W$, then $m^{-1}X \in V$. But

$$\exp X = (\exp m^{-1}X)^m,$$

and exp is differentiable on the sufficiently small neighborhood V of 0. Hence exp is differentiable at Y too.

Since $\text{rank}_0(\exp) = n$ and $\exp 0 = e$, by the inverse function theorem, the map exp is a diffeomorphism from some neighborhood V of 0 in $\mathfrak{g}$ to some neighborhood U of e. Putting these together, we have

THEOREM 1. *Let $\mathfrak{g}$ be the Lie algebra of an n-dimensional Lie group G. The exponential mapping $exp\colon \mathfrak{g} \to G$ is differentiable, and $rank_0(exp) = n$ (0 is the zero vector of $\mathfrak{g}$). There are a neighborhood U of G at e and a neighborhood V of $\mathfrak{g}$ at 0, such that if $g \in U$, then g can be written uniquely as*

$$g = \exp X, \qquad X \in V.^*$$

The map $X \to exp\ X$ is a diffeomorphism from V to U.

Let $\{X_1, \ldots, X_n\}$ be a basis of $\mathfrak{g}$. Choose a positive number c small enough so that the neighborhood V of $\mathfrak{g}$ at 0 in Theorem 1 can be written

---

*Conversely, we sometimes write $X = \log g$. We should note that if $Y \notin V$, it is possible that $g = \exp X = \exp Y$. For example, consider the exponential map from the Lie algebra $\mathfrak{g}$ of T^1 to T^1.

as

$$V = \left\{ \sum_{i=1}^{n} a^i X_i \mid |a^i| < c,\ 1 \le i \le n \right\}.$$

Then the map

$$\exp\left(\sum_{i=1}^{n} a^i X_i \right) \to (a^1, \ldots, a^n)$$

is a diffeomorphism from U to a cube $\{(a^i) \mid |a^i| < c,\ i = 1, \ldots, n\}$ of $\mathbf{R}^n$. Hence we can define a local coordinate system $(x^1, \ldots, x^n)$ on U by

$$x^k\!\left(\exp\!\left(\sum_{i=1}^{n} a^i X_i \right)\right) = a^k \qquad (|a^k| < c,\ k = 1, \ldots, n). \tag{3}$$

DEFINITION 1.   The local coordinate system $(x^1, \ldots, x^n)$ defined by (3) around the identity e is called the *canonical coordinate system* (*of the first kind*) of G with respect to the basis $\{X_1, \ldots, X_n\}$ of $\mathfrak{g}$.

Now let $X \in \mathfrak{g}$, $X = \sum_{i=1}^{n} b^i X_i$. Take $\varepsilon > 0$ sufficiently small so that $|t b^i| < c$ for $|t| < \varepsilon$ $(i = 1, \ldots, n)$. Then, by (3), we have

$$x^i\!\left(\exp t \sum_{k=1}^{n} b^k X_k\right) = b^i t \qquad (i = 1, \ldots, n).$$

That is, for $|t|$ sufficiently small, the one-parameter subgroup $\exp t(\sum_{i=1}^{n} b^i X_i)$ is expressed by the straight line $t \to (b^1 t, \ldots, b^n t)$ with respect to the canonical coordinate system.

Conversely, if $\varphi : t \to \varphi(t)$ is a curve of G such that $\varphi(0) = e$, and $x^i(\varphi(t)) = b^i t\,(i = 1, \ldots, n)$ for $|t| < \varepsilon$,   then   we   have $\varphi(t) = \exp t(\sum_{i=1}^{n} b^i X_i)$ for $|t| < \varepsilon$.

It is often convenient to use a canonical coordinate system to treat a local problem of Lie groups since, as we have seen above, one-parameter subgroups are expressed by straight lines in a canonical coordinate system.

Problem 1.   Let $(x^1, \ldots, x^n)$ be the canonical coordinate system with respect to the basis $\{X_1, \ldots, X_n\}$ of $\mathfrak{g}$. Prove that

$$(X_i)_{\exp t X_i} = \left(\frac{\partial}{\partial x^i} \right)_{\exp t X_i}$$

Problem 2.   Let $(x^1, \ldots, x^n)$ be as in Problem 1, and set $U = \{g \mid |x^i(g)| < c\}$. Since the map $x \to bxb^{-1}$ from G to G is continuous, and $beb^{-1} = e$, we can choose a neighborhood W of e such that $bWb^{-1} \subset U$. Prove that

$$x^i(bgb^{-1}) = \sum_{j=1}^{n} a_j^i(b) x^j(g), \qquad g \in W,\quad i = 1, \ldots, n.$$

where

$$\text{Ad}(b)X_j = \sum_{k=1}^{n} a_j^k(b)X_k \qquad (j = 1, \ldots, n).$$

Problem 3. Let $(x_1, \ldots, x_n)$ and $(\bar{x}_1, \ldots, \bar{x}_n)$ be canonical coordinate systems with respect to bases $\{X_1, \ldots, X_n\}$ and $\{\bar{X}_1, \ldots, \bar{X}_n\}$, respectively. Show that the coordinate transformation between $(x^1, \ldots, x^n)$ and $(\bar{x}^1, \ldots, \bar{x}^n)$ is a linear transformation.

THEOREM 2.   *Let G be a connected Lie group, and $\mathfrak{g}$ its Lie algebra. For any $g \in G$, there are a finite number of elements $X_1, \ldots, X_k$ of $\mathfrak{g}$, such that*

$$g = \exp X_1 \cdots \exp X_k.$$

This is clear from Theorem 1 and Theorem 1 of §4.

DEFINITION 2.   If, for any two elements X, Y in a Lie algebra $\mathfrak{g}$, $[X, Y] = 0$, then $\mathfrak{g}$ is called a *commutative Lie algebra* or an *abelian Lie algebra*.

THEOREM 3.   *Let G be a connected Lie group, and $\mathfrak{g}$ its Lie algebra. Then G is commutative if and only if $\mathfrak{g}$ is commutative.*

Proof. If G is commutative, then A_g $(g \in G)$ is the identity transformation on G. Hence

$$\text{Ad}(g)Y = A_{g*}Y = Y, \qquad Y \in \mathfrak{g}, \quad g \in G.$$

If we set $g = \exp tX$, then by (3) of §8, we obtain $(\exp t\, \text{ad}(X))Y = Y$. Differentiating this relation with respect to t, and then setting $t = 0$, we get $\text{ad}(X)Y = [X, Y] = 0$. Since X, Y are arbitrary elements of $\mathfrak{g}$, this shows that $\mathfrak{g}$ is commutative.

Conversely, if $\mathfrak{g}$ is commutative, by what was said at the end of §8, $\exp X$ and $\exp Y$ commute for arbitrary X, $Y \in \mathfrak{g}$. Hence, by Theorem 2, G is commutative.

From Problem 4 of §8, if X, $Y \in \mathfrak{g}$ are such that $[X, Y] = 0$, then $(\exp tX)(\exp tY) = \exp t(X + Y)$. In particular, setting $t = 1$, we get

$$\exp X \exp Y = \exp(X + Y). \tag{4}$$

In particular, if $\mathfrak{g}$ is commutative, (4) holds for all X, $Y \in \mathfrak{g}$. If G is a a connected commutative Lie group, then, by Theorem 2, any $g \in G$ can be written as $g = \exp X_1 \cdots \exp X_k$, and by (4), $\exp X_1 \cdots \exp X_k = \exp(X_1 + \cdots + X_k)$. Hence an arbitrary element g of G can be written in the simple form

$$g = \exp X.$$

The vector space $\mathfrak{g}$ can be identified with $\mathbf{R}^n$, so that we can consider $\mathfrak{g}$ to be a connected *commutative Lie group* with respect to the addition in $\mathfrak{g}$. If G is a connected commutative Lie group, then (4), and the argument after (4), show that the map $\exp:\mathfrak{g} \to G$ is a homomorphism (§15) from the commutative Lie group $\mathfrak{g}$ onto the commutative Lie group G. We shall use this fact in §16 to investigate the structure of connected commutative Lie groups.

Going back to general Lie groups, by Theorem 1, we can choose a neighborhood V of 0 in $\mathfrak{g}$ and a neighborhood U of e in G so that exp is a diffeomorphism from V onto U. Since the map $(a, b) \to a \cdot b$ is continuous, and $e \cdot e = e$, we can choose a neighborhood U_1 of e such that $U_1 \cdot U_1 \subset U$. Since $U_1 = U_1 \cdot e \subset U_1 \cdot U_1$, we have $U_1 \subset U$. Since exp is a diffeomorphism from V onto U, if we set $\exp^{-1} U_1 = V_1$, then V_1 is a neighborhood of 0 contained in V. If X, $Y \in V_1$, then $\exp X \cdot \exp Y \subset U$, so that there is a unique $Z \in V$ such that

$$\exp X \cdot \exp Y = \exp Z. \tag{5}$$

The element Z is a function of X and Y, so we can write $Z = f(X, Y)$. We would like to know the form of the function f. We can formulate this question in terms of canonical coordinates. Let $\{X_1, \ldots, X_n\}$ be a basis of $\mathfrak{g}$, and write $X = \sum_i a^i X_i$, $Y = \sum_i b^i X_i$, $Z = \sum_i c^i X_i$. Then the canonical coordinates of the points $\exp X$, $\exp Y$, and $\exp Z$, are $(a^1, \ldots, a^n)$, $(b^1, \ldots, b^n)$, and $(c^1, \ldots, c^n)$, respectively. To ask how Z is related to X and Y is the same as to ask how the c^i's are related to (a^i) and (b^i).

Let us prove the following theorem.

THEOREM 4.   *Let $\mathfrak{g}$ be the Lie algebra of a Lie group G. For arbitrary X, $Y \in \mathfrak{g}$, we have*

(a) $\exp tX \cdot \exp tY = \exp\left(t(X + Y) + \dfrac{t^2}{2}[X, Y] + O(t^3)\right)$

(b) $\exp(-tX) \cdot \exp(-tY) \cdot \exp tX \cdot \exp tY = \exp(t^2[X, Y] + O(t^3))$

for sufficiently small $|t|$, where $O(t^3)$ is an element of $\mathfrak{g}$, and $O(t^3)/t^3$ is bounded for sufficiently small $|t|$ and is a C^∞ function of t.*

Proof.   Let D_X, D_Y be the derivations of $C^\infty(G)$ determined by the vector fields X, Y respectively (II, §11). Since $\operatorname{Exp} tX = R_{\exp tX}$ (§8),

---

*By this we mean that the components of the element $O(t^3)/t^3$ of $\mathfrak{g}$, with respect to a basis of $\mathfrak{g}$, are bounded and of class C^∞ as functions of t.

by what was shown at the end of II, §12, for $f \in C^\infty(G)$, we have

$$(D_X^m f)(g \cdot \exp tX) = \frac{d^m}{dt^m} f(g \cdot \exp tX) \qquad (g \in G).$$

Hence we have

$$(D_X^m D_Y^n f)(e) = \left[\frac{\partial^{m+n}}{\partial t^m \partial s^n} f(\exp tX \cdot \exp sY) \right]_{s=t=0},$$

and we get

$f(\exp tX \cdot \exp sY)$
$$= f(e) + t(D_X f)(e) + s(D_Y f)(e) + (t^2(D_X^2 f)(e)$$
$$+ 2ts(D_X D_Y f)(e) + s^2(D_Y^2 f)(e))$$
$$+ \sum_{m+n=3} \frac{t^m s^n}{m! \, n!} (D_X^m D_Y^n f)(\exp \theta t X \cdot \exp \eta s Y)$$

$$(0 < \theta < 1, 0 < \eta < 1). \quad (*)$$

On the other hand, if $|t|$ is sufficiently small, we can set

$$\exp tX \cdot \exp tY = \exp Z(t),$$

where $Z(t) \in \mathfrak{g}$ is a C^∞ function of t, and $Z(0) = 0$. Hence we can write

$$Z(t) = tZ_1 + \frac{t^2}{2} Z_2 + O(t^3).$$

Taking a basis $\{X_1, \ldots, X_n\}$ of $\mathfrak{g}$, let $(x^1, \ldots, x^n)$ be the canonical coordinate system with respect to this basis. By the definition of a canonical coordinate system, and by Taylor's formula (II, §12), we have

$x^i(\exp Z(t))$
$$= \left(\left(tD_{Z_1} + \frac{t^2}{2} D_{Z_2} \right) x^i \right)(e) + \frac{1}{2} \left(\left(tD_{Z_1} + \frac{t^2}{2} D_{Z_2} \right)^2 x^i \right)(e) + O(t^3), \quad (**)$$

where these $O(t^3)$'s denote C^∞ functions of t such that $O(t^3)/t^3$ is bounded for sufficiently small $|t|$. If we let $f = x^i$ and $t = s$ in (*), and compare the coefficients of t and t^2 of the right members of (*) and (**), then we get

$$(D_{Z_1} x^i)(e) = (D_X x^i + D_Y x^i)(e) \qquad (i = 1, \ldots, n),$$
$$(D_{Z_2} x^i)(e) + (D_{Z_1}^2 x^i)(e) = (D_X^2 x^i)(e) + 2(D_X D_Y x^i)(e) + (D_Y^2 x^i)(e)$$
$$(i = 1, \ldots, n).$$

From the first equations, we obtain $(Z_1)_e = X_e + Y_e$. But Z_1, X, Y are all elements of $\mathfrak{g}$, so that we get $Z_1 = X + Y$. Hence $D_{Z_1}^2 x^i = (D_X^2 + D_X D_Y + D_Y D_X + D_Y^2)x^i$. Substituting this in the second equations above, we get

$$(D_{Z_2} x^i)(e) = (D_{[X, Y]} x^i)(e) \qquad (i = 1, \ldots, n).$$

From this we conclude that $Z_2 = [X, Y]$, and thus (a) is proved.

Using (a), the left member of (b) becomes

$$\exp(-t(X + Y) + \frac{t^2}{2} [X, Y] + O(t^3)) \cdot$$

$$\exp(t(X + Y) + \frac{t^2}{2} [X, Y] + O(t^3)).$$

Using (a) once more on this expression, we obtain the right member of (b).

Let us now define a canonical coordinate system of the second kind of the Lie group G. Let $\{X_1, \ldots, X_n\}$ be a basis of $\mathfrak{g}$, and let Φ be the map from $\mathfrak{g}$ to G defined by

$$\Phi\left(\sum_{k=1}^{n} a^k X_k\right) = \exp a^1 X_1 \cdot \exp a^2 X_2 \cdots \exp a^n X_n.$$

The map Φ is clearly differentiable, and it is easy to see that $\mathrm{rank}_0(\Phi) = n$. Hence taking $c > 0$ sufficiently small, and setting $V = \{\sum_{k=1}^{n} a^k X_k | |a^k| < c, k = 1, \ldots, n\}$, Φ is a diffeomorphism from V onto a neighborhood U of e. Hence any element g of U can be written uniquely as

$$g = \exp a^1 X_1 \cdots \exp a^n X_n, \qquad |a^k| < c \quad (k = 1, \ldots, n),$$

and we can define a local coordinate system on U by

$$x^i(g) = a^k \qquad (k = 1, \ldots, n).$$

We call $(x^1, \ldots, x^n)$ the *canonical coordinate system of the second kind* of G with respect to the basis $\{X_1, \ldots, X_n\}$ of $\mathfrak{g}$.

Problem 4. Let $\mathfrak{v}_1, \ldots, \mathfrak{v}_m$ be subspaces of $\mathfrak{g}$, such that $\mathfrak{g}$ is the direct sum of $\mathfrak{v}_1, \ldots, \mathfrak{v}_m$: $\mathfrak{g} = \mathfrak{v}_1 + \cdots + \mathfrak{v}_m$. Define a mapping Φ from $\mathfrak{g}$ to G by $\Phi(\sum_{i=1}^{m} X_i) = \exp X_1 \cdots \exp X_m$ ($X_i \in \mathfrak{v}_i$). Prove that Φ is differentiable, and that $\mathrm{rank}_0(\Phi) = \dim G$.

§11. Complex Lie Groups and Complex Lie Algebras

In II, §16, we have studied complex manifolds. In the theory of Lie groups, just as in the theory of differentiable manifolds, we can also consider complex Lie groups in a natural way.

DEFINITION 1. If a group G has the structure of a complex manifold with a countable base, and if the map $(x, y) \to x \cdot y$ from $G \times G$ to G and the map $x \to x^{-1}$ from G to G are both holomorphic, then we call G a *complex Lie group*. The complex dimension of the complex manifold G is called the complex dimension of the complex Lie group G.

Complex manifolds of complex dimension n are differentiable manifolds of (real) dimension $2n$, and holomorphic maps are differentiable. Hence complex Lie groups are certainly Lie groups in the sense of the definition in §1. Sometimes we shall call a Lie group, in the sense of the definition in §1, a *real Lie group*, as opposed to a complex Lie group. A complex Lie group of complex dimension n is a real Lie group of dimension $2n$.

DEFINITION 2. A Lie algebra over the complex number field $\mathbf{C}$ is called a *complex Lie algebra*.

We sometimes call a Lie algebra over the real number field $\mathbf{R}$ a *real Lie algebra*.

If $\mathfrak{g}$ is an n-dimensional complex Lie algebra, then $\mathfrak{g}$ is also a Lie algebra over the real numbers, and the dimension of $\mathfrak{g}$ over $\mathbf{R}$ is $2n$. That is, an n-dimensional complex Lie algebra can be considered as a $2n$-dimensional real Lie algebra. As in II, §17, consider $\mathfrak{g}$ as a $2n$-dimensional real vector space, and denote the linear transformation of $\mathfrak{g}$, given by $X \to iX$ (i is the imaginary unit), by I. Since $\mathfrak{g}$ is a Lie algebra over $\mathbf{C}$, we have

$$[X, IY] = I[X, Y] \tag{1}$$

for arbitrary $X, Y \in \mathfrak{g}$. That is,

$$\mathrm{ad}(X) \cdot I = I \cdot \mathrm{ad}(X) \tag{1'}$$

holds for all $X \in \mathfrak{g}$.

Conversely, let $\mathfrak{g}$ be a $2n$-dimensional Lie algebra over $\mathbf{R}$. Let I be a complex structure of the real vector space $\mathfrak{g}$(II, §17). For any $X \in \mathfrak{g}$, let $I \cdot X = iX$, and consider $\mathfrak{g}$ as an n-dimensional vector space over $\mathbf{C}$. Suppose I satisfies (1') for all elements X of $\mathfrak{g}$. Then, for an arbitrary complex number $\lambda = a + ib$ ($a, b \in \mathbf{R}$), we have

$$\begin{aligned}
[X, \lambda Y] &= [X, aY + bI \cdot Y] \\
&= a[X, Y] + b[X, I \cdot Y] \\
&= a[X, Y] + bI[X, Y] = \lambda[X, Y].
\end{aligned}$$

We also have $[\lambda X, Y] = -[Y, \lambda X] = -\lambda[Y, X] = \lambda[X, Y]$. Hence $\mathfrak{g}$ is a Lie algebra over **C**, i.e., $\mathfrak{g}$ is a complex Lie algebra. Thus we have shown that a Lie algebra over **R** is a complex Lie algebra if and only if it admits a complex structure satisfying (1').

THEOREM. *If the Lie group G is a complex Lie group, then the Lie algebra $\mathfrak{g}$ of G is a complex Lie algebra. Conversely, if the Lie algebra $\mathfrak{g}$ of a real Lie group G is a complex Lie algebra, then G has the structure of a complex Lie group.*

Proof.   We first show that if the Lie group G is a complex Lie group, then $\mathfrak{g}$ is a complex Lie algebra. Let J be the almost complex structure attached to the complex manifold G (II, §16). The almost complex structure J determines a linear transformation J_s of $T_s(G)$ ($J_s^2 = -1$) at each point s of G. Since G is a complex Lie group, the left translation $x \to L_g(x) = g \cdot x$, and the right translation $x \to R_g(x) = x \cdot g$, are both holomorphic transformations of G. Hence for all $v \in T_x(G)$, we have

$$(L_g)_* J_x v = J_{gx}(L_g)_* v, \qquad (R_g)_* J_x v = J_{xg}(R_g)_* v \qquad (2)$$

for all $g, x \in G$.

If Y is a left invariant vector field on G, then define a vector field JY on G, as in II, §17, by

$$(JY)_x = J_x Y_x.$$

Then, by (2), we have

$$((L_g)_* JY)_x = (L_g)_* J_{g^{-1}x} Y_{g^{-1}x} = J_x((L_g)_* Y)_x = J_x Y_x = (JY)_x,$$

so that JY is left invariant. Hence $Y \to JY$ is a linear transformation of $\mathfrak{g}$, and, from the definition of JY, it is clear that $J^2 Y = -Y$. If we denote this linear transformation on $\mathfrak{g}$ also by J, then J is a complex structure of the real vector space $\mathfrak{g}$. Furthermore, we can show that $(R_{g^{-1}})_*(JY) = J((R_{g^{-1}})_* Y)$, by a similar argument as above, so that by the definition of $\mathrm{Ad}(g)$, we have

$$\mathrm{Ad}(g) JY = J \mathrm{Ad}(g) Y.$$

If we let $g = \exp tX$, then by (3) of §8,

$$\exp(t \, \mathrm{ad}(X)) J = J \exp(t \, \mathrm{ad}(X)).$$

If we differentiate this expression with respect to t, and set $t = 0$, then we get

$$\mathrm{ad}(X) \cdot J = J \cdot \mathrm{ad}(X).$$

Thus $\mathfrak{g}$ is a complex Lie algebra.

Conversely, if $\mathfrak{g}$ is a complex Lie algebra, let us prove that G is a complex Lie group. First, let us suppose that G is connected. The Lie algebra $\mathfrak{g}$ has a complex structure J satisfying (1'). Let us show, that

$$\text{Ad}(g) \cdot J = J \cdot \text{Ad}(g) \tag{3}$$

for all $g \in G$. Since G is connected, each $g \in G$ can be written as $g = \exp Y_1 \cdots \exp Y_k$ for some $Y_1, \ldots, Y_k \in \mathfrak{g}$ (Theorem 2 of §10). Hence it suffices to prove (3) for $g \in G$ of the form $\exp tY$. From §8, (3), we have $\text{Ad}(g) = \exp t \, \text{ad} \, (Y)$. Since $\text{ad} \, Y$ commutes with J by (1'), the exponential function of $\text{ad} \, Y$ also commutes with J, so that $\text{Ad}(g)$ commutes with J, and hence (3) holds.

Using J, let us define an almost complex structure on G, which we shall denote by the same letter J. If $x \in G$ and $v \in T_x(G)$, then there is a unique $X \in \mathfrak{g}$ such that $X_x = v$. So we can define a linear transformation J_x of $T_x(G)$ by

$$J_x v = (JX)_x.$$

Clearly, from the definition, we have $J_x^2 = -1$. Furthermore, it is easy to check that $J : x \to J_x$ is of class C^∞ in the sense of II, §17, and J is an almost complex structure on G.

Let us show that L_g and R_g are "almost complex mappings", i.e., that they satisfy (2). If we let $v = X_x$, where $X \in \mathfrak{g}$, then we have $(R_g)_* J_x v = ((R_g)_* JX)_{xg}$. Since $JX \in \mathfrak{g}$, we have $(R_g)_* JX = \text{Ad}(g^{-1}) JX$. But then, by (3), $\text{Ad}(g^{-1}) JX = J \, \text{Ad}(g^{-1}) X = J((R_g)_* X)$, and hence $((R_g)_* JX)_{xg} = J_{xg}((R_g)_* X)_{xg} = J_{xg}(R_g)_* v$. Hence the second equality in (2) is proved. The proof of the first equality in (2) is direct.

Furthermore, the map $\psi : x \to x^{-1}$ from G to G is also almost complex, i.e.,

$$(\psi_*)_x J_x v = J_{x^{-1}}(\psi_*)_x v, \qquad v \in T_x(G) \tag{4}$$

holds. By the definition of ψ, we have $(\psi_*)_e u = -u$ for an arbitrary $u \in T_e(G)$, so that clearly (4) holds for $x = e$. Since $\psi \cdot L_g = R_{g^{-1}} \cdot \psi$, we have $(\psi_*)_x \cdot ((L_x)_*)_e = ((R_{x^{-1}})_*)_e \cdot (\psi_*)_e$. Hence $(\psi_*)_x = ((R_{x^{-1}})_*)_e \cdot (\psi_*)_e \cdot ((L_{x^{-1}})_*)_x$. From this equality, (4) for $x = e$, and (2), we see that (4) holds for an arbitrary $x \in G$.

Now let us show that J gives a complex structure to G. For this, let us show that we can find a local coordinate system on a neighborhood of the identity element e satisfying the condition of the lemma in II, §17. Let $\{X_1, \ldots, X_n\}$ be a basis of $\mathfrak{g}$ over $\mathbf{C}$, and set $Y_i = JX_i$ ($i = 1$,

$\ldots$, n). Then $\{X_1, Y_1, X_2, Y_2, \ldots, X_n, Y_n\}$ is a basis of $\mathfrak{g}$ over $\mathbf{R}$, and, because of $(1')$, we have

$$[X_i, Y_i] = 0 \qquad (i = 1, \ldots, n). \tag{5}$$

Let $(x_1, y_1, \ldots, x_n, y_n)$ be the canonical coordinate system of the second kind on a neighborhood of e with respect to the basis $\{X_1, Y_1, \ldots, X_n, Y_n\}$ of $\mathfrak{g}$. We shall prove that

$$J\left(\frac{\partial}{\partial x^i}\right) = \frac{\partial}{\partial y^i}, \qquad J\left(\frac{\partial}{\partial y^i}\right) = -\frac{\partial}{\partial x^i} \qquad (i = 1, \ldots, n) \tag{6}$$

holds.

To simplify the notation, set

$$a_i(t) = \exp tX_i, \qquad b_i(t) = \exp tY_i.$$

Since (5) holds, from §8 we have $\mathrm{Ad}(b_i(t))X_i = (\exp t \, \mathrm{ad}(Y_i)) \, X_i = X_i$. Hence we have

$$(R_{b_i(t)})_* X_i = X_i \tag{7}$$

for all t. From (5) and §8, we have also, that $a_i(t)$ and $b_i(s)$ commute. Set

$$g_1 = a_1(t_1)b_1(s_1) \cdots a_{i-1}(t_{i-1})b_{i-1}(s_{i-1}),$$
$$g_2 = a_{i+1}(t_{i+1}) \cdots a_n(t_n)b_n(s_n).$$

Then the point g with coordinates $(t_1, s_1, \ldots, t_n, s_n)$ is equal to $g = g_1 a_i(t_i)b_i(s_i)g_2$. Set $g_3 = a_i(t_i)b_i(s_i)$. Let us show that

$$\left(\frac{\partial}{\partial x^i}\right)_g = (L_{g_1} \cdot R_{g_2})_*(X_i)_{g_3}$$

$$\left(\frac{\partial}{\partial y^i}\right)_g = (L_{g_1} \cdot R_{g_2})_*(Y_i)_{g_3}$$

Since $(\partial/\partial x^i)_g$ is equal to the tangent vector to the curve $t \to g_1 a_i(t)b_i(s_i)g_2$ at $t = t_i$, if we let $a_i'(t_i)$ be the tangent vector to the curve $t \to a_i(t)$ at $t = t_i$, then

$$\left(\frac{\partial}{\partial x^i}\right)_g = (L_{g_1} \cdot R_{b_i(s_i)g_2})_* a_i'(t_i)$$

$$= (L_{g_1} \cdot R_{g_2})_*(R_{b_i(s_i)})_* a_i'(t_i).$$

But $a_i(t) = \exp tX_i$, so that we have $a_i'(t_i) = (X_i)_{a_i(t_i)}$. Hence, by (7), we have $(R_{b_i(s_i)})_* a'(t_i) = (X_i)_{a_i(t_i)b(s_i)} = (X_i)_{g_3}$. Hence we have $(\partial/\partial x_i)_g = (L_{g_1} \cdot R_{g_2})_*(X_i)_{g_3}$. Similarly we can prove $(\partial/\partial y^i)_g = (L_{g_1} \cdot R_{g_2})_*(Y_i)_{g_3}$.

As we have already proved, L_{g_1} and R_{g_2} are almost complex with respect to J. That is, (2) holds for them, so that using (2), from (8) we get

$$J_g \left(\frac{\partial}{\partial x^i} \right)_g = (L_{g_1} \cdot R_{g_2})_* J_{g_3}(X_i)_{g_3}$$

$$= (L_{g_1} \cdot R_{g_2})_*(Y_i)_{g_3}$$

$$= \left(\frac{\partial}{\partial y^i} \right)_g.$$

Similarly we have $J_g(\partial/\partial y^i)_g = -(\partial/\partial x^i)_g$, and (6) is proved.

Now let U_e be a neighborhood of e where $(x^1, y^1, \ldots, x^n, y^n)$ is defined. For an arbitrary element s of G, define a neighborhood U_s of s, and a local coordinate system $(u^1, v^1, \ldots, u^n, v^n)$ on U_s, by

$$U_s = L_s(U_e), \qquad u^i = x^i \circ L_{s^{-1}}, \qquad v^i = y^i \circ L_{s^{-1}}.$$

Since $L_{s^{-1}}$ is almost complex, it is easy to see that $J(\partial/\partial u^i) = \partial/\partial v^i$ and $J(\partial/\partial v^i) = -\partial/\partial u^i$ hold. Hence G has an open covering which satisfies the condition of the lemma in II, §17, so that G is a complex manifold and J is the almost complex structure attached to this complex structure. Hence L_g, R_g, for $g \in G$, and ψ are holomorphic maps of G. From this we see that G becomes a complex Lie group, and the theorem is proved when G is connected.

Now let us consider the case where G is not connected. If we let G_0 be the connected component of G containing the identity, then G_0 is a Lie group whose Lie algebra is $\mathfrak{g}$. From what we have proved already, G_0 is a complex Lie group. Let $\{G_i\}_{i=1, 2, \ldots}$ be the connected components of G other that G_0.* If $g_i \in G_i$, then, since $g_i^{-1} G_i$ is a connected component and contains e, we have $g_i^{-1} G_i = G_0$. The almost complex structure J on G_0 can be extended to an almost complex structure $\tilde{J}$ on G as follows: for $x \in G_i$ and $v \in T_x(G)$, define $\tilde{J}_x$ by

$$\tilde{J}_x v = (L_{g_i})_* J_y (L_{g_i^{-1}})_* v,$$

where $y = g_i^{-1} x \in G_0$. This definition does not depend on the choice of the representative g_i of G_i. Thus we have an almost complex structure

---

*From the definition of a Lie group, G has a countable base, so that the number of connected components is at most countable.

$\tilde{J}$ on G, and we can show that L_g, R_g $(g \in G)$ and ψ are almost complex mappings from G to G. By transforming the local coordinate system $(x^1, y^1, \ldots, x^n, y^n)$ from the neighborhood of e above to a neighborhood of an arbitrary point g of G, we can show, as in the connected case, that $\tilde{J}$ is the almost complex structure attached to the complex structure on G, and that G is a complex Lie group. The theorem is proved.

Remark. Since $[X_k, Y_k] = 0$ holds, we have $\exp t^k X_k \cdot \exp s^k Y_k = \exp (t^k X_k + s^k J X_k)$. Hence, setting $\alpha^k = t^k + is^k$, we can write $\exp t^k X_k \cdot \exp s^k Y_k = \exp \alpha^k X_k$. If we define complex-valued functions z^k by

$$z^k(\exp \alpha^1 X_1 \cdots \exp \alpha^n X_n) = \alpha^k \qquad (k = 1, \ldots, n),$$

then $(z^1, \ldots, z^n)$ is a complex local coordinate system of G at e. We call $(z^1, \ldots, z^n)$ a complex canonical coordinate system of the second kind on G.

Examples of Complex Lie Groups
(1) $GL(n, \mathbf{C})$ is a complex Lie group.
(2) $\mathbf{C}^n$ is a complex Lie group with respect to addition.
(3) Direct products of complex Lie groups are complex Lie groups.
(4) Every even-dimensional commutative Lie group G is a complex Lie group. In fact, since $\mathfrak{g}$ is an even-dimensional real vector space, $\mathfrak{g}$ admits a complex structure J. Since G is commutative, so is $\mathfrak{g}$ (§10). Hence we have $[X, JY] = 0$, $J[X, Y] = 0$, so that $[X, JY] = J[X, Y]$ holds. That is, $\mathfrak{g}$ becomes a complex Lie algebra. Hence, by the theorem, G becomes a complex Lie group.
(5) In particular, an even-dimensional torus becomes a complex Lie group. This complex Lie group is called a *complex torus*.

Problem. If G is a complex Lie group of complex dimension n, prove that $g \to \mathrm{Ad}(g)$ is a holomorphic map from G to $GL(n, \mathbf{C})$.

PROPOSITION. *A compact connected complex Lie group G is commutative.*

Proof. Let $\mathrm{Ad}(g)$ be expressed as a matrix, $\mathrm{Ad}(g) = (a_j^i(g))$. By the problem above, the $a_j^i(g)$'s are holomorphic functions on G. Since a holomorphic function on a compact connected complex manifold is a constant, and since $a_j^i(e) = \delta_j^i$, we have $a_j^i(g) = \delta_j^i$, i.e., $\mathrm{Ad}(g) = 1$ holds for all $g \in G$. Hence we have $\mathrm{Ad}(g)x = gxg^{-1} = x$ for any g and x in G and so G is commutative.

Remark. As we shall see in §16, a compact connected commutative Lie group is necessarily a torus. Hence, by the proposition, a compact connected complex Lie group is a complex torus.

§12. Lie Subgroups of a Lie Group (I)

DEFINITION 1. Let G be a Lie group. A Lie group H is called a *Lie subgroup* of G if it has the following two properties.
(1) H is a submanifold of the manifold G.
(2) H is a subgroup of the group G.
In particular, if H is connected as a manifold, then H is called a *connected Lie subgroup*. If H is a closed submanifold of G, then H is called a *closed Lie subgroup* of G.

DEFINITION 2. Let $\mathfrak{g}$ be a Lie algebra over a field K. A subset $\mathfrak{h}$ of $\mathfrak{g}$ is called a *Lie subalgebra* of $\mathfrak{g}$ if it has the following two properties:

(1) The set $\mathfrak{h}$ is a subspace of $\mathfrak{g}$, i.e., if X, $Y \in \mathfrak{h}$, and λ, $\mu \in K$, then $\lambda X + \mu Y \in \mathfrak{h}$.
(2) If X, $Y \in \mathfrak{h}$, then $[X, Y] \in \mathfrak{h}$.

It is clear that a Lie subalgebra of a Lie algebra is itself a Lie algebra.

We shall show that the Lie algebra $\mathfrak{h}$ of a Lie subgroup H of a Lie group G can be considered as a Lie subalgebra of the Lie algebra $\mathfrak{g}$ of G, and conversely, that to each Lie subalgebra $\mathfrak{h}$ of $\mathfrak{g}$, there corresponds a unique connected Lie subgroup of G.

Let G be a Lie group, H a Lie subgroup of G, and i the injection map from H to G. By condition (1) for a Lie subgroup, i is differentiable, and, at each point h of H, the differential i_* of i is an isomorphism from $T_h(H)$ into $T_{i(h)}(G)$. From condition (2), we also have $i(xy) = i(x)i(y)$ $(x, y \in H)$. Hence, for $h \in H$, we have

$$i \cdot L_h = L_{i(h)} \cdot i. \tag{1}$$

For an element X of the Lie algebra $\mathfrak{h}$ of H, there is a unique element X' of $\mathfrak{g}$ such that $(i_*)_e X_e = X'_e$. Moreover, X and X' are i-related, and we have $i_* X = X'$. In fact, by (1), we have $i_* X_h = i_*(L_h)_* X_e = (L_{i(h)})_* X'_e = X'_{i(h)}$. Conversely, if $X' \in \mathfrak{g}$, and $X'_e \in i_* T_e(H)$, then there is a unique $X \in \mathfrak{h}$ such that $i_* X = X'$. From II, §11, for X, $Y \in \mathfrak{h}$, we have

$$i_*[X, Y] = [i_* X, i_* Y].$$

Hence, if $\mathfrak{h}'$ is the set of all X' of $\mathfrak{g}$ such that $X'_e \in i_* T_e(H)$, then $\mathfrak{h}'$ is a Lie subalgebra of $\mathfrak{g}$, and $\mathfrak{h}$ and $\mathfrak{h}'$ are isomorphic by the map $X \to i_*X$. We identify the Lie algebra $\mathfrak{h}$ of H with a Lie subalgebra of the Lie algebra $\mathfrak{g}$ of G by this isomorphism. We also identify the tangent vector space $T_h(H)$ of H with the subspace $i_* T_h(H)$ of the tangent vector space $T_h(G)$. Then the Lie algebra of H is the set of elements X of $\mathfrak{g}$ satisfying $X_e \in T_e(H)$.

Conversely, let $\mathfrak{h}$ be an arbitrary Lie subalgebra of $\mathfrak{g}$. For an arbitrary element g of G, set $\mathfrak{h}_g = \{X_g | X \in \mathfrak{h}\}$. Then $\mathfrak{D}: g \to \mathfrak{h}_g$ is a completely integrable differential system on G (II, §8). In fact, if $\{Y_1, \ldots, Y_r\}$ is a basis of $\mathfrak{h}$, then $\{Y_1, \ldots, Y_r\}$ is a local basis for this differential system on G. Moreover, since $\mathfrak{h}$ is a Lie algebra, we have

$$[Y_i, Y_j] = \sum_{k=1}^{r} c_{ij}^k Y_k, \qquad c_{ij}^k \in \mathbf{R} \quad (i, j = 1, \ldots, r).$$

Hence, by Theorem 2 of III, §8, the differential system $\mathfrak{D}$ is completely integrable. Let H be the maximal connected integral manifold of $\mathfrak{D}$ passing through e. If $a \in G$, since $(L_a)_* X_g = X_{ag}$ $(X \in \mathfrak{h})$, we have $(L_a)_* \mathfrak{h}_g = \mathfrak{h}_{ag}$. That is, L_a leaves the differential system $\mathfrak{D}$ invariant. Hence L_a transforms a maximal connected integral manifold to a maximal connected integral manifold. Hence, if $h \in H$, then $L_{h^{-1}} H$ is also a maximal connected integral manifold passing through e. That is, we have $L_{h^{-1}} H = H$ $(h \in H)$. From this, we see that H is a subgroup of G. Now let us show that the map from $H \times H$ to H, given by $(a, b) \to a \cdot b$, and the map from H to H, given by $a \to a^{-1}$, are both differentiable. Since H is a submanifold of G, and G is a Lie group, it is clear that $\varphi:(a, b) \to a \cdot b$ is differentiable as a map from $H \times H$ to G. But $\varphi(H \times H) \subset H$, H is a maximal connected integral manifold of $\mathfrak{D}$, and G has a countable base, so, that by Theorem 2 of III, §10, φ is differentiable also as a map from $H \times H$ to H. Similarly, the map $a \to a^{-1}$ from H to H is also differentiable. Thus we have proved that H is a connected Lie subgroup of G, and, from the construction of H, it is clear that the Lie algebra of H coincides with the given Lie subalgebra $\mathfrak{h}$ of $\mathfrak{g}$.

If H is an arbitrary Lie subgroup of G, and if $\mathfrak{h}$ is its Lie algebra, then it is clear that H is an integral manifold of the differential system $\mathfrak{D}: g \to \mathfrak{h}_g$. Hence, if H_0 is the connected component of H containing e, and if H_0' is the maximal connected integral manifold of $\mathfrak{D}$ passing through e, then we have $H_0 \subset H_0'$. But H_0 and H_0' are both connected Lie groups having $\mathfrak{h}$ as their Lie algebra. So, by Theorem 2 of §10, we get $H_0 = H_0'$. The other connected components of H are of the form hH_0, so they are also maximal connected integral manifolds of $\mathfrak{D}$ passing through *h*. Putting together what has been proved above, we have the following theorem.

THEOREM 1. *Let G be a Lie group,* $\mathfrak{g}$ *the Lie algebra of G. If H is a Lie subgroup of G, then the Lie algebra* $\mathfrak{h}$ *of H can be regarded as a Lie subalgebra of* $\mathfrak{g}$. *Conversely, if* $\mathfrak{h}$ *is a Lie subalgebra of* $\mathfrak{h}$, *then there is a unique connected Lie subgroup of G whose Lie algebra is* $\mathfrak{h}$.

Problem 1. Let *H* be a connected Lie subgroup of the Lie group *G*. An element *X* of $\mathfrak{g}$ belongs to the Lie algebra $\mathfrak{h}$ of *H* if and only if exp $tX \in H$ for all $t \in \mathbf{R}$. Prove this. (*Hint*: Apply Theorem 2 of III, §10.)

Now let us consider the case when *G* is a complex Lie group.

DEFINITION 3. Let *G* be a complex Lie group. A complex Lie group *H* is called a *complex Lie subgroup* of *G* if the following two conditions are satisfied:

(1) The complex manifold *H* is a complex submanifold of *G*.
(2) The group *H* is a subgroup of *G*.

A complex Lie subgroup of *G* is of course a Lie subgroup of *G*. We also have the following theorem.

THEOREM 2. *Let* $\mathfrak{g}$ *be the Lie algebra of a complex Lie group G. If H is a Lie subgroup of G, then H is a complex Lie subgroup of G if and only if the Lie algebra* $\mathfrak{h}$ *of H is a complex Lie subalgebra of* $\mathfrak{g}$.

Using the result in §11, we can prove this theorem easily. The proof is left to the reader.

Remark 1. We can also define a connected Lie subgroup *H* of *G* as follows. That is, a subset *H* of *G* is called a connected Lie subgroup of *G* if *H* is a connected submanifold of *G* and a subgroup of *G*. With this definition, it is rather difficult to prove that *H* itself is actually a Lie group.

Problem 2. If *H* is a subset of *G* satisfying (1) *H* is a regular submanifold of *G*, and (2) *H* is a subgroup of *G*, then show that *H* is a Lie subgroup of *G*.

Remark 2. If a subset *H* of *G* satisfies the two conditions (1) *H* is a closed subset of *G*, and (2) *H* is a subgroup of *G*, then it will be shown in §20 that *H* is a closed Lie subgroup of *G* (Theorem 2 of §20).

DEFINITION 4. Let *H* be a Lie subgroup of a Lie group *G*. If the group *H* is a normal subgroup of *G*, i.e., if $gHg^{-1} = H$ holds for all $g \in G$, then *H* is called a *Lie normal subgroup* of *G*.

DEFINITION 5.   Let $\mathfrak{h}$ be a Lie subalgebra of the Lie algebra $\mathfrak{g}$. If, for all $X \in \mathfrak{g}$, $Y \in \mathfrak{h}$, we have $[X, Y] \in \mathfrak{h}$, then $\mathfrak{h}$ is called an *ideal* of $\mathfrak{g}$.

THEOREM 3.   *If H is a connected normal Lie subgroup of a Lie group G, then the Lie algebra $\mathfrak{h}$ of H is an ideal of the Lie algebra $\mathfrak{g}$ of G. Conversely, if G is a connected Lie group with Lie algebra $\mathfrak{g}$, and if $\mathfrak{h}$ is an ideal of $\mathfrak{g}$, then the connected Lie subgroup H of G corresponding to $\mathfrak{h}$ is a normal subgroup of G.*

Proof.   (1) Let H be a connected Lie normal subgroup of G, and $\mathfrak{h}$ the Lie algebra of H. If $Y \in \mathfrak{h}$, then $\exp tY$ is a one-parameter subgroup of H. Since H is normal, $g(\exp tY)g^{-1}$ belongs to H for an arbitrary $g \in G$. By (4) of §8, we have $g(\exp tY)g^{-1} = \exp t \, \mathrm{Ad}(g)Y$, so that $\exp t \, \mathrm{Ad}(g)Y$ belongs to H, and hence by Problem 1, $\mathrm{Ad}(g)Y \in \mathfrak{h}$. Here, if we set $g = \exp(-tX)$, $X \in \mathfrak{g}$, then $\mathrm{Ad}(\exp(-tX)) \in \mathfrak{h}$. But

$$[X, Y] = \lim_{t \to 0} \frac{1}{t} [Y - \mathrm{Ad}(\exp(-tX))Y],$$

and since the right member belongs to $\mathfrak{h}$, we get $[X, Y] \in \mathfrak{h}$. Here X is an arbitrary element of $\mathfrak{g}$, Y an arbitrary element of $\mathfrak{h}$, and hence $\mathfrak{h}$ is an ideal of $\mathfrak{g}$.

(2) Let $\mathfrak{h}$ be an ideal of $\mathfrak{g}$, and assume that the Lie group G is connected. Let H be the connected Lie subgroup of G whose Lie algebra is $\mathfrak{h}$. For $X \in \mathfrak{g}$, $Y \in \mathfrak{h}$, we have $\mathrm{Ad}(\exp X) \cdot Y = \sum_{n=0}^{\infty}(1/n!)(\mathrm{ad}\ X)^n \cdot Y$ ((3) of §8). Since $\mathfrak{h}$ is an ideal of $\mathfrak{g}$, we have $(\mathrm{ad}\ X)^n \cdot Y \in \mathfrak{h}$ $(n = 0, 1, 2, \ldots)$. Hence we have $\mathrm{Ad}(\exp X) \cdot Y \in \mathfrak{h}$. But, by (4) of §8, $(\exp X)(\exp Y)$ $(\exp X)^{-1} = \exp(\mathrm{Ad}(\exp X) \cdot Y)$. Hence $(\exp X)(\exp Y)(\exp X)^{-1} \in H$. If h is an arbitrary element of H, then, since H is connected, by Theorem 2 of §10, there are elements $Y_1, \ldots, Y_k \in \mathfrak{h}$ such that

$$h = \exp Y_1 \cdots \exp Y_k.$$

Then

$$(\exp X)h(\exp X)^{-1}$$
$$= (\exp X)(\exp Y_1)(\exp X)^{-1} \cdots (\exp X)(\exp Y_k)(\exp X)^{-1}.$$

As we have proved above, $(\exp X)(\exp Y_i)(\exp X)^{-1} \in H$, so this shows that we also have $(\exp X)h(\exp X)^{-1} \in H$. Now if g is an arbitrary element of G, again by §10, Theorem 2, we have elements $X_1, \ldots, X_m \in \mathfrak{g}$ such that $g = g_1 \cdots g_m$, where $g_i = \exp X_i$. Then, by induction on m, we can prove that $ghg^{-1} \in H$. Hence H is a normal subgroup of G.

§13. Linear Lie Groups

A Lie subgroup of $GL(n, \mathbf{R})$ or $GL(n, \mathbf{C})$ is called a *linear Lie group*. We shall study various examples of linear Lie groups in this section.

(1) The *special linear groups* $SL(n, \mathbf{R})$, $SL(n, \mathbf{C})$. We denote the set of all $n \times n$ real[complex] matrices of determinant 1 by $SL(n, \mathbf{R})[SL(n, \mathbf{C})]$, and call it the *real [complex] special linear group*. The group $SL(n, \mathbf{R})[SL(n, \mathbf{C})]$ is a subgroup of $GL(n, \mathbf{R})[GL(n, \mathbf{C})]$. For $x \in GL(n, \mathbf{R})$, set $f(x) = \det x$. Then $SL(n, \mathbf{R})$ is the set of all $x \in GL(n, \mathbf{R})$ satisfying $f(x) - 1 = 0$. We have

$$df = \sum_{k=1}^{n} \begin{vmatrix} x_1^1 & \cdots & dx_k^1 & \cdots & x_n^1 \\ x_1^2 & \cdots & dx_k^2 & \cdots & x_n^2 \\ \cdot & & & & \\ \cdot & & & & \\ \cdot & & & & \\ x_1^n & \cdots & dx_k^n & \cdots & x_n^n \end{vmatrix}$$

Hence $(df)_x$ is not 0 at any $x \in SL(n, \mathbf{R})$. Hence, by the corollary to Theorem 1 of II, §10, $SL(n, \mathbf{R})$ is an $(n^2 - 1)$-dimensional closed submanifold of $GL(n, \mathbf{R})$. Hence, by the proposition of II, §10, the maps $(a, b) \to a \cdot b$ and $b \to b^{-1}$ are differentiable. Hence $SL(n, \mathbf{R})$ is a closed Lie subgroup of $GL(n, \mathbf{R})$. Now let $\exp tX$ ($X \in \mathfrak{gl}(n, \mathbf{R})$) be a one-parameter subgroup of $GL(n, \mathbf{R})$. Then we have

$$\det(\exp tX) = \exp t(\operatorname{tr} X).$$

Hence $\exp tX$ is a one-parameter subgroup of $SL(n, \mathbf{R})$ if and only if $\operatorname{tr} X = 0$. Set

$$\mathfrak{sl}(n, \mathbf{R}) = \{X \in \mathfrak{gl}(n, \mathbf{R}) | \operatorname{tr} X = 0\}.$$

Then $\mathfrak{sl}(n, \mathbf{R})$ is an $(n^2 - 1)$-dimensional subalgebra of $\mathfrak{gl}(n, \mathbf{R})$, and $\mathfrak{sl}(n, \mathbf{R})$ is the Lie algebra of $SL(n, \mathbf{R})$.

Let $\mathbf{R}^+$ be the Lie group formed by the set of all positive real numbers with respect to multiplication. Define a map φ from $\mathbf{R}^+ \times SL(n, \mathbf{R})$ to $GL^+(n, \mathbf{R})$ by $\varphi(\lambda, a) = \lambda \cdot a (\lambda \in \mathbf{R}^+, a \in SL(n, \mathbf{R}))$. Then φ is a 1:1 homomorphism from $\mathbf{R}^+ \times SL(n, \mathbf{R})$ onto $GL^+(n, \mathbf{R})$. Hence, by the corollary to Theorem 2 of §5, φ is a homeomorphism from $\mathbf{R}^+ \times SL(n, \mathbf{R})$ onto $GL^+(n, \mathbf{R})$. By the proposition of §4, $GL^+(n, \mathbf{R})$ is connected, and hence $\mathbf{R}^+ \times SL(n, \mathbf{R})$ is also connected. The group $SL(n, \mathbf{R})$ is a continuous image of $\mathbf{R}^+ \times SL(n, \mathbf{R})$, and hence $SL(n, \mathbf{R})$ is connected.

Similarly, we can prove that $SL(n, \mathbf{C})$ is a closed complex Lie subgroup of $GL(n, \mathbf{C})$ of complex dimension $n^2 - 1$, and that its Lie algebra is

$$\mathfrak{sl}(n, \mathbf{C}) = \{X \in \mathfrak{gl}(n, \mathbf{C}) | \operatorname{tr} X = 0\}.$$

The group $SL(n, \mathbf{C})$ is also connected.

Remark. The left invariant vector field $X = \sum_{i,j} (\sum_k a_k^i x_j^k) \, \partial/\partial x_j^i$ of $GL(n, \mathbf{R})$ belongs to the Lie algebra of $SL(n, \mathbf{R})$ if and only if X_e (where e is the unit matrix) is a tangent vector to $SL(n, \mathbf{R})$. Since $SL(n, \mathbf{R})$ is defined by $f - 1 = 0$, the condition that X_e is tangent to $SL(n, \mathbf{R})$ is $(df)_e(X_e) = 0$. However, we have $(df)_e = \sum (dx_k^k)_e$. Hence $(df)_e(X_e) = \sum a_k^k = \operatorname{tr} a$. Thus X belongs to the Lie algebra of $SL(n, \mathbf{R})$ if and only if $\operatorname{tr} a = 0$. This is another way to show that the Lie algebra of $SL(n, \mathbf{R})$ is $\mathfrak{sl}(n, \mathbf{R})$.

(2) *The orthogonal group $O(n)$ and the special orthogonal group $SO(n)$*

Let $O(n)$ be the set of all $n \times n$ orthogonal matrices, and set $SO(n) = O(n) \cap SL(n, \mathbf{R})$. The sets $O(n)$ and $SO(n)$ are both subgroups of $GL(n, \mathbf{R})$. We call $O(n)$ the *orthogonal group*, and $SO(n)$ the *special orthogonal group* (§5, Example 2).

If $x = (x_j^i)$ is an element of $GL(n, \mathbf{R})$, define the $\frac{1}{2}n(n+1)$ functions f_{ij} $(1 \leq i \leq k \leq n)$ by

$$f_{ik}(x) = \sum_{j=1} x_i^j x_k^j \qquad (1 \leq i \leq k \leq n).$$

Then we have

$$O(n) = \{a \in GL(n, \mathbf{R}) | f_{ik}(a) - \delta_{ik} = 0, 1 \leq i \leq k \leq n\}.$$

We can easily check by computation that the df_{ik}'s are linearly independent at each point of $O(n)$. Hence, by the corollary to Theorem 1 of II, §10, $O(n)$ is a $\frac{1}{2} n(n-1)$-dimensional closed submanifold of $GL(n, \mathbf{R})$. Hence $O(n)$ is a closed Lie subgroup of $GL(n, \mathbf{R})$ (cf. Problem 2 of §12 and the case of $SL(n, \mathbf{R})$).*

If $\exp tX(X \in \mathfrak{gl}(n, \mathbf{R}))$ is a one-parameter subgroup of $GL(n, \mathbf{R})$, then, since ${}^t(\exp tX) = \exp t{}^tX$ and $(\exp tX)^{-1} = \exp(-tX)$, $\exp tX$ belongs to $O(n)$ if and only if

$${}^tX + X = 0.\dagger$$

---

*A quicker way to get to this conclusion is to use Theorem 2 of §20, because it is clear that $O(n)$ is a closed subgroup of $GL(n, \mathbf{R})$. This way, we can also prove directly that $SO(n)$ is a closed Lie subgroup of $GL(n, \mathbf{R})$.

†By ${}^tX$ we mean the transpose matrix of X.

An $n \times n$ matrix satisfying this condition is called a *skew-symmetric matrix*. The set of all $n \times n$ skew-symmetric matrices is denoted by $\mathfrak{o}(n)$. It is a subalgebra of $\mathfrak{gl}(n, \mathbf{R})$ and the Lie algebra of $O(n)$ is $\mathfrak{o}(n)$.

Problem 1. Verify by direct computation that $\mathfrak{sl}(n, \mathbf{R})$ and $\mathfrak{o}(n)$ are Lie subalgebras of $\mathfrak{gl}(n, \mathbf{R})$.

Problem 2. A left invariant vector field X of $GL(n, \mathbf{R})$ belongs to the Lie algebra of $O(n)$ if and only if $(df_{ik})_e(X_e) = 0$ (e is the unit matrix) for all $1 \leq i \leq k \leq n$. Use this to show that the Lie algebra of $O(n)$ is isomorphic to $\mathfrak{o}(n)$.

Problem 3. Show that the set of all real matrices of the form (triangular matrices)

$$\begin{bmatrix} a_1^1 & 0 & \cdots & 0 \\ & a_2^2 & 0 & 0 \\ & & \cdot & 0 \\ & * & & \cdot \\ & & & a_n^n \end{bmatrix}, \quad a_k^k > 0 \quad (k = 1, \ldots, n).$$

is a connected closed Lie subgroup of $GL(n, \mathbf{R})$, and find its Lie algebra.

(3) *The complex orthogonal group* $O(n, \mathbf{C})$. The set of all $n \times n$ complex matrices satisfying ${}^t a\, a = 1_n$ (1_n is the $n \times n$ unit matrix) is denoted by $O(n, \mathbf{C})$. The set $O(n, \mathbf{C})$ is a complex Lie subgroup of $GL(n, \mathbf{C})$, and has complex dimension $\frac{1}{2} n(n - 1)$. The Lie algebra of $O(n, \mathbf{C})$ is the subalgebra $\mathfrak{o}(n, \mathbf{C})$ of $\mathfrak{gl}(n, \mathbf{C})$ formed by all $n \times n$ skew symmetric complex matrices.

These facts can be all verified using arguments similar to those used for $O(n)$. The group $O(n, \mathbf{C})$ is called the *complex orthogonal group*.

(4) *The unitary group* $U(n)$ *and the special unitary group* $SU(n)$. Let $U(n)$ denote the set of all $n \times n$ unitary matrices, and let $SU(n)$ denote the set of all $n \times n$ unitary matrices with determinant 1. The sets $U(n)$ and $SU(n)$ are subgroups of $GL(n, \mathbf{C})$. Since, for $a \in GL(n, \mathbf{C})$,

$$a \in U(n) \Leftrightarrow {}^t \bar{a} a = 1_n,$$

the entries of the matrices belonging to $U(n)$ are bounded, and $U(n)$ is a closed set of $GL(n, \mathbf{C})$. Hence $U(n)$ is compact. Similarly, $SU(n)$ is also compact. Hence these are closed subgroups of $GL(n, \mathbf{C})$. Invoking Theorem 2 of §20, we see that $U(n)$ and $SU(n)$ are closed Lie subgroups of $GL(n, \mathbf{C})$. The groups $U(n)$ and $SU(n)$ are called the *unitary group* and the *special unitary group*, respectively.

Now let $\exp tX$ ($X \in \mathfrak{gl}(n, \mathbf{C})$) be a one-parameter subgroup of $GL(n, \mathbf{C})$. Since ${}^t(\overline{\exp tX}) = \exp t{}^t\bar{X}$ and $(\exp tX)^{-1} = \exp(-tX)$, we have $\exp tX \in U(n)$ if and only if

$$ {}^t\bar{X} + X = 0.$$

An $n \times n$ complex matrix X satisfying this condition is called an $n \times n$ *skew-Hermitian matrix*. The set of all $n \times n$ skew-Hermitian matrices is denoted by $\mathfrak{u}(n)$. The set $\mathfrak{u}(n)$ is a Lie subalgebra of $\mathfrak{gl}(n, \mathbf{C})$, and the Lie algebra of $U(n)$ is $\mathfrak{u}(n)$. Similarly, the set $\mathfrak{su}(n)$ of all $n \times n$ skew-Hermitian matrices whose traces are 0 is also a Lie subalgebra of $\mathfrak{gl}(n, \mathbf{C})$, and the Lie algebra of $SU(n)$ is $\mathfrak{su}(n)$.

(5) Let $\mathfrak{a}$ be an n-dimensiona algebra over $\mathbf{R}$.* If a nonsingular linear transformation θ of $\mathfrak{a}$ satisfies $\theta(a \circ b) = (\theta a) \circ (\theta b)$ for any $a, b \in \mathfrak{a}$, then θ is called an automorphism of $\mathfrak{a}$. Since products of automorphisms, and inverses of automorphisms, are automorphisms, the set G of all automorphisms of $\mathfrak{a}$ is a group of linear transformations of $\mathfrak{a}$. Taking a basis $\{u_1, \ldots, u_n\}$ of $\mathfrak{a}$, and writing θ as an $n \times n$ matrix, we can consider G as a subgroup of $GL(n, \mathbf{R})$.

Suppose the multiplication of elements of the basis $\{u_1, \ldots, u_n\}$ is given by

$$u_i \circ u_j = \sum_{k=1}^{n} \alpha_{ij}^k u_k, \qquad \alpha_{ij}^k \in \mathbf{R}.$$

Let θ be a nonsingular linear transformation of $\mathfrak{a}$, and write

$$\theta u_i = \sum_{j=1}^{n} a_i^j u_j.$$

The transformation θ is an automorphism of $\mathfrak{a}$ if and only if $\theta(u_i \circ u_j) = (\theta u_i) \circ (\theta u_j)$ holds for all $i, j = 1, \ldots, n$. But

$$\theta(u_i \circ u_j) = \sum_{k, m} \alpha_{ij}^k a_k^m u_m, \qquad (\theta u_i) \circ (\theta u_j) = \sum_{k, l, m} a_i^k a_j^l \alpha_{kl}^m u_m.$$

Hence θ is an automorphism of $\mathfrak{a}$ if and only if the following holds:

$$\sum_{k, l=1}^{n} \alpha_{kl}^m a_i^k a_j^l = \sum_{k=1}^{n} a_{ij}^k \alpha_k^m \qquad (i, j, m = 1, \ldots, n). \tag{1}$$

That is, $a = (a_j^i) \in GL(n, \mathbf{R})$ belongs to G if and only if (1) holds. Hence the group G of all automorphisms of $\mathfrak{a}$ is a closed subgroup of $GL(n, \mathbf{R})$, and hence, invoking Theorem 2 of §20 again, we conclude that G is a closed Lie subgroup of $GL(n, \mathbf{R})$.

Let $D \in \mathfrak{gl}(n, \mathbf{R})$. Let us find a condition for the one-parameter subgroup $\exp tD = a(t)$ of $GL(n, \mathbf{R})$ to belong to G. Let $a(t) \subset G$ and let the components of $a(t)$ be $a_j^i(t)$. Then $a_j^i(t)$ satisfies (1). Setting $D = (d_j^i)$, we have $(da_j^i(t)/dt)_{t=0} = d_j^i$. If we substitute $a_j^i = a_j^i(t)$ in (1), and dif-

*Cf. II, §11.

ferentiate with respect to t and set $t = 0$, then we get

$$\sum_{k=1}^{n} (\alpha_{kj}^m d_i^k + \alpha_{ik}^m d_j^k) = \sum_{k=1}^{n} \alpha_{ij}^k d_k^m \qquad (i, j, m = 1, \ldots, n). \qquad (2)$$

Now, if we define a linear transformation D on $\mathfrak{a}$ by

$$Du_i = \sum_{j=1}^{n} d_i^j u_j,$$

then condition (2) is equivalent to $Du_i \circ u_j + u_i \circ Du_j = D(u_i \circ u_j)$
$(i, j = 1, \ldots, n)$. The latter means that D is a derivation of $\mathfrak{a}$. Hence,
if $\exp tD$ is in G, then D is a derivation of $\mathfrak{a}$. Conversely, we shall show
that if D is a derivation of $\mathfrak{a}$, then $\exp tD$ is in G. By induction on n,
we can easily show that

$$D^n(u \circ v) = \sum_{r=0}^{n} \binom{n}{r} D^{n-r} u \cdot D^r v \qquad (u, v \in \mathfrak{a}).$$

Hence we get

$$\sum_{n=0}^{k} \frac{t^n}{n!} D^n(u \circ v) = \sum_{i+j \leq k} \frac{t^i}{i!} D^i u \circ \frac{t^j}{j!} D^j v.$$

From this, we easily obtain $(\exp tD)(u \circ v) = ((\exp tD)u) \circ ((\exp tD)v)$.
That is, $\exp tD$ is an automorphism of $\mathfrak{a}$. We thus have the following
theorem.

THEOREM. *The automorphism group of a finite-dimensional algebra
$\mathfrak{a}$ is a Lie group, and its Lie algebra is the Lie algebra formed by all the
derivations of $\mathfrak{a}$.*

§14. Quotient Spaces and Quotient Groups of Lie Groups

Let G be a Lie group, and let H be a closed Lie subgroup. If we give
a topology to the set of left cosets G/H as in §2, then, since H is a closed
subgroup, G/H becomes a Hausdorff space. Furthermore, since G has a
countable base, we can easily see that G/H has a countable base. Let
π be the natural map from G to G/H. Let φ be the map from $G \times (G/H)$
to G/H defined by $(g, hH) \to ghH$, and let t_g be the homeomorphism
from G/H to itself given by $t_g(hH) = ghH$. Let us show that we can
introduce a differentiable structure in G/H so that π and φ become
differentiable maps, and t_g becomes a diffeomorphism from G/H to itself.

(1) Choose a basis $\{X_1, \ldots, X_m, Y_1, \ldots, Y_r\}$ of the Lie algebra $\mathfrak{g}$ of G (where $m + r = n = \dim G$) so that $\{Y_1, \ldots, Y_r\}$ is a basis of the Lie algebra $\mathfrak{h}$ of H. Let $(y^1, \ldots, y^r)$ be the canonical coordinate system of H of the second kind with respect to $\{Y_1, \ldots, Y_r\}$, and set $U_2 = \{h \mid |y^i(h)| < c_2, i = 1, \ldots, r\}$. Since H is a closed submanifold of G, we can choose a neighborhood V of e in G such that $V \cap H = U_2$. Let $(x^1, \ldots, x^m, x^{m+1}, \ldots, x^{m+r})$ be the canonical coordinate system of G of the second kind with respect to $\{X_1, \ldots, X_m, Y_1, \ldots, Y_r\}$. Let i be the injection map from H into G. Then we have $x^{m+j} \circ i = y^j$ $(j = 1, \ldots, r)$. We can choose a neighborhood U of e in G by taking c_1 sufficiently small, and defining U by

$$U = \{g \mid |x^i(g)| < c_1 (1 \leq i \leq m), |x^{m+j}(g)| < c_2 (1 \leq j \leq r)\},$$

so that U is contained in V. Then we have

$$U_2 = \{g \in U \mid x^i(g) = 0 \ (1 \leq i \leq m)\}.$$

Setting $U_1 = \{g \in U \mid x^{m+j}(g) = 0 \ (1 \leq j \leq r)\}$, we have $U = U_1 \cdot U_2$, $U_1 \cap U_2 = \{e\}$, $U \cap H = U_2$, and $U \cdot H = U_1 \cdot H$.

Now define a map σ from the neighborhood $\pi(U)$ of the point $\pi(e) = H$ in G/H, to G, as follows: if $g \in U$, then g can be written uniquely as $g = g_1 g_2$, where $g_1 \in U_1, g_2 \in U_2$, and we have $gH = g_1 H$; set $\sigma(gH) = g_1$. Then σ is a $1:1$ map from $\pi(U)$ onto U_1, and we have $\pi(\sigma(gH)) = gH$. That is, the restriction of π to U_1 is the inverse of σ. The map σ is continuous. In fact, if W is an open set of G, then $\sigma^{-1}(W) = \pi(W \cap U_1)$. Since $W \cap U_1$ is open in U_1, $(W \cap U_1)U_2$ is open in U. Hence $\sigma^{-1}(W)$ is open in $\pi(U)$, and σ is continuous. Thus σ is a homeomorphism from $\pi(U)$ onto the subspace U_1 of G.

We define a map $\sigma_{\pi(g)}$ from the neighborhood $\pi(gU)$ of an arbitrary point $\pi(g) = gH$ of G/H, to G, by $\sigma_{\pi(g)} = L_g \cdot \sigma \cdot t_{g^{-1}}$. The map $\sigma_{\pi(g)}$ is a homeomorphism from $\pi(gU)$ onto the subspace gU_1 of G, and the restriction of π to gU_1 is the inverse of $\sigma_{\pi(g)}$.

(2) Let us define a differentiable structure on G/H. The set U_1 is homeomorphic to the cube $Q = \{(v^1, \ldots, v^m) \mid |v^i| < c_1 (1 \leq i \leq m)\}$ in $\mathbf{R}^m$. Since U_1 is homeomorphic to gU_1 by L_g, and $\pi(gU)$ is homeomorphic to gU_1 by $\sigma_{\pi(g)}$, we see that $\pi(gU)$ is homeomorphic to Q, and that G/H is an m-dimensional topological manifold with a local coordinate system $x_{\bar{g}}^1, \ldots, x_{\bar{g}}^m$ (where $\pi(g) = \bar{g}$) given by

$$x_{\bar{g}}^i(\pi(gu)) = x^i(u_1) \qquad (i = 1, \ldots, m),$$

where for $u \in U$, we write $u = u_1 \cdot u_2, u_1 \in U_1, u_2 \in U_2$. Since $x^i(u_1) =$

$x^i(u)$, for $i = 1, \ldots, m$, we can define $x_{\bar{g}}^i$ by

$$x_{\bar{g}}^i(\pi(gu)) = x^i(u) \qquad (i = 1, \ldots, m).$$

We can also define it using $\sigma_{\bar{g}} : \pi(gU) \to G$ as

$$x_{\bar{g}}^i = x^i \circ L_{g^{-1}} \circ \sigma_{\bar{g}} \qquad (i = 1, \ldots, m).$$

Suppose $\pi(gU) \cap \pi(g'U) \neq \varnothing$. If $\bar{a} = \pi(ga_1) = \pi(g'a_1')\,(a_1, a_1' \in U_1)$, then, setting $g'^{-1}g = b$, there is an $h \in H$ such that $ba_1 h = a_1'$. Hence, by taking a sufficiently small neighborhood V of a_1 such that $V \subset U$, we have $bVh \subset U$. Here we can assume that V is of the form $V = V_1 \cdot V_2$, where $V_1 = V \cap U_1$, $V_2 = V \cap U_2$. Since h and V_2 are both contained in H, we have $\pi(g'bVh) = \pi(g'bV_1)$. On the other hand, since $g'b = g$, we have $\pi(g'bVh) = \pi(gV)$. Hence $\pi(g'bV_1) = \pi(gV)$ is a neighborhood of $\bar{a}$ contained in $\pi(gU) \cap \pi(g'U)$. Since $x \to bxh$ is differentiable, for $v_1 \in V_1$, we have

$$x^i(bv_1 h) = F^i(x^1(v_1), \ldots, x^m(v_1), x^{m+1}(v_1), \ldots, x^{m+r}(v_1)),$$

where F^i is a C^∞ function on $m + r$ variables.

Now, since $v_1 \in V_1 \subset U$, we have $x^{m+1}(v_1) = \cdots = x^{m+r}(v_1) = 0$, and hence $x_{\bar{g}}^i(\pi(g'bv_1 h)) = x^i(bv_1 h)$. Furthermore, since $\pi(g'bv_1 h) = \pi(gv_1)$, we have $x_{\bar{g}}^i(\pi(g'bv_1 h)) = x^i(v_1)$. Hence the coordinate transformation between $(x_{\bar{g}}^1, \ldots, x_{\bar{g}}^m)$ and $(x_{\bar{g}'}^1, \ldots, x_{\bar{g}'}^m)$ is of class C in a neighborhood of $\bar{a}$. Hence G/H is a C^∞ manifold.

It is easy to prove that the maps $\pi : G \to G/H$ and $\varphi : G \times G/H \to G/H$ are differentiable, and that t_g is a diffeomorphism of G/H. We leave the details of the proof as an exercise for the reader.

It is also clear that the cross section $\sigma_{\bar{g}} : \pi(gU) \to G$ is differentiable.

(3) If G is a complex Lie group, and H a complex closed Lie subgroup of G, then, using a complex canonical coordinate system of the second kind, we can prove, just as before, the following: The quotient G/H is a complex manifold; the maps $\pi : G \to G/H$ and $\varphi : G \times G/H \to G/H$ are holomorphic maps, and t_g is a holomorphic isomorphism of G/H. The map $\sigma_{\bar{g}}$ is also holomorphic.

The manifold G/H, as defined above, is called the *quotient manifold* or the *quotient space* of the Lie group G by the closed Lie subgroup H. Examples of various quotient spaces will be given in §18.

If H is a normal closed Lie subgroup of G, then G/H is a group. We shall now prove that the quotient manifold G/H becomes a Lie group with respect to this group structure.

The product of two elements aH, bH of G/H is abH, and the inverse of aH is $a^{-1}H$. Let $\bar{\varphi}$ denote the map $G/H \times G/H \to G/H$ given by $(aH, bH) \to abH$, and let $\bar{\psi}$ denote the map $G/H \to G/H$ given by $aH \to a^{-1}H$. If φ is the map $G \times G/H \to G/H$ defined before, then $\bar{\varphi}(p, q)$ is equal to $\varphi(a, q)$, where $p, q \in G/H$, and $a \in G$ is a representative of p. Fix $p_0 \in G/H$, and set $p_0 = g_0 H$, $\sigma_{\pi(g_0)} = \sigma_{p_0}$, $\pi(g_0 U) = U_{p_0}$. The map $\pi \cdot \sigma_{p_0}$ is the identity map on U_{p_0}. If $p \in U_{p_0}$, then $\bar{\varphi}(p, q) = \varphi(\sigma_{p_0}(p), q)$, so that $\bar{\varphi}$ is differentiable on $U_{p_0} \times G/H$, and hence $\bar{\varphi}$ is differentiable. Since $\bar{\psi}(p) = \pi(a^{-1})$, if $p \in U_{p_0}$, then $\bar{\psi}(p) = \pi(\sigma_{p_0}(p)^{-1})$, and hence $\bar{\psi}$ is differentiable. Since $\bar{\varphi}$ and $\bar{\psi}$ are both differentiable, G/H is a Lie group. It is called the *quotient group* of G by H.

If G is a complex Lie group, H a complex normal closed Lie subgroup of G. then the quotient group G/H is a complex Lie group.

Let $\mathfrak{g}$ be a Lie algebra over a field K, and let $\mathfrak{h}$ be an ideal of $\mathfrak{g}$. Consider the quotient vector space $\mathfrak{g}/\mathfrak{h}$. Define the product $[\bar{X}, \bar{Y}]$ of two elements $\bar{X}, \bar{Y}$ of $\mathfrak{g}/\mathfrak{h}$, as follows. Let $X, Y \in \mathfrak{g}$ be representatives of $\bar{X}$, $\bar{Y}$, respectively, and let

$$[\bar{X}, \bar{Y}] = \text{the element of } \mathfrak{g}/\mathfrak{h} \text{ represented by } [X, Y].$$

Since $\mathfrak{h}$ is an ideal, this definition does not depend on the choice of the representatives for $\bar{X}, \bar{Y}$. We can prove that $\mathfrak{g}/\mathfrak{h}$ becomes a Lie algebra with respect to this product. The Lie algebra $\mathfrak{g}/\mathfrak{h}$ is called the quotient algebra of $\mathfrak{g}$ by its ideal $\mathfrak{h}$.

Let G be a Lie group, H a normal closed Lie subgroup of G. If $\mathfrak{g}$, $\mathfrak{h}$ are the Lie algebras of G, H, respectively, then by Theorem 3 of §12, $\mathfrak{h}$ is an ideal of $\mathfrak{g}$. We shall show in §15 that the Lie algebra of the quotient group G/H is isomorphic to the Lie algebra $\mathfrak{g}/\mathfrak{h}$.

§15. Isomorphisms and Homomorphisms of Lie Groups; Representations of Lie Groups

DEFINITION 1. Let G and G' be Lie groups. If a differentiable map φ from G to G' satisfies

$$\varphi(a \cdot b) = \varphi(a) \cdot \varphi(b)$$

for any two elements a, b of G, then φ is called a *homomorphism* from G into G'. If, furthermore, $\varphi(G) = G'$, then φ is called a homomorphism from G onto G'. If φ is a diffeomorphism from G onto G', and at the same time is a homomorphism, then φ is called an *isomorphism* from G onto G'.

If there is an isomorphism from G onto G', then we say that G is isomorphic to G', and write $G \cong G'$. The relation of isomorphism of Lie groups is an equivalence relation.

We have defined homomorphisms of Lie algebras in Definition 3 of §6. If α is a homomorphism from a Lie algebra $\mathfrak{g}_1$ into a Lie algebra $\mathfrak{g}_2$, then the image $\alpha(\mathfrak{g}_1)$ of $\mathfrak{g}_1$ is a subalgebra of $\mathfrak{g}_2$.

If we set

$$\mathfrak{n} = \{X \in \mathfrak{g}_1 | \alpha(X) = 0\},$$

then $\mathfrak{n}$ is an ideal of $\mathfrak{g}_1$. In fact, if X, $Y \in \mathfrak{n}$, and λ, $\mu \in \mathbf{R}$, then, since $\alpha(\lambda X + \mu Y) = \lambda\alpha(X) + \mu\alpha(Y) = 0$, we have $\lambda X + \mu Y \in \mathfrak{n}$, and hence $\mathfrak{n}$ is a subspace of $\mathfrak{g}_1$. For $X \in \mathfrak{g}_1$, $Y \in \mathfrak{n}$, we have $\alpha[X, Y] = [\alpha(X), \alpha(Y)] = [\alpha(X), 0] = 0$. Hence $[X, Y] \in \mathfrak{n}$, and so $\mathfrak{n}$ is an ideal of $\mathfrak{g}_1$. The ideal $\mathfrak{n}$ is called the kernel of the homomorphism α. We can define a map $\bar{\alpha}$ from the quotient algebra $\mathfrak{g}_1/\mathfrak{n}$ to $\alpha(\mathfrak{g}_1)$ by setting $\bar{\alpha}(X + \mathfrak{n}) = \alpha(X)$. The map $\bar{\alpha}$ is an isomorphism of $\mathfrak{g}_1/\mathfrak{n}$ to $\alpha(\mathfrak{g}_1)$.

Now let φ be a homomorphism from a Lie group G into a Lie group G'. Let $\mathfrak{g}$, $\mathfrak{g}'$ be the Lie algebras of G, G', respectively.

The image $\varphi(\exp tX)$ of a one-parameter subgroup of G corresponding to $X \in \mathfrak{g}$ is a one-parameter subgroup of G', since φ is a homomorphism. Hence there is an $X' \in \mathfrak{g}'$ such that

$$\varphi(\exp tX) = \exp tX'. \tag{1}$$

Let us prove that the map $X \to X'$ is a homomorphism from $\mathfrak{g}$ to $\mathfrak{g}'$. Both sides of (1) are curves in G'. Considering the tangent vectors to these curves at $t = 0$, we get

$$\varphi_* X_e = X'_e. \tag{2}$$

Hence we see that the map $X \to X'$ is a linear mapping from $\mathfrak{g}$ to $\mathfrak{g}'$. We shall now show that X and X' are φ-related (II, §11), i.e., that $\varphi_* X_g = X'_{\varphi(g)}$ for any $g \in G$. Since X and X' are left invariant, we have $X_g = (L_g)_* X_e$ and $X'_{\varphi(g)} = (L_{\varphi(g)})_* X'_{e'}$, and since $\varphi(g \cdot x) = \varphi(g) \cdot \varphi(x)$, we have $\varphi \cdot L_g = L_{\varphi(g)} \cdot \varphi$. Hence $\varphi_* X_g = \varphi_*(L_g)_* X_e = (L_{\varphi(g)})_* \cdot \varphi_* X_e = X'_{\varphi(g)}$. Hence X and X' are φ-related, and by II, §11, we write $X' = \varphi_* X$, and we have

$$\varphi(\exp tX) = \exp t\varphi_* X. \tag{1'}$$

By II, §11, we have $\varphi_*[X, Y] = [\varphi_* X, \varphi_* Y]$, and hence φ_* is a homomorphism from $\mathfrak{g}$ to $\mathfrak{g}'$ the *differential* of the homomorphism φ from G to G'.

Let G be connected and let φ and ψ be homomorphisms from G to G'. If the differentials of φ and ψ and equal, i.e., if $\varphi_* = \psi_*$, then φ and ψ coincide. In fact, since G is connected, by Theorem 2 of §10, any $g \in G$ can be written as $g = \exp X_1 \cdots \exp X_k$ $(X_i \in \mathfrak{g})$. Hence we have $\varphi(g) = \exp \varphi_*(X_1) \cdots \exp \varphi_*(X_k)$ and $\psi(g) = \exp \psi_*(X_1) \cdots \exp \psi_*(X_k)$. So if $\varphi_* = \psi_*$, then we have $\varphi_*(X_i) = \psi_*(X_i)$, and hence $\varphi(g) = \psi(g)$.

Again supposing G connected, let us prove that the image $\varphi(G)$ of G is a connected Lie subgroup of G'. It is easy to see that the image $\varphi_*(\mathfrak{g})$ of $\mathfrak{g}$ is a subalgebra of $\mathfrak{g}'$. Let G'' be the connected Lie subgroup of G' corresponding to $\varphi_*(\mathfrak{g})$. Any element g'' in G'' can be written as $g'' = \exp \varphi_*(X_1) \cdots \exp \varphi_*(X_k)(X_i \in \mathfrak{g})$. Hence, if we set $g = \exp X_1 \cdots \exp X_k$, then we have $\varphi(g) = g''$. That is, $G'' \subset \varphi(G)$. Since $\varphi(G) \subset G''$ can be proved in a similar manner, we conclude that $\varphi(G) = G''$. Hence $\varphi(G)$ is equal to the connected Lie subgroup of G' corresponding to the Lie subalgebra $\varphi_*(\mathfrak{g})$ of $\mathfrak{g}'$.

Problem 1.   If $g \in G$, $X \in \mathfrak{g}$, then prove that $\varphi_*(\mathrm{Ad}(g)X) = \mathrm{Ad}(\varphi(g))\varphi_*X$. Using this, prove that $\varphi_*[X, Y] = [\varphi_*X, \varphi_*Y]$.

DEFINITION 2.   Let G and G' be complex Lie groups. If a holomorphic map φ from G to G' satisfies $\varphi(ab) = \varphi(a)\varphi(b)$ for any two elements a, b of G, then φ is called a *holomorphic homomorphism* from the complex Lie group G to the complex Lie group G'. If φ is a holomorphism from the complex manifold G onto the complex manifold G', and a homomorphism between groups, then φ is called a *holomorphic isomorphism* from the complex Lie group G onto the complex Lie group G'.

Let φ be a homomorphism from a complex Lie group G to a complex Lie group G'. That is, suppose φ is a homomorphism from G to G', when G and G' are considered as real Lie groups. The differential φ_* of φ is a homomorphism from the real Lie algebra $\mathfrak{g}$ into the real Lie algebra $\mathfrak{g}'$. But $\mathfrak{g}$ and $\mathfrak{g}'$ are both complex Lie algebras. The map φ_* is a homomorphism of the complex Lie algebra $\mathfrak{g}$ into the complex Lie algebra $\mathfrak{g}'$ (i.e., $\varphi_*(\lambda X) = \lambda\varphi_*(X)$ for all $\lambda \in \mathbf{C}$ and all $X \in \mathfrak{g}$) if and only if φ is a holomorphic homomorphism. In fact, let J and J' be the almost complex structures attached to the complex structures of G and G', respectively. The complex structure of $\mathfrak{g}$, $\mathfrak{g}'$ are defined by $iX = JX$ $(X \in \mathfrak{g})$, $iX' = J'X'$ $(X' \in \mathfrak{g}')$, respectively, and φ_* is a homomorphism of complex Lie algebras if and only if $\varphi_*(JX) = J'\varphi_*(X)$ holds for all $X \in \mathfrak{g}$. But this is precisely the condition for φ to be holomorphic, by (12) of II, §16.

Let G be a connected complex Lie group, G' a complex Lie group, and φ a holomorphic homomorphism from G to G', then $\varphi_*(\mathfrak{g})$ is a complex Lie subalgebra of $\mathfrak{g}'$. Hence the connected Lie subgroup $\varphi(G)$ of G corresponding to $\varphi_*(\mathfrak{g})$ is a complex Lie subgroup of G'.

We put together the results above and state it as a theorem.

THEOREM 1. *Let φ be a homomorphism from a Lie group G to a Lie group G'. The homomorphism φ induces a homomorphism φ_* from the Lie algebra $\mathfrak{g}$ of G into the Lie algebra $\mathfrak{g}'$ of G', called the differential of φ. If $X \in \mathfrak{g}$, then*

$$(\varphi_* X)_{e'} = \varphi_* X_e,$$

where e and e' are the identity elements of G and G', respectively. If G is connected, then the image $\varphi(G)$ of G is a connected Lie subgroup of G', and its Lie algebra is equal to $\varphi_*(\mathfrak{g})$. If G and G' are complex Lie groups, then φ is a holomorphic homomorphism if and only if the differential φ_* of φ is a homomorphism from the complex Lie algebra $\mathfrak{g}$ into the complex Lie algebra $\mathfrak{g}'$. Furthermore, if G is connected and φ is holomorphic, then $\varphi(G)$ is a connected complex Lie subgroup of G'.

Problem 2. If $a = (a_j^i)$ is a complex $n \times n$ matrix, then let $\bar{a} = (\bar{a}_j^i)$ (where $\bar{\lambda}$ denotes the complex conjugate of the complex number λ). If we define a map φ from $GL(n, \mathbf{C})$ by $\varphi(a) = \bar{a}$, then show that φ is an isomorphism from the Lie group $GL(n, \mathbf{C})$ onto itself, but not a holomorphic isomorphism from the complex Lie group $GL(n, \mathbf{C})$ onto itself.

Problem 3. Let G and G' be two connected Lie groups, and φ a homomorphism from G to G'. If φ is injective, then prove that φ_* is injective. Give an example to show that even if φ_* is injective, it is not necessarily true that φ is injective.

Let N be a normal closed Lie subgroup of a Lie group G. The natural map π from G onto G/N is a homomorphism. If we denote the Lie algebras of G, N, G/N by $\mathfrak{g}$, $\mathfrak{n}$, $\bar{\mathfrak{g}}$, respectively, then the kernel of the homomorphism π_* from $\mathfrak{g}$ onto $\bar{\mathfrak{g}}$ is equal to $\mathfrak{n}$. In fact, if $X \in \mathfrak{n}$, then $\pi(\exp tX) = \bar{e}$, where $\bar{e}$ is the identity element of G/N, so that we have $\pi_*(X) = 0$, and X is contained in the kernel $\mathfrak{n}'$ of π_*. Hence $\mathfrak{n} \subset \mathfrak{n}'$. Since $\mathfrak{g}/\mathfrak{n}' \cong \bar{\mathfrak{g}}$, we have $\dim G - \dim N = \dim G/N = \dim \bar{\mathfrak{g}} = \dim \mathfrak{g} - \dim \mathfrak{n}'$. Hence $\dim \mathfrak{n} = \dim \mathfrak{n}'$, and we get $\mathfrak{n} = \mathfrak{n}'$. Hence we have proved that the Lie algebra $\bar{\mathfrak{g}}$ of the quotient group G/N is isomorphic to the quotient Lie algebra $\mathfrak{g}/\mathfrak{n}$.

Let φ be a homomorphism from a Lie group G onto a Lie group G'. The groups G, G' are locally compact groups, and φ is of course a homomorphism of locally compact groups. If N is the kernel of φ, then N is a closed subgroup of G. Using Theorem 2 of §20, we see that N is a normal closed Lie subgroup of G. We shall also prove in §19 that the quotient group G/N and G' are isomorphic as Lie groups.

Problem 4.   Prove directly, that the kernel N of φ is a closed Lie subgroup of G, following the indicated steps: Show that $\mathrm{rank}_g(\varphi) = \dim G'$ for all points $g \in G$, and using this, show that $\varphi^{-1}(h)$ is a closed submanifold of G for $h \in G'$.

DEFINITION 3.   A homomorphism ρ of a Lie group G into $GL(n, \mathbf{C})$ or $GL(n, \mathbf{R})$ is called an n-dimensional *representation* of G. If G is a complex Lie group, a holomorphic homomorphism from G into $GL(n, \mathbf{C})$ is called an n-dimensional *holomorphic representation* of the complex Lie group G. Similarly a homomorphism of a Lie algebra g over $\mathbf{R}$ into $\mathfrak{gl}(n, \mathbf{C})$ or $\mathfrak{gl}(n, \mathbf{R})$ is called an n-dimensional representation of g, and a homomorphism of a complex Lie algebra g into $\mathfrak{gl}(n, \mathbf{C})$ is called an n-dimensional *complex representation* of the complex Lie algebra g.

By Theorem 1, for a representation ρ of G, we obtain a representation ρ_* of the Lie algebra g as the differential of ρ, and we have

$$\rho(\exp tX) = \exp t\rho_*(X),$$

for any X of g. Here the right member is the exponential function of the matrix $\rho_*(X)$.

If G is a complex Lie group, then ρ is a holomorphic representation of G if and only if ρ_* is a complex representation of g.

Example.   If G has dimension n, then the adjoint representation Ad of G is an n-dimensional representation of G. The differential of Ad is the adjoint representation ad of the Lie algebra of G. If G is a complex Lie group, then Ad is a holomorphic representation.

Remark.   Let G, G' be connected Lie groups and g, g' their Lie algebras, respectively. Suppose α is a homomorphism from g onto g'. Then we do not always have a homomorphism φ from G onto G' such that $\varphi_* = \alpha$. For example, if G is the n-dimensional torus, and $G' = \mathbf{R}^n$, then g and g' are both n-dimensional commutative Lie algebras, so that any linear map α from g onto g' is an isomorphism of Lie algebras. However, there is no homomorphism from G onto G'. However, if G is "*simply connected*," then for a homomorphism α from g onto g', there is a homomorphism φ from G onto G', such that $\varphi_* = \alpha$. Since we have not included discussions on covering spaces and covering groups in this book, we cannot prove the last statement above, and we refer the reader to Pontryagin, *Topological Groups*, 2nd ed., Gordon and Breach, New York, 1966.

§16. The Structure of Connected Commutative Lie Groups

Let G be an n-dimensional connected commutative Lie group, and let $\mathfrak{g}$ be its Lie algebra. Consider the vector space $\mathfrak{g}$ as a Lie group with respect to addition, and write this Lie group as $\mathbf{R}^n$. Then the exponential mapping exp from $\mathfrak{g}$ to G is a homomorphism from $\mathbf{R}^n$ to G (cf. §10). Let Γ be the kernel of the homomorphism exp. Since exp is $1:1$ on a neighborhood of 0, there is a neighborhood U of 0 such that $\Gamma \cap U = \{0\}$. We shall now give a general definition.

DEFINITION. A subgroup Γ of a Lie group G is called a *discrete subgroup* if Γ is a closed subset of G, and if for some neighborhood U of the identity element e of G, $\Gamma \cap U = \{e\}$ holds.

If Γ is a discrete subgroup of G, and $a \in \Gamma$, then there is a neighborhood U_a of a in G such that $\Gamma \cap U_a = \{a\}$. In fact, if U is a neighborhood of e satisfying $\Gamma \cap U = \{e\}$, then letting $aU = U_a$, we have a neighborhood U_a of a such that $\Gamma \cap U_a = \{a\}$.

The kernel Γ of the exponential mapping exp is a discrete subgroup of $\mathbf{R}^n$.

Let us prove the following theorem.

THEOREM 1. *If Γ is a discrete subgroup of $\mathbf{R}^n$, then there exists a set of elements $u_1, \ldots, u_d$ of Γ such that*

(1) *$u_1, \ldots, u_d$ are linearly independent over $\mathbf{R}$.*
(2) *Every element of Γ can be written uniquely as $m_1 u_1 + \cdots + m_d u_d$, where $m_1, \ldots, m_d$ are integers.*
The cardinality d of a set of elements of Γ with the properties (1) and (2) is unique, and is called the rank of Γ.

Proof. Let $\{e_1, e_2, \ldots, e_n\}$ be a basis of $\mathbf{R}^n$. For an element $a = \alpha_1 e_1 + \cdots + \alpha_n e_n$ of $\mathbf{R}^n$, set $|a| = \sum_{i=1}^n |\alpha_i|$. Using the fact that Γ is a discrete subgroup, we can prove easily that the set $\{|a| \,|\, a \in \Gamma, a \neq 0\}$ of positive numbers has a minimum $m_0 > 0$.

Let e_1' be an element of Γ satisfying $|e_1'| = m_0$, and set $\Gamma_1 = \{me_1' | m:$ integer$\}$. If $\Gamma_1 = \Gamma$, then we are through, so suppose $\Gamma_1 \neq \Gamma$. Adjoin $(n-1)$ elements $e_2', \ldots, e_n'$ of $\mathbf{R}^n$ to e_1', so that they form a basis of $\mathbf{R}^n$ together. Let $a = \sum_{i=1}^n \alpha_i e_i'$ be an element of Γ and set $|a| = \sum_{i=1}^n |\alpha_i|$ again and set $|a|' = \sum_{i=2}^n |\alpha_i|$. Then $|a|' > 0$ if and only if $a \notin \Gamma_1$. Let us show that there is a minimum in the set $\{|a|' \,|\, a \in \Gamma, a \notin \Gamma_1\}$. For this, we shall show that if we assume that for each positive integer m

we have $a_m \in \Gamma$, $a_m \notin \Gamma_1$ such that $|a_m|' < 1/m$, then we arrive at a contradiction. To each a_m we can add an integral multiple of e_1', so that when the resulting element is written as a linear combination of $e_1', \ldots,$ e_n', the coefficient of e_1' has absolute value less than 1. By this process, the properties $a_m \in \Gamma$, $a_m \notin \Gamma_1$, $|a_m|' < 1/m$ are not disturbed. So we assume that the coefficient of e_1' of each a_m has absolute value less than 1. Then, since $|a_m| \leqq 1 + 1/m$, the sequence $\{a_m\}$ is bounded, so that there is a convergent subsequence. Since $a_m \in \Gamma$ and Γ is closed, the limit belongs to Γ, and since Γ is discrete, we may assume that there is a sufficiently large N such that $a_m = a_n$ for m, $n > N$. Hence $|a_m|' = 0$ for $m > N$, i.e., $a_m \in \Gamma_1$ for $m > N$, a contradiction to the choice of the sequence $\{a_m\}$. Therefore $m_0' = \inf\{|a|' \mid a \in \Gamma, a \notin \Gamma_1\}$ is positive, but using the fact that Γ is closed and discrete again, the same argument just employed shows us that the infimum m_0' is actually attained by some element in Γ. Choose $e_2'' \in \Gamma$ such that $e_2'' \notin \Gamma_1$ and $|e_2''|' = m_0'$, and set $\Gamma_2 = \{m_1 e_1' + m_2 e'' \mid m_i : \text{integer}\}$. If $\Gamma_2 = \Gamma$ we are through. If $\Gamma_2 \neq \Gamma$, pick $(n-2)$ elements $e_3'', \ldots, e_n''$ of Γ so that $e_1''(=e_1'')$, $e_2'', \ldots, e_n''$ form a basis of $\mathbf{R}^n$, and construct Γ_3 as we have constructed Γ_2. Repeating this process, we find that there is some integer $d (1 \leqq d \leqq n)$ such that $\Gamma_d = \Gamma$. If $\{v_1, \ldots, v_r\}$ is another set of elements of Γ with the properties (1) and (2), then $\{u_1, \ldots, u_d\}$ and $\{v_1, \ldots, v_r\}$ span the same subspace W of $\mathbf{R}^n$, and they are both bases of W. Hence we have $r = d$.

Let Γ be a discrete susbgroup of $\mathbf{R}^n$ of rank d. The group $\mathbf{R}^n/\Gamma$ and $T^d \times \mathbf{R}^{n-d}$ are isomorphic as topological groups. In particular, if $\mathbf{R}^n/\Gamma$ is compact, then $\mathbf{R}^n/\Gamma$ is isomorphic to T^n, and $\mathbf{R}^n/\Gamma$ is compact if and only if the rank of Γ is n.

Thus we have the following theorem:

THEOREM 2.   *If G is an n-dimensional connected commutative Lie group, then G is isomorphic to $T^d \times \mathbf{R}^{n-d}$ for some d, $0 \leqq d \leqq n$. In particular, if G is compact, then G is isomorphic to the n-dimensional torus T^n.*

Problem.   Determine all closed Lie subgroups of $\mathbf{R}^n$. (*Suggestion*: First, using the exponential map, show that each connected Lie subgroup of $\mathbf{R}^n$ is a vector subspace of $\mathbf{R}^n$, and hence closed.)

§17. Lie Transformation Groups and Homogeneous Spaces of Lie Groups

DEFINITION 1.   Let G be a Lie group, and let M be a manifold. The group G is called a *Lie transformation group* of M if there is a differentiable map φ from $G \times M$ to M such that, setting $\varphi(g, p) = g \cdot p$,

(1) $(g \cdot h) \cdot p = g \cdot (h \cdot p)$, for g, $h \in G$ and $p \in M$,
(2) $e \cdot p = p$ for the identity element e of G and $p \in M$, are satisfied.

If this is the case, then the map $p \to g \cdot p$ is a differentiable map from M to M, and by (1) and (2) of the definition, we have $g^{-1} \cdot (g \cdot p) = g \cdot (g^{-1} \cdot p) = e \cdot p = p$, so that $p \to g \cdot p$ is a diffeomorphism of M.

It is clear, from the definition above, that a Lie transformation group G of M is also a topological transformation group of M (cf. §5). If a Lie transformation group G acts transitively on M (cf. §5), then M is called a *homogeneous space* of G.

If N is the set of all elements g of G such that $g \cdot p = p$ for all points p of M, then N is a closed normal subgroup of G (cf. §5). Hence by Theorem 2 of §20, N is a normal Lie subgroup of G. In §5 we said that G acts *effectively* on M if $N = \{e\}$. We now say that G acts *almost effectively* on M if the connected component N_0 of N containing the identity e consists of e only, i.e., if N is a discrete normal subgroup of G.

If G is a Lie transformation group of M, then the *isotropy subgroup* $H = \{g \in G \,|\, g \cdot p = p\}$ of G at a point p of M is a closed subgroup of G. Hence, by Theorem 2 of §20, H is a closed Lie subgroup of G. Furthermore, if M is a homogeneous space of G, then M and G/H are not only homeomorphic (Theorem 1 of §5), but we shall prove that they are also diffeomorphic.

We shall denote the diffeomorphism $p \to g \cdot p$ of M by t_g. Let h be an element of the isotropy subgroup H at p. Since t_h leaves p fixed, the differential t_{h*} of t_h at p is a nonsingular linear transformation of the tangent space $T_p(M)$. This gives us a group $\overline{H} = \{t_{h*} \,|\, h \in H\}$ of linear transformations of $T_p(M)$, which we call the *linear isotropy group* of G at p.

Let G be a Lie group, and H a closed Lie subgroup of G. The quotient space G/H is a homogeneous space of G. In this case N is a closed normal subgroup of G contained in H, and every normal subgroup of G contained in H is a subgroup of N. Hence G acts effectively on G/H if and only if the only normal subgroup of G contained in H is $\{e\}$. Furthermore, G acts on G/H almost effectively if and only if the Lie subalgebra $\mathfrak{h}$ of the Lie algebra $\mathfrak{g}$ of G corresponding to H contains no nonzero ideal of $\mathfrak{g}$.

Let π be the natural map from G to G/H. The isotropy subgroup of G at the point $\pi(g) = gH$ of G/H is equal to gHg^{-1}. Let us set $\pi(e) = o$

and study the isotropy subgroup at o. For $g \in G$ we have $t_g \cdot \pi = \pi \cdot L_g$, and for $h \in H$ we have $\pi \cdot R_d = \pi$. Hence for $h \in H$ we have

$$t_h \cdot \pi = \pi \cdot A_h. \tag{1}$$

If $X \in \mathfrak{g}$, then $(A_h)_* X = \mathrm{Ad}(h)X$ and $(A_h)_* X_e = (\mathrm{Ad}(h)X)_e$, where $(A_h)_*$ is the differential of A_h at e. If we let the differential of π at e be denoted by π_*, then, by (1), we have $(t_h)_*(\pi_* X_e) = \pi_*(\mathrm{Ad}(h)X)_e$ $(X \in \mathfrak{g})$. But $T_e(G) = \{X_e | X \in \mathfrak{g}\}$, and it is easy to see that the kernel of the linear map $\pi_* : T_e(G) \to T_o(G/H)$ is $\{X_e | X \in \mathfrak{h}\}$. Hence, if we identify $\mathfrak{g}$ with $T_e(G)$ by the correspondence $X \to X_e$, then $\mathfrak{g}/\mathfrak{h}$ is identified with $T_o(G/H)$, and π_* is identified with the natural map from $\mathfrak{g}$ onto $\mathfrak{g}/\mathfrak{h}$. If $h \in H$, then $\mathrm{Ad}(h)$ maps $\mathfrak{h}$ into $\mathfrak{h}$, so that $\mathrm{Ad}(h)$ induces a linear transformation on $\mathfrak{g}/\mathfrak{h}$. If we denote this linear transformation by $\mathrm{Ad}_{\mathfrak{g}/\mathfrak{h}}(h)$, then

$$\mathrm{Ad}_{\mathfrak{g}/\mathfrak{h}}(h)\pi_* X = \pi_* \mathrm{Ad}(h)X$$

holds for all $h \in H$, $X \in \mathfrak{g}$. Hence, if we identify $T_o(G/H)$ with $\mathfrak{g}/\mathfrak{h}$, then we have $(t_h)_* = \mathrm{Ad}_{\mathfrak{g}/\mathfrak{h}}(h)(h \in H)$. That is, the linear isotropy group at o is equal to the group of all linear transformations on $\mathfrak{g}/\mathfrak{h}$ induced by $\mathrm{Ad}(h)$ $(h \in H)$. The map $h \to \mathrm{Ad}_{\mathfrak{g}/\mathfrak{h}}(h)$ is a representation of the Lie group H by linear transformations on $\mathfrak{g}/\mathfrak{h}$. The representation of the Lie algebra $\mathfrak{h}$, obtained as the differential of the representation of H above, is denoted by $\mathrm{ad}_{\mathfrak{g}/\mathfrak{h}}(Y)$ $(Y \in \mathfrak{h})$. If $Y \in \mathfrak{h}$, then $\mathrm{ad}(Y)$ maps $\mathfrak{h}$ to $\mathfrak{h}$, and hence $\mathrm{ad}(Y)$ induces a linear transformation on $\mathfrak{g}/\mathfrak{h}$. This linear transformation is $\mathrm{ad}_{\mathfrak{g}/\mathfrak{h}}(Y)$.

Problem 1. Let ω be a differential form of order p on G/H. If $t_g^* \omega = \omega$ holds for all $g \in G$, then ω is said to be G-invariant. If ω is G-invariant, then for tangent vectors $v_1, \ldots, v_p$ at o, show that $\omega_o(\mathrm{Ad}_{\mathfrak{g}/\mathfrak{h}}(h)v_1, \ldots, \mathrm{Ad}_{\mathfrak{g}/\mathfrak{h}}(h)v_p) = \omega_o(v_1, \ldots, v_p)$ holds for all $h \in H$.

Now suppose a Lie transformation group G is acting on a manifold M. For an arbitrary element X of the Lie algebra $\mathfrak{g}$ of G, $t_{\exp sX}$ is a one-parameter group of transformations of M. If the infinitesimal transformation of the one-parameter group of transformations $s \to t_{\exp(-sX)}$ is denoted by X^*, then we have a map $\alpha : X \to X^*$ from $\mathfrak{g}$ into the Lie algebra $\mathfrak{X}(M)$ of all vector fields on M. Let us prove that α is a Lie algebra homomorphism from $\mathfrak{g}$ into $\mathfrak{X}(M)$. Fix a $p \in M$, and set $S_p(g) = t_g(p)$. By the condition on Lie transformation groups, the map $g \to S_p(g)$ is a differentiable map from G to M. The vector X^*_p is the tangent vector at $s = 0$ to the curve $s \to S_p(\exp(-sX))$. Hence we have $X^*_p = -S_{p*}X_e$. Hence $(X + Y)^*_p = -S_{p*}(X + Y)_e = X^*_p + Y^*_p$, and similarly $(\lambda X)^*_p = \lambda X^*_p$. Since these hold for each point p of M, we have $(X + Y)^* = X^* +$

Y^* and $(\lambda X)^* = \lambda X^*$. Hence α is a linear map. Let us now prove $[X, Y]^* = [X^*, Y^*]$. From II, §12, we have

$$[X^*, Y^*]_p = \lim_{s \to 0} \frac{1}{s} \{ Y_p^* - (t_{a_s*} Y^*)_p \}, \qquad a_s = \exp(-sX).$$

But $(t_{a_s*} Y^*)_p = t_{a_s*} Y^*_{a_s^{-1}p} = -t_{a_s*} S_{a_s^{-1}p*} Y_e$. On the other hand, we have $(t_{a_s} S_{a_s^{-1}p})(g) = a_s g a_s^{-1} p = A_{a_s}(g)p = (S_p A_{a_s})(g)$. Hence we have

$$(t_{a_s*} Y^*)_p = -S_{p*}(A_{a_s*} Y)_e = -S_{p*} R_{a_s^{-1}*} Y_e,$$

and thus we get

$$[X^*, Y^*]_p = -S_{p*} \left\{ \lim_{s \to 0} \frac{1}{s} \{ Y_e - (R_{a_s^{-1}*} Y)_e \} \right\}$$
$$= -S_{p*} [X, Y]_e = [X, Y]_p^*.$$

Since this holds for each point p, we get $[X^*, Y^*] = [X, Y]^*$. That is, $X \to X^*$ is a homomorphism from $\mathfrak{g}$ to $\mathfrak{X}(M)$. Now, if $X^* = 0$, then $t_{\exp sX}(p) = p$ holds for each point p of M. Letting N denote, as before, the normal subgroup of G consisting of all elements g of G such that $t_g p = p$ for all points p of M, we have $\exp sX \in N$ for all $s \in \mathbf{R}$. Conversely, if $\exp sX \in N$, then $X^* = 0$. Hence the kernel of the homomorphism $X \to X^*$ is equal to the Lie algebra $\mathfrak{n}$ of N. In particular, if G acts almost effectively on M, then $\mathfrak{n} = \{0\}$, so that $X \to X^*$ is an isomorphism.

Note that, by the definition of X^*, we have

$$\mathrm{Exp}\, sX^* = t_{\exp(-sX)}.$$

We call the subalgebra $\mathfrak{g}^* = \{X^* | X \in \mathfrak{g}\}$ of $\mathfrak{X}(M)$ the *Lie algebra formed by the infinitesimal transformations of the Lie transformation group G*. If G acts almost effectively, then $\mathfrak{g}^*$ is isomorphic to $\mathfrak{g}$.

Problem 2. Let $M = G/H$. For $X \in \mathfrak{g}$, let X' be a right invariant vector field on G such that $X_e = X'_e$. Prove that $-X'$ and X^* are π-related, i.e., that $\pi_*(-X') = X^*$.

THEOREM. *Let M be a manifold which is a homogeneous space of a Lie group G, and let H be the isotropy subgroup of G at a point p_0 of M. Then the map α given by*

$$\alpha : gH \to t_g p_0 = g p_0$$

is a diffeomorphism from G/H onto M, and $\alpha(t_g \xi) = t_g \alpha(\xi)$ holds for all $\xi \in G/H$ and $g \in G$.

Proof. By Theorem 1 of §5, α is a homeomorphism from G/H onto

M. As before, set $\pi(e) = o$, and let S denote the map from G to M given by $g \to t_g p_0$. Then, in Fig. 4.2, we have $\alpha \cdot \pi = S$. First let us show that

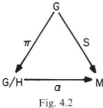

Fig. 4.2

α is differentiable on a neighborhood of o. For this, choose a basis $\{X_1, \ldots, X_m, Y_1, \ldots, Y_r\}$ $(m + r = n = \dim G)$ of the Lie algebra $\mathfrak{g}$ of G so that $\{Y_1, \ldots, Y_r\}$ is a basis of the Lie algebra $\mathfrak{h}$ of H, and let $\{x^1, \ldots, x^m, x^{m+1}, \ldots, x^n\}$ be a canonical coordinate system of the second kind of G with respect to this basis. We have shown, in §14, that we can choose a local coordinate system $\{y^1, \ldots, y^m\}$ on a neighborhood of o such that $y^i \circ \pi = x^i$ $(1 \leqq i \leqq m)$ holds. If f is a C^∞ function on M, then, since α is continuous, $f \circ \alpha$ is a continuous function of G/H. Hence we can write $(f \circ \alpha)(\xi) = F(y^1(\xi), \ldots, y^m(\xi))$ on a neighborhood of o. Here $F(u^1, \ldots, u^m)$ is a continuous function defined on a neighborhood of $(0, \ldots, 0)$. Hence, for g in a neighborhood of e in G, we have $(f \circ \alpha)(\pi(g)) = F(y^1(\pi(g)), \ldots, y^m(\pi(g)))$. Since $\alpha \circ \pi = S$ and $y^i \circ \pi = x^i$, we have $(f \circ S)(g) = F(x^1(g), \ldots, x^m(g))$. But, since S is differentiable, $f \circ S$ is a C^∞ function on G. Hence $F(u^1, \ldots, u^m)$ is of class C^∞, and hence α is differentiable in a neighborhood of o. Next we shall show that, for a sufficiently small neighborhood of o, at each point ξ of this neighborhood the differential α_* of α is 1:1. The vectors of $T_o(G/H)$ are of the form $\pi_* X_e$ $(X \in \mathfrak{g})$. If $\alpha_*(\pi_* X_e) = 0$, then, since $\alpha_* \cdot \pi_* = S_*$, we have $S_* X_e = 0$. But $S_* X_e = X_{p_0}^*$, so that we have $X_{p_0}^* = 0$, and hence $(\operatorname{Exp} sX^*)(p_0) = p_0$ for all s. But $\operatorname{Exp} sX^* = t_{\exp(-sX)}$, so that $\exp(-sX)$ is contained in the isotropy subgroup H at p_0. Hence $X \in \mathfrak{h}$ and $\pi_* X_e = 0$. Hence the kernel of the linear map α_* from $T_o(G/H)$ to $T_{p_0}(M)$ is $\{0\}$, and α_* is 1:1. By the theorem of II, §7, we can choose local coordinate systems $(y^1, \ldots, y^m)$ and $(z^1, \ldots, z^l)$ at o and $p_0 = \alpha(0)$, respectively, so that $z^i \circ \alpha = y^i$ $(i = 1, \ldots, m)$ and $z^j \circ \alpha = 0$ $(j = m + 1, \ldots, l)$ hold. But α is a homeomorphism from G/H onto M, so that we must have $m = l$. Then, by the inverse function theorem, α^{-1} is also differentiable on a neighborhood of p_0.

Now, if $\xi = \pi(g)$ is an arbitrary point of G/H, then for each point η in a neighborhood of ξ, we have $\alpha(\eta) = t_g(\alpha(t_{g^{-1}}(\eta)))$. Since the map

α is differentiable in a neighborhood of o, and t_g is a diffeomorphism of G/H and of M, we conclude from the last equality that α is a diffeomorphism on a neighborhood of ξ. Similarly α^{-1} is a diffeomorphism at each point of M. This concludes the proof.

DEFINITION 2.   Let M be a homogeneous space of a Lie group G. If M is a complex manifold, and if for each $g \in G$, t_g is a holomorphism of M, then we call M a *complex homogeneous space* of the (real) Lie group G. If G is a complex Lie group, and $G \times M \to M$ is a holomorphic map, then we call M a complex homogeneous space of the complex Lie group G.

Remark.   In the case of a complex homogeneous space M of a complex Lie group G, we can show that the isotropy subgroup H of G at a point p_0 of M is a closed complex Lie subgroup of G, and that M and G/H are holomorphically isomorphic. The proof is left to the reader.

Problem 3.   Let a connected manifold M be a homogeneous space of a Lie group G. Prove that the connected component G_0 of G containing the identity also acts transitively on M, and hence that M is a homogeneous space for the connected Lie group G_0.

§18. Examples of Homogeneous Spaces

(1) The affine space $\mathbf{A}^n$ is a homogeneous space of the affine transformation group, and the euclidean space $\mathbf{E}^n$ is a homogeneous space of the euclidean motion group.

(2) The unit sphere $S^n = \{(x^1, \ldots, x^{n+1}) | \sum_{i=1}^{n+1} (x^i)^2 = 1\}$ in $\mathbf{R}^{n+1}$ is a homogeneous space of the compact Lie groups $O(n+1)$ and $SO(n+1)$. In fact the map $\tilde{\varphi}$ from $GL(n+1, R) \times \mathbf{R}^{n+1}$ to $\mathbf{R}^{n+1}$, defined by $(a, x) \to ax$ $(a \in GL(n+1, \mathbf{R}), x \in \mathbf{R}^{n+1})$, is clearly differentiable, and $O(n+1) \times S^n$ and S^n are closed submanifolds of $GL(n+1, \mathbf{R}) \times \mathbf{R}^{n+1}$ and $\mathbf{R}^{n+1}$, respectively. Furthermore $\tilde{\varphi}$ maps $O(n+1) \times S^n$ into S^n. Hence, by the proposition of II, §10, the map $O(n+1) \times S^n \to S^n$ is differentiable.

The isotropy subgroup of $O(n+1)$(resp. $SO(n+1)$) at the point $(1, 0, \ldots, 0)$ of S^n is identified with $O(n)$(resp. $SO(n)$)(cf. §5). Hence, by the theorem in §17, $O(n+1)/O(n)$ and $SO(n+1)/SO(n)$ are diffeomorphic to S^n.

(3) Let p^n be the real projective space (cf. II, §2), and let π be the projection from $\mathbf{R}^{n+1} - \{0\}$ onto p^n. For a matrix a in $GL(n+1, \mathbf{R})$, set

$a \cdot \pi(x) = \pi(a \cdot x)$ $(x \in \mathbf{R}^{n+1}, x \neq 0)$. Then the map $(a, \pi(x)) \rightarrow a \cdot \pi(x)$ is a differentiable map from $GL(n + 1, \mathbf{R}) \times P^n$ to P^n, and $GL(n + 1, \mathbf{R})$ becomes a Lie transformation group on P^n, acting transitively on P^n. The isotropy subgroup of $GL(n + 1, \mathbf{R})$ at a point $\pi(x_0)$ $(x_0 = (1, 0, \ldots, 0))$ of P^n is a closed subgroup $\tilde{H}$ of $GL(n + 1, \mathbf{R})$ consisting of all the matrices of the form

$$
\begin{bmatrix}
a & * & \cdots & * \\
0 & & & \\
\cdot & & A & \\
\cdot & & & \\
\cdot & & & \\
0 & & &
\end{bmatrix}, \quad a \neq 0, \quad A \in GL(n, \mathbf{R}). \tag{1}
$$

Similarly, $SL(n + 1, \mathbf{R})$ and $SO(n + 1)$ act transitively on P^n. The isotropy subgroup of $SO(n + 1)$ at $\pi(x_0)$ is the closed subgroup H of $SO(n + 1)$ consisting of all of the orthogonal matrices of the form

$$
\begin{bmatrix}
a & 0 & \cdots & 0 \\
0 & & & \\
\cdot & & A & \\
\cdot & & & \\
\cdot & & & \\
0 & & &
\end{bmatrix}, \quad a = \pm 1, \quad A \in O(n), \quad a \cdot \det A = 1. \tag{2}
$$

Hence the quotient space of $GL(n + 1, \mathbf{R})$ by H, and the quotient space of $SO(n + 1)$ by H, are both diffeomorphic to P^n.

Problem 1. The action of $GL(n + 1, \mathbf{R})$ on P^n is not almost effective. The action of $SL(n + 1, \mathbf{R})$ on P^n and the action of $SO(n + 1)$ on P^n are both almost effective. Check these statements.

(4) Similarly, the complex projective space $P^n(C)$ (cf. II, §16, **K**) is a complex homogeneous space of the complex Lie groups $GL(n + 1, \mathbf{C})$ and $SL(n + 1, \mathbf{C})$. As in the case of (3), $SU(n + 1)$ also acts transitively on $P^n(C)$.

Problem 2. Find the isotropy subgroups of $GL(n + 1, \mathbf{C})$ and $SU(n + 1)$ at the point $\pi(x_0)$ $(x_0 = (1, 0, \ldots, 0)$ of $P^n(\mathbf{C}))$.

(5) The upper half-plane $H = \{z = x + iy \in \mathbf{C} \mid y \geq 0\}$ is a complex homogeneous space of $SL(2, \mathbf{R})$ (cf. II, §9). An element $\begin{pmatrix} a & b \\ c & d \end{pmatrix}$ of $SL(2, \mathbf{R})$ acts transitively on H by the action $z \rightarrow (az + b)/(cz + d)$.

Problem 3.   Show that the action of $SL(2, \mathbf{R})$ on H is almost effective. Find the isotropy subgroup of $SL(2, \mathbf{R})$ at the point i of H.

(6) Let V be an N-dimensional real vector space, and let $G_{N, n}$ be the set of all n-dimensional subspaces of V. Fix a basis $\{e_1, \ldots, e_N\}$ of V and identify V with $\mathbf{R}^N$. Also identify the group of all nonsingular linear transformations of V with $GL(N, \mathbf{R})$. If W is an n-dimensional subspace of V, and $a \in GL(N, \mathbf{R})$, then $aW = \{a \cdot x | x \in W\}$ is also an n-dimensional subspace of V. If W and W' are two n-dimensional subspaces of V, then there is an $a \in GL(N, \mathbf{R})$ such that $aW = W'$. Hence $GL(N, \mathbf{R})$ acts transitively on $G_{N, n}$. If W_0 is the n-dimensional subspace of V spanned by $e_1, \ldots, e_n$, then $b \in GL(N, \mathbf{R})$ satisfying $bW_0 = W_0$ is a matrix of the form

$$\begin{bmatrix} A & * \\ 0 & B \end{bmatrix}, \qquad A \in GL(n, \mathbf{R}), \quad B \in GL(N - n, \mathbf{R}),$$

and these matrices form a closed Lie subgroup $\tilde{H}$ of $GL(N, \mathbf{R})$. By the map $\alpha : a\tilde{H} \to aW_0$, we have a $1:1$ correspondence from the set $GL(N, \mathbf{R})/\tilde{H}$ onto $G_{N, n}$. Since $GL(N, \mathbf{R})/\tilde{H}$ is a differentiable manifold, we can put a differentiable structure on $G_{N, n}$ so that α is a diffeomorphism, and $G_{N, n}$ becomes a homogeneous space of $GL(N, \mathbf{R})$. The manifold $G_{N, n}$ is called the *Grassmann manifold* of n-dimensional subspaces in an N-dimensional real vector space. If $n = 1$, then $G_{N, 1}$ is the real projective space P^{N-1}.

Now define a positive inner product in V so that $\{e_1, \ldots, e_N\}$ is an orthonormal basis. Let W and W' be n-dimensional subspaces of V, and let $\{u_1, \ldots, u_n\}$ and $\{v_1, \ldots, v_n\}$ be orthonormal bases of W and W', respectively. Renumbering the subscripts of $v_1, \ldots, v_n$ if necessary, we can find an $a \in SO(N)$ such that $au_i = v_i$ $(i = 1, \ldots, n)$. Then we have $aW = W'$, so that $SO(N)$ also acts transitively on $G_{N, n} = GL(N, \mathbf{R})/\tilde{H}$. The isotropy subgroup of $SO(N)$ at W_0 is the closed subgroup H of $SO(N)$ consisting of all the orthogonal matrices of the form

$$\begin{bmatrix} A & 0 \\ 0 & B \end{bmatrix}, \qquad A \in O(N), \quad B \in O(N - n), \quad \det A \cdot \det B = 1.$$

Hence, by the theorem of §17, $G_{N, n}$ is also diffeomorphic to $SO(N)/H$. Since $SO(N)$ is compact and connected, so is $SO(N)/H$. Hence $G_{N, n}$ is also compact and connected.

(7) Let V be an N-dimensional complex vector space. Let $G_{N, n}(\mathbf{C})$ be

the set of all n-dimensional (complex) subspaces of V. Choose a basis $\{e_1, \ldots, e_N\}$ of V, identify V with $\mathbf{C}^N$, and let W_0 be the n-dimensional subspace spanned by $e_1, \ldots, e_n$. As in (6), $GL(N, \mathbf{C})$ acts transitively on $G_{N,n}(\mathbf{C})$. The elements of $GL(N, \mathbf{C})$ leaving W_0 fixed are the matrices of the form

$$\begin{bmatrix} A & * \\ 0 & B \end{bmatrix}, \qquad A \in GL(n, \mathbf{C}), \quad B \in GL(N - n, \mathbf{C}),$$

and they form a closed complex Lie subgroup $\tilde{H}$ of $GL(N, \mathbf{C})$. Hence, as in (6), we can define a complex structure on $G_{N,n}(\mathbf{C})$ which is holomorphically isomorphic to the complex structure on $GL(N, \mathbf{C})/\tilde{H}$, and $G_{N,n}(\mathbf{C})$ becomes a complex homogeneous space of the complex Lie group $GL(N, \mathbf{C})$. We call $G_{N,n}(\mathbf{C})$ the complex *Grassmann manifold* of n-dimensional subspaces in an N-dimensional complex vector space. We can also show, as in (6), that $SU(N)$ acts transitively on $G_{N,n}(\mathbf{C})$, and the isotropy subgroup of $SU(N)$ at W_0 is a closed subgroup H of $SU(N)$ consisting of all the matrices of the form

$$\begin{bmatrix} A & 0 \\ 0 & B \end{bmatrix}, \qquad A \in U(n), \quad B \in U(N - n), \quad \det A \cdot \det B = 1.$$

Hence $G_{N,n}(\mathbf{C})$ is diffeomorphic to $SU(N)/H$, so it is compact and connected. On the other hand we can define a complex structure on $SU(N)/H$, which is holomorphically isomorphic to the complex structure on $G_{N,n}(\mathbf{C})$. Hence $SU(N)/H$ is a complex homogeneous space of $SU(N)$. If $n = 1$, then $G_{N,1}(\mathbf{C})$ is nothing but the complex projective space P^{N-1}.

(8) Let V be an N-dimensional complex vector space. A sequence $(V_N, V_{N-1}, \ldots, V_1)$ of subspaces of V is called a *flag* of V if

(1) $\dim_{\mathbf{C}} V_k = k \ (k = 1, \ldots, N)$
(2) $V = V_N \supset V_{N-1} \supset \cdots \supset V_1$

are satisfied. This flag is denoted by $F = (V_N, V_{N-1}, \ldots, V_1)$. The set of all flags of V is denoted by $\mathfrak{F}$. As in (7), choose a basis $\{e_1, \ldots, e_N\}$ of V, identify V with $\mathbf{C}^N$, and the group of all nonsingular linear transformations of V with $GL(N, \mathbf{C})$. If $F = (V_N, V_{N-1}, \ldots, V_1)$ is a flag, $a \in GL(N, \mathbf{C})$, then $(aV_N, aV_{N-1}, \ldots, aV_1)$ is also a flag. For any two flags $F = (V_N, V_{N-1}, \ldots, V_1)$ and $F' = (V_N', V_{N-1}', \ldots, V_1')$, there is an $a \in GL(N, \mathbf{C})$ such that $aF = F'$. Choose a basis $\{u_1, \ldots, u_N\}$ of V so that for each $k = 1, \ldots, N$, the subspace spanned by $\{u_1, \ldots, u_k\}$ is equal to V_k. Similarly, choose $\{u_1', u_2', \ldots, u_N'\}$ so that for each $k = 1, \ldots, N$, $\{u_1', \ldots, u_k'\}$ spans V_k'. There exists an $a \in GL(N, \mathbf{C})$ such that $a \cdot u_k = u_k'$ $(k = 1, \ldots, N)$. This a clearly satisfies $aF = F'$. Hence $GL(n, \mathbf{C})$ acts transitively on the set $\mathfrak{F}$ of flags. Using the basis $\{e_1, \ldots, e_N\}$ we have chosen at the beginning, we define a flag $F^0 = (V_N^0, V_{N-1}^0, \ldots, V_N^0)$ by letting V_k^0 be the k-dimensional subspace of V spanned by $\{e_1, \ldots, e_k\}$. The elements b satisfying $bF^0 = F^0$ are the matrices of the form

$$\begin{bmatrix} a_1^1 & a_2^1 & \cdots & a_N^1 \\ 0 & a_2^2 & \cdots & a_N^2 \\ \cdot & & & \\ \cdot & & & \\ \cdot & & & \\ 0 & 0 & \cdots & a_N^N \end{bmatrix}$$

(which are called upper triangular matrices), and they form a closed complex Lie subgroup B of $GL(N, \mathbf{C})$. Hence, by the map $\alpha : aB \to aF^0$, we have a $1:1$ correspondence from $GL(N, \mathbf{C})/B$ onto the set $\mathfrak{F}$ of all flags of V. Since $GL(N, \mathbf{C})/B$ is a complex manifold, we can define a complex manifold structure on $\mathfrak{F}$ so that α is a holomorphic isomorphism from $GL(N, \mathbf{C})/B$ onto $\mathfrak{F}$. Then $\mathfrak{F}$ becomes a complex homogeneous space of the complex Lie group $GL(N, \mathbf{C})$. This complex homogeneous space is called a *flag manifold*. As in (7), $U(N)$ (and $SU(N)$) acts transitively on $\mathfrak{F}$, and the isotropy subgroup of $U(N)$ at F^0 is a closed subgroup T of $U(N)$, which consists of all the diagonal matrices of the form

$$\begin{bmatrix} a_1^1 & & & & \\ & a_2^2 & & & \\ & & \cdot & & \\ & & & \cdot & \\ & & & & a_N^N \end{bmatrix}, \quad |a_i^i| = 1, \quad (i = 1, \ldots, N).$$

Hence, by the theorem in §17, $\mathfrak{F}$ is diffeomorphic to $U(N)/T$, and hence is compact and connected. We can also give a complex structure to $U(N)/T$ holomorphically isomorphic to the complex structure on $\mathfrak{F}$, and the quotient space $U(N)/T$ is a complex homogeneous space of $U(N)$.

Problem 4. Show that the map, which assigns to each n-dimensional subspace W of an N-dimensional real vector space V its orthogonal complement $W^\perp$, is a diffeomorphism from $G_{N,n}$ to $G_{N,N-n}$.

Problem 5. Find the dimensions of $G_{N,n}$, $G_{N,n}(\mathbf{C})$, and $\mathfrak{F}$.

§19. The Differentiability of One-Parameter Subgroups

We have defined a one-parameter subgroup of a Lie group G to be a differentiable curve a of G, defined on $(-\infty, +\infty)$, such that

$$a(s)a(t) = a(s + t) \tag{1}$$

holds. However, we can prove that a continuous curve of G, defined on $(-\infty, +\infty)$ and satisfying (1), is indeed differentiable.

We shall prove this now. Let $(x^1, \ldots, x^n)$ be a canonical coordinate system of G with respect to a basis $\{X_1, \ldots, X_n\}$ of the Lie algebra $\mathfrak{g}$ of G. For a sufficiently small positive number c, let $U_c = \{g \mid |x^k(g)| < c, k = 1, \ldots, n\}$.

LEMMA. *Let $d = c/2$. If for an element g of U_d, g, $g^2, \ldots, g^m$ all belong to U_d, then $g \in U_{d/m}$, and we have $x^k(g^l) = lx^k(g)(k = 1, \ldots, n; l = 1, \ldots, m)$.*

Proof. Let $X = \sum_{k=1}^n x^k(g)X_k$. If $|tx^k(g)| < c$ $(k = 1, \ldots, n)$, then $\exp tX \in U_c$. If $g = e$, then the lemma is clearly true, so we assume $g \neq e$. Define t_0 by $t_0 \cdot \max_k(|x^k(g)|) = c$. For the positive integer m_0 such that $m_0 < t_0 \leq m_0 + 1$, we have $|m_0 x^k(g)| < c$ $(k = 1, \ldots, n)$, so $\exp m_0 X = g^{m_0} \in U_c$. On the other hand, for the index k' such that $t_0|x^{k'}(g)| = c$, we have $(m_0 + 1)|x^{k'}(g)| \geq t_0|x^{k'}(g)|$, so $m_0|x^{k'}(g)| \geq c - |x^{k'}(g)| \geq c - d = d$, and hence $g^{m_0} \notin U_d$. Thus we have $t_0 > m_0 > m$. Hence, if $l \leq m$, then $l|x^k(g)| < c(k = 1, \ldots, n)$ holds, and we have $\exp lX = g^l$ and $x^k(g^l) = lx^k(g)$. If we set $l = m$, then we have $x^k(g^m) = mx^k(g)$. But since $|x^k(g^m)| < d$, we get $|x^k(g)| < d/m$, that is, $g \in U_{d/m}$.

Now let a be a continuous curve in G satisfying condition (1). From (1), it is immediate that $a(0) = e$. Hence, for $d = c/2$, there is an $\varepsilon > 0$ such that $a(t) \in U_d$ for $|t| < \varepsilon$. Set

$$a^k(t) = x^k(a(t)) \qquad (|t| < \varepsilon) \quad (k = 1, \ldots, n).$$

Fix an ε_0 such that $0 < \varepsilon_0 < \varepsilon$. We shall prove that

$$a^k(t\varepsilon_0) = ta^k(\varepsilon_0) \qquad (k = 1, \ldots, n) \tag{2}$$

holds for $|t| \leq 1$. Both members of (2) are continuous functions of t, and for the left member we have $a^k(-t\varepsilon_0) = -a^k(t\varepsilon_0)$. Hence, it suffices to prove (2) for positive rational values of t. Let $t = s/r$, where r, s are integers and $0 \leq s \leq r$. If we set $g = a(\varepsilon_0/r)$, then for integers l, $1 \leq l \leq r$, we have $g^l = a(l\varepsilon_0/r)$, $|l\varepsilon_0/r| \leq \varepsilon_0$, so we get $g^l \in U_d$. Hence, by the lemma, $a^k(s\varepsilon_0/r) = sa^k(\varepsilon_0/r)$, and $a^k(\varepsilon_0) = ra^k(\varepsilon_0/r)$. Hence we have $a^k(s\varepsilon_0/r) = (s/r)a^k(\varepsilon_0)$, and (2) is proved. From (2) it is clear that $a(t)$ is differentiable in a neighborhood of $t = 0$ given by $|t| < \varepsilon_0$. Now for an arbitrary real number t_0 we have $a(t) = L_{a(t_0)}a(t - t_0)$. Hence a is differentiable on a neighborhood of t_0 given by $|t - t_0| < \varepsilon_0$. Thus we have proved the following theorem.

THEOREM 1. *Let a be a continuous curve of G defined on $(-\infty, +\infty)$ and satisfying (1). Then a is differentiable, and is a one-parameter subgroup of G.*

Let us give some applications of Theorem 1. Let φ be a continuous map from a Lie group G to a Lie group G', satisfying $\varphi(a \cdot b) = \varphi(a) \cdot \varphi(b)$. Then we call φ a continuous homomorphism from G to G'. That is, a homomorphism from G to G', considered as locally compact groups, is called a continuous homomorphism.

THEOREM 2. *A continuous homomorphism φ from a Lie group G to a Lie group G' is differentiable. Hence φ is a homomorphism of Lie groups. In particular, a continuous homomorphism from G into $GL(n, \mathbf{C})$ (or $GL(n, \mathbf{R})$) is a representation of G.*

Proof. If we can prove that φ is differentiable on a neighborhood of the identity element e, then it is easy to show that φ is differentiable at all points of G. Let $\{X_1, \ldots, X_n\}$ be a basis of the Lie algebra $\mathfrak{g}$ of G, and let $(x^1, \ldots, x^n)$ be the canonical coordinate system of the second kind with respect to this basis on a neighborhood U of e. Let V be a neighborhood of the identity element e' of G', and $(y^1, \ldots, y^m)$ a local coordinate system on V. Choose U sufficiently small so that $\varphi(U) \subset V$. Since φ is a continuous homomorphism, $\varphi(\exp tX_i)$ is a continuous curve in G' and satisfies (1). Hence, by Theorem 1, $\varphi(\exp tX_i)$ is a one-parameter subgroup of G', and we can write $\varphi(\exp tX_i) = \exp tY_i$ ($Y_i \in \mathfrak{g}'$, $i = 1, \ldots, n$), where $\mathfrak{g}'$ is the Lie algebra of G'. Hence we have

$$\varphi(\exp x^1X_1 \ldots \exp x^nX_n) = \exp x^1 Y_1 \ldots \exp x^n Y_n.$$

From this we see easily that $y^i \circ \varphi$ is a C^∞ function of $x^1, \ldots, x^n$. Hence φ is differentiable on U.

Now if φ is a continuous homomorphism from G onto G', then by

Theorem 2 of §5, φ is an open homomorphism. Hence, if N is the kernel of φ, then by the homomorphism theorem in §3, we have $G/N \cong G'$ as locally compact groups. But as we have remarked in §15, N is a closed normal Lie subgroup of G, so that G/N is a Lie group. Hence, by Theorem 2, G/N and G' are isomorphic as Lie groups. In particular, if φ is a $1:1$ continuous homomorphism from G onto G', then since $N = \{e\}$, G and G' are isomorphic by φ, and φ^{-1} is also differentiable. We have proved the following theorem.

THEOREM 3.   *Let φ be a continuous homomorphism from a Lie group G onto a Lie group G', and let N be the kernel of φ. Then N is a closed normal Lie subgroup of G, and the quotient Lie group G/N is isomorphic to G'. In particular, if φ is $1:1$, then φ is an isomorphism from G onto G'.*

§20. Lie Subgroups of a Lie Group (II)

In this section we identify the Lie algebra $\mathfrak{g}$ of a Lie group G with the tangent space $T_e(G)$ of G at the identity element e by the identification map $X \to X_e$.

PROPOSITION 1.   *Let G be a Lie group, and H a subgroup of G as an abstract group. Let $\mathfrak{h}$ be the subset of $T_e(G)$ consisting of all $u \in T_e(G)$ satisfying the property: "there exists a differentiable curve a of G such that $a(0) = e$, $a(t) \in H$ for sufficiently small $|t|$, and $\dot{a}(0) = u$."* Then $\mathfrak{h}$ is a subalgebra of $\mathfrak{g}$.*

To prove this proposition we need the following two lemmas.

LEMMA 1.   *Let a and b be differentiable curves in G such that $a(0) = b(0) = e$. Let c be the curve in G defined by $c(t) = a(t) \cdot b(t)$. Then $c(0) = e$ and $\dot{c}(0) = \dot{a}(0) + \dot{b}(0)$.*

This lemma can be easily proved by using canonical coordinates and Theorem 4 of §10. We leave the proof as an exercise for the reader.

LEMMA 2.   *Let a and b be differentiable curves in G such that $a(0) = b(0) = e$. Let*

$$c(t) = a(t^{1/2})^{-1}b(t^{1/2})^{-1}a(t^{1/2})b(t^{1/2}) \qquad \text{for} \quad t \geq 0,$$
$$c(t) = c(-t)^{-1} \qquad\qquad\qquad\qquad\qquad \text{for} \quad t \leq 0.$$

---

*The tangent vector to a differentiable curve $a(t)$(cf. II, §5) at $t = t_1$ is denoted by $\dot{a}(t_1)$.

Then c is a differentiable curve in G such that $c(0) = e$ and $\dot{c}(0) = [\dot{a}(0), \dot{b}(0)]$.

Proof. Let $\{X_1, \ldots, X_n\}$ be a basis of $\mathfrak{g}$, and let $\{x^1, \ldots, x^n\}$ be the corresponding canonical coordinate system in a neighborhood U of e. Let V be a neighborhood of e such that $V^{-2}V^2 \subset U$, and let $a, b \in V$. Then, by Theorem 4 of §10, we have

$$x^k(a^{-1}b^{-1}ab) = \sum_{i,j} c^k_{ij}a^i b^j + h^k(a, b) \qquad (k = 1, \ldots, n),$$

where $a^i = x^i(a)$, $b^j = x^j(b)$, $[X_i, X_j] = \sum_k c^k_{ij}X_k$, and $h^k(a, b)$ denotes a higher order term.

Let $a = a(s)$, $b = b(s)$ for $|s|$ sufficiently small. Then we have

$$x^k(a^{-1}(s)b^{-1}(s)a(s)b(s)) = \sum c^k_{ij}a'^i(0)b'^j(0)s^2 + O(s^3).^*$$

From this, we see easily that

$$c'^k(0) = (dx^k(c(t))/dt)_{t=0} = \sum_{i,j} c^k_{ij}a'^i(0)b'^j(0).$$

Since $\dot{a}(0) = \sum_i a'^i(0)X_i$ and $\dot{b}(0) = \sum_j b'^j(0)X_j$, we obtain

$$\dot{c}(0) = \sum_k c'^k(0)X_k = [\dot{a}(0), \dot{b}(0)].$$

Proof of Proposition 1. We show first that $\mathfrak{h}$ is a subspace of $\mathfrak{g}$. Let $u, v \in \mathfrak{h}$. There exist two curves a and b such that $a(t), b(t) \in H$ for $|t| < \varepsilon$, $a(0) = b(0) = e$ and $\dot{a}(0) = u$, $\dot{b}(0) = v$. Let c be the curve defined by $c(t) = a(t)b(t)$. Then we have $c(0) = e$, $c(t) \in H$ for $|t| < \varepsilon$, and hence $\dot{c}(0)$ belongs to $\mathfrak{h}$. By Lemma 1, we have $\dot{c}(0) = \dot{a}(0) + \dot{b}(0) = u + v$. Thus $u + v \in \mathfrak{h}$ for any $u, v \in \mathfrak{h}$. Let now λ be any real number, and let a be the curve defined by $a_\lambda(t) = a(\lambda t)$. Then $a_\lambda(0) = e$, and $a_\lambda(t) \in H$ for $|t|$ sufficiently small. Hence $\dot{a}_\lambda(0) \in \mathfrak{h}$. But $\dot{a}_\lambda(0) = \lambda\dot{a}(0) = \lambda u$, so that λu belongs to $\mathfrak{h}$. Thus $\mathfrak{h}$ is a subspace of $\mathfrak{g}$. Using Lemma 2, an analogous argument shows us that $[u, v]$ also belongs to $\mathfrak{h}$. Hence $\mathfrak{h}$ is a subalgebra of $\mathfrak{g}$.

In the following, we shall always denote by H a subgroup of a Lie group G, and by $\mathfrak{h}$ the subalgebra of $\mathfrak{g}$ defined in Proposition 1.

DEFINITION.   The arcwise connected component H_c of e in H is the subset of H consisting of all elements h in H with the following property: there exists a differentiable curve a in G such that $a(0) = e$, $a(1) = h$, and $a(t) \in H$ for all t.

$^*a'^i(0)$ means $(da^i/ds)_{s=0}$, and similarly for the others.

We can see easily that H_c *is a normal subgroup of H.*

Problem 1. Verify this property of H_c.

In the following we shall show that H_c is the connected Lie subgroup of G corresponding to the subalgebra $\mathfrak{h}$ of $\mathfrak{g}$.

To see this, we denote by H_0 the connected Lie subgroup of G corresponding to $\mathfrak{h}$. We shall denote by the same letter $\mathfrak{h}$ the completely integrable differential system on G defined by $\mathfrak{h}$(cf. §12). Then H_0 is the leaf of $\mathfrak{h}$ passing through e. If a is a differentiable curve in G such that $a(0) = e$ and $a(t) \in H$ for all t, then a is an integral curve of the differential system $\mathfrak{h}$, that is, $\dot{a}(t_0) \in \mathfrak{h}(a(t_0))$ for all t_0. In fact, let

$$b(t) = a(t_0)^{-1}a(t).$$

Then $b(t_0) = e$, $b(t) \in H$, and hence $\dot{b}(t_0) \in \mathfrak{h}(e) = \mathfrak{h}$. On the other hand, $\dot{b}(t_0) = (L_{a(t_0)^{-1}})_*\dot{a}(t_0)$. Hence $\dot{a}(t_0) = (L_{a(t_0)})_*\dot{b}(t_0)$, and so $\dot{a}(t_0) \in \mathfrak{h}(a(t_0))$.

Now each integral curve of a completely integrable differential system is contained in a leaf, and since $a(0) = e$, $a(t)$ is contained in H_0 for all t. This implies that H_c is contained in H_0.

We show now that H_0 is contained in H_c. To show this, let $\{X_1, \ldots, X_r\}$ be a basis of $\mathfrak{h}$. Then there exists a differentiable curve a_i in G, for each i, such that $a_i(0) = e$, $a_i(t) \in H$ for $|t| < \varepsilon$, and $\dot{a}_i(0) = X_i$. Let φ be the map from the cube $Q = \{t = (t^1, \ldots, t^r) \in \mathbf{R}^r| \ |t^i| < \varepsilon, i = 1, \ldots, r\}$ into G defined by

$$\varphi(t) = a_1(t^1) \ldots a_r(t^r).$$

Then $\varphi(t) \in H_c$ since each $a_i(t^i)$ belongs to H_c, and H_c is a subgroup. The map φ is differentiable, and the image $\varphi(Q)$ is contained in H_c, and so in H_0. Since H_0 is a leaf of the differential system $\mathfrak{h}$, φ is a differentiable map from Q into H_0, by Theorem 2 of III, §10. On the other hand, the the differential of φ at 0 is of rank r, since $X_1, \ldots, X_r$ are linearly independent. By the inverse function theorem, φ defines a diffeomorphism of a neighborhood of 0 in $\mathbf{R}^r$ onto a neighborhood U of e in H_0. The neighborhood U of e is contained in $\varphi(Q)$, and hence in H_c. Since H_0 is connected, H_0 is generated by $U \cap U^{-1}$ (Theorem 1 of §4), and since $U \cap U^{-1} \subset H_c$ and H_c is a subgroup, this shows that $H_0 \subset H_c$. Since we have already proved that $H_c \subset H_0$, we obtain $H_c = H_0$.

We can now put the structure of a Lie group on H_c, so that the identity map from H_c onto H_0 becomes an isomorphism of Lie groups, and we have proved the following theorem of Freudenthal and Yamabe.

THEOREM 1.   *Let H be a subgroup of a Lie group G, and H_c the arcwise connected component of H containing e. Then H_c is a normal subgroup of H, and H_c is the connected Lie subgroup of G whose Lie algebra is the one defined in Proposition 1.*

COROLLARY 1.   *Every arcwise connected subgroup* H of a Lie group is a connected Lie subgroup.*

COROLLARY 2.   *Let H be a subgroup of G. Assume that H admits a structure of a connected Lie subgroup of G. Then such a structure is unique.*

Proof. Let H_1 denote a connected Lie subgroup of G, whose underlying abstract subgroup is H, and let $\mathfrak{h}_1$ be the subalgebra of the Lie algebra $\mathfrak{g}$ of G corresponding to H_1. Then we see readily that $\mathfrak{h}_1$ is equal to $\mathfrak{h}$ given in Proposition 1, and hence $H_1 = H_c$ as Lie groups.

Problem 2. Let H be a Lie subgroup of G, and H_0 the connected component of H containing the identity. Show that $H_0 = H_c$.

Let H be a subgroup of a Lie group G, and suppose that H is a closed set of G. Then, with respect to the topology induced from G, H is a locally compact group with a countable base. From now on, we regard H as a locally compact group with respect to the induced topology, and let H_0 be the connected component of H containing the identity. Then H_c is contained in H_0, because if a is a differentiable curve in G such that $a(0) = e$ and $a(t) \in H$, we may regard a as a continuous curve in H. The image of the closed interval $[0, t]$ by a is connected, so it is contained in the connected component H_0 of H containing e.

Now we shall show that $H_0 \subset H_c$. To show this, it is enough to show that a neighborhood of e in H_0 is contained in H_c, because H_0 is generated by a neighborhood of e. Let $\mathfrak{h}$ be the Lie subalgebra of $\mathfrak{g}$ corresponding to the Lie subgroup H_c of G (cf. Theorem 1), and let $\mathfrak{m}$ be a subspace of $\mathfrak{g}$ such that

$$\mathfrak{g} = \mathfrak{m} + \mathfrak{h}, \qquad \mathfrak{m} \cap \mathfrak{h} = \{0\}.$$

Let V_1 and V_2 be sufficiently small neighborhoods of 0 in the vector spaces $\mathfrak{m}$ and $\mathfrak{h}$, respectively, and let $\varphi: V_1 \times V_2 \to G$ be the map which assigns to each $(X, Y) \in V_1 \times V_2$ the element $\exp X \cdot \exp Y$ of G. This map φ is a diffeomorphism from $V_1 \times V_2$ onto a neighborhood U of e, if we choose V_1 and V_2 sufficiently small (cf. Problem 4 of §10). Moreover φ defines a diffeomorphism from $\{0\} \times V_2$ onto a neighborhood

---

*This means that $H = H_c$.

U_2 of e in H_c. We show that U_2 contains a neighborhood of e in H. In fact, if this were not the case, then there would be a sequence of points $\{g_n\}$, $g_n \in H$, tending to e, and $g_n \notin U_2$ for each n. Then since $\varphi(V_1 \times V_2) = U$, there would be sequences $\{X_n\}$, $X_n \in V_1$, and $\{Y_n\}$, $Y_n \in V_2$, both tending to 0, and such that

$$g_n = \varphi(X_n, Y_n) = \exp X_n \cdot \exp Y_n.$$

Since g_n is not in U_2, X_n is different from 0 for each n. Let

$$a_n = \exp X_n = g_n(\exp Y_n)^{-1}.$$

Then, since g_n and $\exp Y_n$ belong to H, a_n also belongs to H, and a_n tends to e. Let ε be a sufficiently small positive real number, and let

$$Z_n = \frac{\varepsilon X_n}{|X_n|},$$

where $|X_n|$ denotes the euclidean length of the vector X_n in the real vector space $\mathfrak{g}$. Then Z_n belongs to the compact set $\{X \in \quad | \; |X| = \varepsilon\}$, and $Z_n \in \mathfrak{m}$. Hence Z_n converges to a vector such that $|Z| = \varepsilon$, $Z \in \mathfrak{m}$. Choosing ε sufficiently small, we may assume that Z belongs to V_1. For each real number t, let r_n be the integer defined by

$$r_n < \frac{t\varepsilon}{|X_n|} \leq r_n + 1,$$

and let

$$s_n = \frac{t\varepsilon}{|X_n|} - r_n.$$

Then

$$tZ_n = \frac{t\varepsilon}{|X_n|} X_n = r_n X_n + s_n X_n.$$

Since $0 < s_n \leq 1$ for each n, $s_n X_n$ tends to 0. On the other hand, tZ_n tends to tZ, and hence $r_n X_n \to tZ$ as $n \to +\infty$. Therefore, $\exp r_n X_n = (\exp X_n)^{r_n}$ tends to $\exp tZ$. Since $\exp X_n \in H$, we have $(\exp X_n)^{r_n} \in H$, and since H is closed in G, we get $\exp tZ \in H$ for each $t \in \mathbf{R}$. Therefore, $\exp tZ \in H_c$, by the definition of H_c, so $Z \in \mathfrak{h}$. On the other hand Z is an element of $\mathfrak{m}$. But $\mathfrak{h} \cap \mathfrak{m} = \{0\}$, so that we have $Z = 0$. This, however, contradicts $|Z| = \varepsilon > 0$. The contradiction shows that there must be a neighborhood V of e in H contained in U_2, and we conclude that $H_0 \subset H_c$.

We have shown that $H_0 = H_c$ as subsets of G. The identity map i from H_c onto H_0 is a continuous map, because H_c is a submanifold of G and H_0 has the induced topology from G. Since H_c and H_0 are locally compact topological groups with countable bases, the continuous bijective homomorphism from H_c onto H_0 is an isomorphism of topological groups (Corollarly 1 of Theorem 2 of §5), so $H_c = H_0$ as topological groups. (It follows, in particular, that the topology of H_c is the induced topology from G.) We can put a structure of a connected Lie subgroup of G on H_0 by demanding that $i : H_c \to H_0$ be an isomorphism of Lie groups. On each coset hH_0 of H by H_0, we can introduce a submanifold structure, by asking the map $x \to hx$ from H_0 to hH_0 to be diffeomorphic. In this way we can introduce a submanifold structure of G on H, and it is easy to see that H then becomes a closed Lie subgroup of G.

We state the result proved above in the following form.

THEOREM 2 (E. Cartan). *Every closed subgroup of a Lie group is a closed Lie subgroup.*

We prove now the following theorem.

THEOREM 3. *Let H be a Lie subgroup of a Lie group G. If the topology of H is the induced topology, then H is closed.*

To prove this theorem, we need the following elementary lemma.

LEMMA 3. *Let M be a C^∞ manifold, and N a regular submanifold of M such that $\dim N < \dim M$. Then the closure $\bar{N}$ of N is different from M.*

Proof. For each point p of N, there exist a neighborhood U of p in M, and r functions $f_1, \ldots, f_r$ on U ($r = \dim M - \dim N$), such that $df_1, \ldots, df_r$ are linearly independent at each point of $N \cap U$, and such that

$$N \cap U = \{q \in U | f_1(q) = \cdots = f_r(q) = 0\}$$

(cf. II, §10). If $\bar{N} = M$, then we have $f_1(q) = \cdots = f_r(q) = 0$ for all $q \in U$, and this is impossible because $df_1, \ldots, df_r$ are linearly independent at q. Therefore $\bar{N} \neq M$.

Proof of Theorem 3. Let $\bar{H}$ be the closure of H. Then $\bar{H}$ is a closed Lie subgroup of G, by Theorem 2. The identity map $i : H \to G$ is differentiable, and $i(H) \subset \bar{H}$. Since $\bar{H}$ is a regular submanifold of G, by the proposition of II, §10, i is a differentiable map from H into $\bar{H}$. Since

H also has the topology induced from G by the hypothesis, this shows that H is a regular submanifold of $\bar{H}$. But then, from Lemma 3, we get that dim H = dim $\bar{H}$, and we conclude that $H = \bar{H}$.

Problem 3. Let H and H' be Lie subgroups of a Lie group G such that $H' \subset H$. Show that H' is a Lie subgroup of H.

Integration of Differential Forms
and Their Applications

§1. Orientation of Manifolds

Just as we distinguish between positive and negative directions on a straight line, or as we decide on the positive or negative direction of a rotation, or a positive or negative measurement of an angle, it is necessary to decide on an "orientation" of a manifold. On a staight line or in a plane, the concept of "orientation" has been introduced naively, relying on intuitive perception. However, in the case of a manifold, we will have to define "positive orientation" and "negative orientation" more abstractly.

First, let us define the *orientation* of an n-dimensional real vector space. An ordered set of n linearly independent vectors of V is called an *ordered basis* of V. Let $e_1, \ldots, e_n$ be a set of n linearly independent vectors of V. If we order $e_1, \ldots, e_n$ in two different ways, for example, $(e_1, e_2, \ldots, e_n)$ and $(e_2, e_1, \ldots, e_n)$, then we consider the two sets to be distinct as ordered bases of V. We denote the set of all ordered bases of V by $\mathfrak{F}$ and define an equivalence relation in $\mathfrak{F}$ as follows. Let $(e_1, \ldots, e_n), (e_1', \ldots, e_n') \in \mathfrak{F}$ and set

$$e_i' = \sum_{j=1}^{n} a_i^j e_j \qquad (i = 1, \ldots, n) \tag{1}$$

Since $\{e_1, \ldots, e_n\}, \{e_1', \ldots, e_n'\}$ are bases of V, we have $\det(a_i^j) \neq 0$. If $\det(a_i^j) > 0$, then we define $(e_1, \ldots, e_n)$ and $(e_1', \ldots, e_n')$ to be equivalent. It is easy to see that this defines an equivalence relation. There are only two equivalence classes with respect to this equivalence relation. To prove this, it suffices to show that two elements $(e_1', \ldots, e_n')$ and $(e_1'', \ldots, e_n'')$ of $\mathfrak{F}$ are equivalent if they are both not equivalent to a third element $(e_1, \ldots, e_n)$ of $\mathfrak{F}$. Suppose $\{e_i\}$ and $\{e_i'\}$ are related by (1) and $\{e_i''\}$ and $\{e_i\}$ are related by

$$e_i = \sum_{j=1}^{n} b_i^j e_j''.$$

Then we have

$$e_i' = \sum_{j=1}^{n} c_i^j e_j'', \qquad c_i^j = \sum_{k=1}^{n} a_i^k b_k^j.$$

Hence $\det(c_i^j) = \det(a_i^j)\det(b_i^j)$. But $\det(a_i^j)$ and $\det(b_i^j)$ are both negative, so we have $\det(c_i^j) > 0$, and hence $(e_1', \ldots, e_n')$ and $(e_1'', \ldots, e_n'')$ are equivalent. The equivalence classes $\mathfrak{F}_1$, $\mathfrak{F}_2$ of $\mathfrak{F}$ are called the *orientations* of the real vector space V. Hence, V has two orientations. We pick one of them and call it the *positive orientation*, and call the other the *negative orientation*. From the logical point of view, it does not matter whether we choose $\mathfrak{F}_1$ or $\mathfrak{F}_2$ as the positive orientation. The pair $(V, \mathfrak{F}_i)$ of V and a positive orientation $\mathfrak{F}_i$ is called an *oriented vector space*. To simplify the notation, instead of $(V, \mathfrak{F}_i)$, we shall simply write V. We shall call the bases belonging to the positive orientation of V the positive bases of the oriented vector space V, or simply the positive bases of V, and we shall call the bases belonging to the negative orientation the negative bases.

Example 1. Let V be a one-dimensional real vector space. Choose a nonzero vector e of V and let $\mathfrak{F}_1 = \{(\lambda e)|\lambda > 0\}$ and $\mathfrak{F}_2 = \{(\lambda e)|\lambda < 0\}$. If we identify V with the real line, then choosing e as the unit vector from 0 to 1, it is customary to call $\mathfrak{F}_1$ the positive direction and $\mathfrak{F}_2$ the negative direction. This choice comes from the fact that most people are right-handed, and there is no logical reason behind the choice (Fig. 5.1).

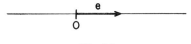

Fig. 5.1

Example 2. Let V be a two-dimensional real vector space. Identify V with $\mathbf{R}^2$ and let e_1 and e_2 be two linearly independent unit vectors in

V. Let the equivalence class containing (e_1, e_2) be denoted by $\mathfrak{F}_1$. Then $\mathfrak{F}_2$ is represented by (e_2, e_1). It is customary to call $\mathfrak{F}_1$ the positive orientation and $\mathfrak{F}_2$ the negative orientation (Fig. 5.2).

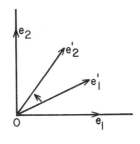

Fig. 5.2

Problem 1. Let $(e_1, \ldots, e_n)$ be a positive basis and σ a permutation of $1, \ldots, n$. Then $(e_{\sigma(1)}, \ldots, e_{\sigma(n)})$ is a positive basis if σ is an even permutation, and is a negative basis if σ is an odd permutation. Show this.

Problem 2. In Example 2, let (u_1, u_2) be an ordered basis of $V = \mathbf{R}^2$, and suppose u_1 and u_2 are of unit length. Letting

$$u_1 = \begin{pmatrix} a \\ b \end{pmatrix}, \qquad u_2 = \begin{pmatrix} c \\ d \end{pmatrix},$$

we can write

$$\begin{pmatrix} c \\ d \end{pmatrix} = \begin{pmatrix} \cos\theta & -\sin\theta \\ \sin\theta & \cos\theta \end{pmatrix} \begin{pmatrix} a \\ b \end{pmatrix} \qquad (-\pi < \theta < \pi)$$

uniquely, and if (u_1, u_2) is positive, then $\theta > 0$, while if (u_1, u_2) is negative, then $\theta < 0$. Prove this.

Now let us give an alternate definition of the orientation of an n-dimensional real vector space V which is convenient for applications. The vector space $\wedge^n V^*$ formed by the alternating n-linear functions on V is one dimensional. Hence, if $\omega, \theta \in \wedge^n V^*$ and $\omega \neq 0$, then we can write $\theta = \lambda\omega$ $(\lambda \in \mathbf{R})$. We denote the set of all *nonzero* alternating n-linear functions on V by $\mathfrak{F}^*$, and define an equivalence relation in $\mathfrak{F}^*$ as follows. For $\theta, \omega \in \mathfrak{F}^*$, we define θ and ω to be equivalent if $\theta = \lambda\omega$ with $\lambda > 0$. Then there are exactly two equivalence classes in $\mathfrak{F}^*$. Now let $(e_1, \ldots, e_n)$ and $(e_1', \ldots, e_n')$ be two ordered bases of V, and let $\omega \in \mathfrak{F}^*$. Then $(e_1, \ldots, e_n)$ and $(e_1', \ldots, e_n')$ determine the same orientation if and only if $\omega(e_1, \ldots, e_n)$ and $\omega(e_1', \ldots, e_n')$ are both positive or both negative. In fact, if $(e_1, \ldots, e_n)$ and $(e_1', \ldots, e_n')$ are related by (1), then we have

$$\omega(e_1', \ldots, e_n') = \sum_{j_1, \ldots, j_n = 1}^{n} a_1^{j_1} \cdots a_n^{j_n} \omega(e_{j_1}, \ldots, e_{j_n})$$

$$= \sum_{\sigma} \varepsilon(\sigma) a_1^{\sigma(1)} \cdots a_n^{\sigma(n)} \omega(e_1, \ldots, e_n)$$

$$= \det(a_i^j) \cdot \omega(e_1, \ldots, e_n)$$

On the other hand, if ω, $\theta \in \mathfrak{F}^*$, and if $(e_1, \ldots, e_n)$ is an ordered basis, then ω and θ are equivalent if and only if $\omega(e_1, \ldots, e_n)$ and $\theta(e_1, \ldots, e_n)$ have the same signature. Hence, by a suitable indexing of the orientations $\mathfrak{F}_1$, $\mathfrak{F}_2$ of V and the equivalence classes $\mathfrak{F}_1^*$, $\mathfrak{F}_2^*$ of $\mathfrak{F}^*$, we have

$$\mathfrak{F}_i^* = \{\omega \in \mathfrak{F}^* | \omega(e_1, \ldots, e_n) > 0, (e_1, \ldots, e_n) \in \mathfrak{F}_i\}$$
$$\mathfrak{F}_i = \{(e_1, \ldots, e_n) \in \mathfrak{F} | \omega(e_1, \ldots, e_n) > 0, \omega \in \mathfrak{F}_i^*\}$$
$$(i = 1, 2).$$

Hence we can consider $\mathfrak{F}_1^*$, $\mathfrak{F}_2^*$ to be the orientations of V, and we can call one of them the positive orientation and the other the negative orientation. The alternating n-linear functions belonging to the positive orientation are called *positive alternating n-linear functions*, and the alternating n-linear functions belonging to the negative orientation are called *negative alternating n-linear functions*. If we agree to call $\mathfrak{F}_i$(resp. $\mathfrak{F}_i^*$) positive when $\mathfrak{F}_i^*$(resp. $\mathfrak{F}_i$) is the positive orientation, then we have the relations

$(e_1, \ldots, e_n)$ is a positive basis $\leftrightarrow$ $\begin{cases} \omega(e_1, \ldots, e_n) > 0 \text{ for any positive} \\ \text{alternating } n\text{-linear function} \end{cases}$

$\left. \begin{array}{l} \omega \text{ is a positive alternating} \\ n\text{-linear function} \end{array} \right\} \leftrightarrow \begin{cases} \omega(e_1, \ldots, e_n) > 0 \text{ for any positive} \\ \text{basis } (e_1, \ldots, e_n) \end{cases}$

Problem 3. Let $\{e_1, \ldots, e_n\}$ and $\{f^1, \ldots, f^n\}$ be bases of V and V^* dual to each other. Show that $(e_1, \ldots, e_n)$ is positive if and only if $f^1 \wedge \cdots \wedge f^n$ is positive.

DEFINITION 1. An n-dimensional manifold M is said to be *orientable* if there is a continuous differential form of order n which does not vanish at any point of M.

Let us define the "orientation of M" when M is orientable. We have two cases:

(1) *M is connected.* Let $\mathfrak{F}^*$ be the set of all continuous differential forms of order n which do not vanish at any point of M. If ω and θ are

two elements of $\mathfrak{F}^*$, then ω and θ are related by

$$\theta = f \cdot \omega,$$

where f is a continuous function on M which does not vanish at any point of M. Since we are assuming M to be connected, f is positive for all points of M or negative for all points of M. We define ω and θ to be equivalent if $f > 0$. This gives an equivalence relation in $\mathfrak{F}^*$, and there are two equivalence classes. An equivalence class of $\mathfrak{F}^*$ is called an *orientation* of M. We choose one of the equivalence classes to be the *positive orientation* and the other to be the *negative orientation*. A manifold with a chosen positive orientation is called an *oriented manifold*. A differential form of order n belonging to the positive orientation is called a *positive differential form of order n*, or a *volume element*, of an oriented manifold.

(2) *M is not connected.* Let M_α ($\alpha \in I$) be the connected components of M, and let ω_α be the restriction to M_α of a continuous differential form ω of order n on M which does not vanish at any point of M. Since ω_α is a differential form of order n on M_α which does not vanish at any point of M_α, M_α is orientable. We determine a positive orientation of M by choosing a positive orientation for each M_α. As in the connected case, a manifold with a chosen positive orientation is called an *oriented manifold*. If, for a differential form ω on M of order n, its restriction ω_α to M_α is positive for each M_α, then ω is called a *positive differential form of order n*, or a *volume element*, of the oriented manifold M. If ω is a differential form of order n on M which does not vanish at any point of M, then we can choose a positive orientation of M so that ω is a volume element of M. In fact, one only has to choose ω_α to be in the positive orientation of M_α.

Now let ω be a volume element of an oriented n-dimensional manifold M. By choosing ω_p to be a positive alternating linear function on the tangent space $T_p(M)$ for each $p \in M$, we can determine a positive orientation on $T_p(M)$. If $(x^1, \ldots, x^n)$ is a local coordinate system on an open set U of M, and if, for each point p of U, $((\partial/\partial x^1)_p, \ldots, (\partial/\partial x^n)_p)$ is a positive basis of $T_p(M)$, i.e., if

$$\omega_p((\partial/\partial x^1)_p, \ldots, (\partial/\partial x^n)_p) > 0 \qquad (\forall p \in U)$$

holds, then $(x^1, \ldots, x^n)$ is called a *positive local coordinate system* of M. If, at each point p of U, we have

$$\omega_p((\partial/\partial x^1)_p, \ldots, (\partial/\partial x^n)_p) < 0,$$

then $(x^1, \ldots, x^n)$ is called a negative local coordinate system of M. If $(x^1, \ldots, x^n)$ is positive and σ is a permutation of $(1, 2, \ldots, n)$, then $(x^{\sigma(1)}, \ldots, x^{\sigma(n)})$ is positive if σ is an even permutation, and negative if σ is an odd permutation.

If $(x^1, \ldots, x^n)$ is a positive local coordinate system on U, then we can write

$$\omega = a(x) \, dx^1 \wedge \cdots \wedge dx^n, \qquad a(x) > 0.$$

If U is a connected open set of M, and $(x^1, \ldots, x^n)$ is an arbitrary local coordinate system on U, then we have $\omega = a(x) \, dx^1 \wedge \cdots \wedge dx^n$ on U. Since U is connected, $a(x)$ is positive at all points of U or negative at all points of U. If $a(x) > 0$, then $(x^1, \ldots, x^n)$ is positive, and if $a(x) < 0$, then $(x^1, \ldots, x^n)$ is negative. That is, a local coordinate system on a connected open set U is either positive or negative.

If $(x^1, \ldots, x^n)$ and $(y^1, \ldots, y^n)$ are positive local coordinate systems on U and V, respectively, and if $U \cap V \neq \varnothing$, then it is easy to see that

$$\det \left(\frac{\partial y^i}{\partial x^j} \right) > 0$$

holds on $U \cap V$.

Problem 4.   Prove the last statement above.

THEOREM 1.   *Let M be an n-dimensional oriented manifold. We can choose a coordinate neighborhood system $\{U_\alpha, \psi_\alpha\}_{\alpha \in A}$ on M satisfying the following condition: If $(x_\alpha^1, \ldots, x_\alpha^n)$ is the local coordinate system on $\{U_\alpha, \psi_\alpha\}$, and $U_\alpha \cap U_\beta \neq \varnothing$, then*

$$\det \left(\frac{\partial x_\alpha^i}{\partial x_\beta^j} \right) > 0$$

holds on $U_\alpha \cap U_\beta$. Conversely, if M is paracompact, and if we can choose a coordinate neighborhood system on M satisfying the above condition, then M is orientable. Moreover, then, there exists a positive differential form of order n and of class C^∞ on M.

Proof.   The first half of the theorem is clear from what has been said already, so we shall prove the converse, assuming M to be paracompact. Choose a locally finite refinement $\{V_\lambda\}_{\lambda \in \Lambda}$ of $\{U_\alpha\}_{\alpha \in A}$ such that $\overline{V}_\lambda$ is compact for each $\lambda \in \Lambda$. Let $\{f_\lambda\}_{\lambda \in \Lambda}$ be a partition of unity subordinate to $\{V_\lambda\}_{\lambda \in \Lambda}$. For each V_λ, choose an α such that $V_\lambda \subset U_\alpha$, and let x_λ^i be the restriction of x_α^i to V_λ. Then $(x_\lambda^1, \ldots, x_\lambda^n)$ is a local coordinate system on V_λ, and if $V_\lambda \cap V_\mu \neq \emptyset$, then

$$\det\left(\frac{\partial x_\lambda^i}{\partial x_\mu^j}\right) > 0$$

holds on $V_\lambda \cap V_\mu$. Define a differential form ω^λ of order n on V_λ by

$$\omega^\lambda = dx_\lambda^1 \wedge \cdots \wedge dx_\lambda^n.$$

Then we can define a differential form ω on M of order n and of class C^∞ by

$$\omega_p = \sum_{\lambda \in \Lambda} f_\lambda(p) \cdot (\omega^\lambda)_p.$$

We have $\omega_p \neq 0$ at each point p. In fact, we can assume that $1, \ldots, s$ are the indices λ such that $f_\lambda(p) \neq 0$. Then, since

$$(dx_\alpha^1 \wedge \cdots \wedge dx_\alpha^n)_p = \det\left(\frac{\partial x_\alpha^i}{\partial x_1^j}\right)_p (dx_1^1 \wedge \cdots \wedge dx_1^n)_p \quad (\alpha = 1, \ldots, s),$$

we have

$$\omega_p = \left(\sum_{\alpha=1}^s f_\alpha(p) \det\left(\frac{\partial x_\alpha^i}{\partial x_1^j}\right)_p\right)(dx_1^1 \wedge \cdots \wedge dx_1^n)_p,$$

and the coefficient of $dx_1^1 \wedge \cdots \wedge dx_1^n$ is positive. Hence M is orientable, an the C^∞ differential form ω of order n does not vanish at any point of M. If we choose ω to be positive, then it is clear from the definition of ω that $(x_\alpha^1, \ldots, x_\alpha^n)$ is a positive coordinate system for each $\alpha \in A$.

COROLLARY. *If an n-dimensional paracompact manifold M is orientable, then there is a differential form of order n and of class C^∞ on M that does not vanish at any point of M.*

THEOREM 2. *Let M be an n-dimensional paracompact manifold. Suppose there are an open covering $\{U_\alpha\}_{\alpha \in A}$ of M and a differential form ω_α of order n on each U_α, such that if $U_\alpha \cap U_\beta \neq \emptyset$, then $\omega_\beta = f_{\beta\alpha}\omega_\alpha$, $f_{\beta\alpha} > 0$ on $U_\alpha \cap U_\beta$. Then M is orientable.*

Proof. Let $(x^1, \ldots, x^n)$ be a local coordinate system on a neighborhood U of a point. Taking U sufficiently small, we choose α such that $U \subset U_\alpha$. If we have $(\omega_\alpha)_q((\partial/\partial x^1)_q, \ldots, (\partial/\partial x^n)_q) > 0$ for each point q of U, then we say that $(x^1, \ldots, x^n)$ is positive. From the assumption on $\{\omega_\alpha\}_{\alpha \in A}$, we see that this definition does not depnd on the choice of U_α containing U. It is also easy to see that $\{U, (x^1, \ldots, x^n)\}$ satisfies the condition of Theorem 1. Hence M is orientable.

Remark. Let M be a connected and nonorientable manifold. There are a connected manifold $\tilde{M}$ and a differentiable map π from $\tilde{M}$ to M with the following properties:

(1) The manifold $\tilde{M}$ is orientable, and $\pi^{-1}(p)$ consists of two points of $\tilde{M}$ for each $p \in M$.

(2) For each point p of M, there is a connected neighborhood U of p such that $\pi^{-1}(U)$ consists of two connected components U_+ and U_-, and π restricted to U_+(resp. U_-) is a diffeomorphism from U_+(resp. U_-) onto U.

Let $\mathbf{E}^{n+1}$ be an $(n + 1)$-dimensional Euclidean space, and let M be an m-dimensiona submanifold of $\mathbf{E}^{n+1}$. We identify the tangent vector space at each point p of M with a subspace of $T_p(\mathbf{E}^{n+1})$. A vector in $T_p(\mathbf{E}^{n+1})$ which is orthogonal to $T_p(M)$ is called a *normal vector* of M at p. A *normal vector field* N defined on a subset A of M is a function which assigns to each point $p \in A$ a normal vector N_p of M at p. We can define continuous (resp. differentiable) normal vector fields on A as in the case of (tangent) vector fields.

Now let $(x^1, \ldots, x^{n+1})$ be the canonical coordinates in $\mathbf{E}^{n+1}$. Then we can identify $T_p(\mathbf{E}^{n+1})$ with the underlying vector space of $\mathbf{E}^{n+1}$ by the identification map $\sum_i \xi^i(\partial/\partial x^i)_p \rightarrow (\xi^1, \ldots, \xi^{n+1})$ for each p. We shall denote by e_i $(i = 1, \ldots, n + 1)$ the $(n + 1)$-dimensional vector whose ith component is 1 and whose other components are 0.

Now we assume that M is an n-dimensional submanifold of $\mathbf{E}^{n+1}$ (we call M a *hypersurface* of $\mathbf{E}^{n+1}$), and shall show that there is a $1:1$ correspondence between an everywhere nonvanishing continuous (resp. differentiable) differential form of order n on M and an everywhere nonvanishing continuous (resp. differentiable) normal vector field on M.

LEMMA. *Let V be an $(n + 1)$-dimensional real vector space with an inner product. Let F be a nonzero alternating $(n + 1)$-linear function on V, and let a be a nonzero alternating n-linear function on an n-dimensional subspace W of V. Then there exists a unique vector w' satisfying $w' \perp W$ and $a(w_1, \ldots, w_n) = F(w', w_1, \ldots, w_n)$ for any $w_1, \ldots, w_n \in W$.*

Proof. For any $v \in V$, $i(v)F:(w_1, \ldots, w_n) \rightarrow F(v, w_1, \ldots, w_n)$ is an alternating n-linear function on W. Since the space of all alternating n-linear functions on W is 1-dimensional, and $a \neq 0$, there exists $\lambda_v \in \mathbf{R}$ such that $i(v)F = \lambda_v a$. Moreover, if $v \in W$, then $i(v)F = 0$, and since $F \neq 0$, there exists an element $v \in V$ such that $i(v)F \neq 0$. Let $v = w + v'$, $w \in W$ and $v' \perp W$. Then $i(v)F = i(v')F = \lambda_v a$ and $\lambda_v \neq 0$. Let $w' = v'/\lambda_v$. Then w' satisfies the conditions of the lemma. Moreover, if

w'' is orthogonal to W, and $i(w'')F = 0$, then we see that $w'' = 0$. Hence the uniqueness of w'.

Now let M be a hypersurface in $\mathbf{E}^{n+1}$, and let ω be an everywhere nonvanishing continuous (resp. differentiable) form of order n on M. Let F be the alternating $(n+1)$-linear function on the underlying vector space V of $\mathbf{E}^{n+1}$ satisfying $F(e_1, \ldots, e_{n+1}) = 1$. Since we have identified V with $T_p(\mathbf{E}^{n+1})$ for each p, we may consider F as an alternating $(n+1)$-linear function on $T_p(\mathbf{E}^{n+1})$. Now the value ω_p of ω at p is a nonzero alternating n-linear function on the n-dimensional subspace $T_p(M)$ of $T_p(\mathbf{E}^{n+1})$. Hence, by the lemma, there is a unique normal vector N_p, $N_p \neq 0$, at p, such that $\omega_p(v_1, \ldots, v_n) = F(N_p, v_1, \ldots, v_n)$ for any $v_1, \ldots, v_n \in T_p(M)$. Then $N: p \to N_p$ defines an everywhere nonvanishing normal vector field on M, and the continuity (resp. differentiability) of N follows from that of ω. Conversely, let N be an everywhere non-vanishing continuous (resp. differentiable) normal vector field on M. Then we can define an everywhere nonvanishing differential form ω of order n on M by the condition $\omega_p(v_1, \ldots, v_n) = F(N_p, v_1, \ldots, v_n)$ for all $v_1, \ldots, v_n \in T_p(M)$ at each $p \in M$. Then ω is continuous or differentiable according as N is continuous or differentiable. Thus we have proved the following:

THEOREM 3. *Let M be a hypersurface in $\mathbf{E}^{n+1}$. Then M is orientable if and only if M admits an everywhere nonvanishing continuous normal vector field. Moreover, if M is connected and orientable, then M admits an everywhere nonvanishing differentiable normal vector field.*

Proof. The second statement is proved by observing that a connected submanifold of $\mathbf{E}^{n+1}$ is paracompact (II, §15), and applying the Corollary to Theorem 1.

Let M be orientable, and let N be an everywhere nonvanishing differentiable normal vector field on M. Then the length $\|N_p\|$ of the vector $N_p \in T_p(\mathbf{E}^{n+1})$ is nonzero, and if we put $L_p = N_p/\|N_p\|$ at each $p \in M$, then $L: p \to L_p$ is also a differentiable normal vector field on M, and we have $\|L_p\| = 1$ at each p. If we identify $T_p(\mathbf{E}^{n+1})$ with $\mathbf{E}^{n+1}$, then $p \to L_p$ defines a map from M into the unit sphere S^n in $\mathbf{E}^{n+1}$, which we call the *Gauss map* from the hypersurface M into S^n.

As an example of a nonorientable surface, we shall describe the Möbius band. Take a rectangle $abdc$ as in Fig. 5.3, give it a 180° twist along the

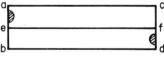

Fig. 5.3

center line ef and paste the edges ab and cd together, so that a matches d and b matches c. Then delete the original edges ac and bd so that the resulting object is a 2-dimensional manifold. We now have a (hyper) surface M in E^3 as shown in Fig. 5.4. This surface is called the Möbius band. There is no continuous vector field on M which is not 0 at each point of M. (Check this by the figure.) That is, the Möbius band is not orientable.

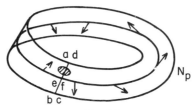

Fig. 5.4

Problem 5. Show that 2-dimensional projective space is not orientable. In general, n-dimensional projective space is orientable if n is odd, and is nonorientable if n is even (cf. §4).

Example 3. Any paracompact almost complex manifold M is orientable. In particular, any paracompact complex manifold is orientable.

Proof. Let J be the almost complex structure of M. On a sufficiently small neighborhood U of an arbitrary point p of M, we can choose complex differential forms ω_α of order 1 $(\alpha = 1, \ldots, n; \dim M = 2n)$ such that $\omega_1, \ldots, \omega_n$ are linearly independent at each point q of U and ${}^t J_q(\omega_\alpha)_q = \sqrt{-1}(\omega_\alpha)_q$ $(\alpha = 1, \ldots, n)$ holds. Set

$$\Omega = (\sqrt{-1})^n \omega_1 \wedge \bar\omega_1 \cdots \omega_n \wedge \bar\omega_n.$$

Since $\bar\Omega = \Omega$, Ω is a real differential form of order $2n$. Moreover, since $\{\omega_1, \bar\omega_1, \ldots, \omega_n, \bar\omega_n\}$ forms a basis of $T_q^{*C}(M)$ at each point q of U, we have $\Omega_q \neq 0$ for each q of U. Now, if θ_α $(\alpha = 1, \ldots, n)$ are also complex differential forms of order 1 which have the same properties as ω_α $(\alpha = 1, \ldots, n)$, but defined on a neighborhood V, then if $U \cap V \neq \varnothing$, on $U \cap V$ we have

$$\theta_\alpha = \sum_\beta a_\alpha^\beta \omega_\beta.$$

Hence, setting

$$\theta = (\sqrt{-1})^n \theta_1 \wedge \bar\theta_1 \cdots \theta_n \wedge \bar\theta_n,$$

we have, on $U \cap V$,

$$\theta = |\det(a_\alpha^\beta)|^2 \Omega.$$

Hence M is orientable by Theorem 2.

Example 4. If G is a Lie group, then G is orientable. In fact, if ω^1, ..., ω^n are n linearly independent left invariant differential forms of order 1, then set $\omega = \omega^1 \wedge \cdots \wedge \omega^n$. ω is a left invariant differential form of order n which does not vanish at any point of G. Hence G is orientable.

Example 5. Let G be a connected Lie group, H a closed subgroup of G, and let $\mathfrak{g}$ and $\mathfrak{h}$ be the Lie algebras of G and H, respectively. Let the adjoint representation of H on $\mathfrak{g}/\mathfrak{h}^*$ be denoted by $\mathrm{Ad}_{\mathfrak{g}/\mathfrak{h}}(h)$ $(h \in H)$. If

$$\det(\mathrm{Ad}_{\mathfrak{g}/\mathfrak{h}}(h)) = 1 \tag{2}$$

holds for all $h \in H$, then the homogeneous space G/H is orientable.†

Proof. Let the projection from G to G/H be π, and let $\pi(e) = O$ (e is the identity element of G). As we have done before (cf. IV §17), we identify $\mathfrak{g}/\mathfrak{h}$ with the tangent vector space of G/H at O, and we identify $\{\mathrm{Ad}_{\mathfrak{g}/\mathfrak{h}}(h)|h \in H\}$ with the linear isotropy group at O. For simplicity of notation, let us write $\mathrm{Ad}_{\mathfrak{g}/\mathfrak{h}}(h) = \alpha(h)$. For a basis $\{e_1, \ldots, e_n\}$ of $T_O(G/H)$, we set

$$\alpha(h)e_i = \sum_{j=1}^{n} \alpha_i^j(h)e_j.$$

Then, by (1), we have

$$\det(\alpha_i^j(h)) = 1$$

for all $h \in H$. Let ω_0 be a nonzero alternating n-linear function on $T_O(G/H)$. We define a differential form ω of order n on G/H as follows. For each p of G/H, there is an element g of G such that $g(O) = p$. In general, for $g' \in G$, we denote the transformation $x \to g' \cdot x$ of G/H by $t_{g'}$. We define ω_p by

$$\omega_p(u_1, \ldots, u_n) = \omega_0(t_{g^{-1}*}u_1, \ldots, t_{g^{-1}*}u_n) \qquad (u_i \in T_p(G/H)).$$

The choice of an element g such that $g \cdot O = p$ is not unique. To show that the above definition of ω_p does not depend on the choice of g, we suppose $g_1 \cdot O = g_2 \cdot O = p$. Since $g_2^{-1}g_1 = h \in H$, we have

---

*Cf. IV, §17.

†Conversely, we shall show later that if G is connected and H is compact, and if G/H is orientable, then (1) holds for all $h \in H$.

$(t_{g_2^{-1}*})_p \cdot (t_{g_1^*})_0 = \alpha(h)$, and since $(t_{g_1^*})_0 = (t_{g_1^{-1}*})_p^{-1}$, we get $(t_{g_2^{-1}*})_p = \alpha(h)(r_{g_1^{-1}*})_p$. Hence

$$\omega_0(t_{g_2^{-1}*}u_1, \ldots, t_{g_2^{-1}*}u_n) = \omega(\alpha(h)(t_{g_1^{-1}*}u_1), \ldots, \alpha(h)(t_{g_1^{-1}*}u_n))$$

holds. But we have

$$\omega_0(\alpha(h)e_1, \ldots, \alpha(h)e_n) = \omega_0\left(\sum_{j_1=1}^n \alpha_1^{j_1}e_{j_1}, \ldots, \sum_{j_n=1}^n \alpha_n^{j_n}e_{j_n}\right)$$

$$= \sum_{j_1, \ldots, j_n=1}^n \alpha_1^{j_1} \cdots \alpha_n^{j_n}\omega_0(e_{j_1}, \ldots, e_{j_n})$$

$$= \sum_\sigma \varepsilon(\sigma)\alpha_1^{\sigma(1)} \cdots \alpha_n^{\sigma(n)}\omega_0(e_1, \ldots, e_n)$$

$$= \det(\alpha_i^j)\omega_0(e_1, \ldots, e_n)$$

$$= \omega_0(e_1, \ldots, e_n).$$

From this, we see that for n arbitrary vectors $v_1, \ldots, v_n$ of $T_0(G/H)$, we have

$$\omega_0(\alpha(h)v_1, \ldots, \alpha(h)v_n) = \omega_0(v_1, \ldots, v_n).$$

Hence, in particular, we have

$$\omega_0(t_{g_2^{-1}*}u_1, \ldots, t_{g_2^{-1}*}u_n) = \omega_0(t_{g_1^{-1}*}u_1, \ldots, t_{g_1^{-1}*}u_n),$$

and we see that the definition of ω_p does not depend on the choice of an element g of G such that $g \cdot O = p$. Clearly, at each point p, we have $\omega_p \neq 0$. Hence we have defined a differential form of order n which does not vanish at any point of G/H. By the definition of ω, we have $t_g^*\omega = \omega$ for an arbitrary $g \in G$, that is, ω is G-invariant. In general, using the differentiable structure of G/H, we can show that a G-invariant differential form on the quotient space G/H is always of class C^∞. Hence, in particular, ω is of class C^∞, and G/H is orientable.

Problem 6. Prove that any G-invariant differential form on the quotient space G/H is of class C^∞.

Problem 7. Using Example 3, show that a sphere is orientable.

§2. Integration of Differential Forms

Let M be an oriented paracompact manifold of dimension n. We shall now define the integral

$$\int_M \theta \tag{1}$$

of a continuous differential form θ of order n on M with compact support.*

Let $(x^1, \ldots, x^n)$ be a local coordinate system on an open set U of M, $a \in U$, $\varepsilon > 0$ and let

$$Q_\varepsilon(a) = \{p \in U \mid |x^i(a) - x^i(p)| < \varepsilon, \, i = 1, \ldots, n\} \qquad (2)$$

be a cubic coordinate neighborhood centered at a and of half-width ε. Throughout this section we want to consider only positive local coordinate systems of M, so that *any local coordinate system mentioned in this section is assumed to be positive*. Let θ be a continuous differential form of order n on M with compact support.

(1) First we shall consider the case where the support of θ is contained in a cubic coordinate neighborhood. Let us assume that this cube is defined by (2). Let

$$\theta = f \, dx^1 \wedge \cdots \wedge dx^n,$$

and define the integral (1) by

$$\int_M \theta = \int_{a^1-\varepsilon}^{a^1+\varepsilon} \cdots \int_{a^n-\varepsilon}^{a^n+\varepsilon} f(u^1, \ldots, u^n) \, du^1 \cdots du^n \qquad (a^i = x^i(a)) \quad (3)$$

We remind the reader that $(x^1, \ldots, x^n)$ is a positive local coordinate system by our assumption. The integral of the right member is an ordinary n-fold integral. We shall show that this integral does not depend on the choice of the cubic coordinate neighborhood which contains the support of θ. For this, we let

$$Q_\eta(b) = \{q \in V \mid |y^i(b) - y^i(q)| < \eta, \, i = 1, \ldots, n\}$$

be another cubic coordinate neighborhood of M containing the support of θ. On $Q_\eta(b)$ we set

$$\theta = g \, dy^1 \wedge \cdots \wedge dy^n$$

and

$$I(\theta) = \int_{b^1-\eta}^{b^1+\eta} \cdots \int_{b^n-\eta}^{b^n+\eta} g(v^1, \ldots, v^n) \, dv^1 \cdots dv^n, \qquad (b^i = y^i(b)), \quad (4)$$

and we shall show that the right members of (3) and (4) have the same value. Let $p \to (x^1(p), \ldots, x^n(p))$ (resp. $q \to (y^1(q), \ldots, y^n(q))$) be the diffeomorphism φ (resp. ψ) from $Q_\varepsilon(a)$ (resp. $Q_\eta(b)$) onto a cube in $\mathbf{R}^n$.

*The closure of the subset $\{p \in M \mid \theta_p \neq 0\}$ of M is called the support of θ.

Since the support of θ is contained in $Q_\varepsilon(a) \cap Q_\eta(b)$, f (resp. g) vanishes outside of $\varphi(Q_\varepsilon(a) \cap Q_\eta(b))$ (resp. $\psi(Q_\varepsilon(a) \cap Q_\eta(b))$). Hence if we set $W = Q_\varepsilon(a) \cap Q_\eta(b)$, then the right member of (3) (resp. (4)) is equal to

$$\int_{\varphi(W)} f \, du^1 \cdots du^n \text{ (resp. } \int_{\psi(W)} g \, dv^1 \cdots dv^n). \tag{5}$$

If we denote the diffeomorphism $\varphi \circ \psi^{-1}$ from $\psi(W)$ to $\varphi(W)$ by $(v^1, \ldots, v^n) \to (h^1(v), \ldots, h^n(v))$, then by change of variables in multiple integration we have

$$\int_{\varphi(W)} f(u) \, du^1 \cdots du^n = \int_{\psi(W)} f(h(v)) \left| \det\left(\frac{\partial h^i}{\partial v^j}\right) \right| dv^1 \cdots dv^n. \tag{6}$$

But on W we have

$$\theta = f \, dx^1 \wedge \cdots \wedge dx^n = g \, dy \wedge \cdots \wedge dy^n,$$

from which it follows that

$$f \cdot \det\left(\frac{\partial x^i}{\partial y^j}\right) = g.$$

From the definition of φ, ψ, and h, we have $(\partial x^i / \partial y^j)_p = (\partial h^i / \partial v^j)_{v = \psi(p)}$. Hence we have

$$f(h(v)) \cdot \det\left(\frac{\partial h^i}{\partial v^j}\right) = g(v). \tag{7}$$

Since $(x^1, \ldots, x^n)$ and $(y^1, \ldots, y^n)$ are *both positive*, we get $\det (\partial x^i / \partial y^j) > 0$. Hence $\det (\partial h^i / \partial v^j) > 0$, and so by (6) and (7) the two integrals in (5) have the same value. Hence the right members in (3) and (4) are equal, and we have proved that definition (3) of the integral of θ on M does not depend on the choice of the cubic neighborhood containing the support of θ. If the supports of θ and η are both contained in the same cubic coordinate neighborhood, then by definition (3), it is clear that $\int_M(\theta + \eta) = \int_M \theta + \int_M \eta$ holds.

(2) We shall now consider the general case of a continuous differential form θ of order n with compact support. For each point x of M, we can take a cubic coordinate neighborhood $\tilde{Q}_x$ and get an open covering $\{\tilde{Q}_x\}_{x \in M}$ of M. Since M is paracompact, we can pick a locally finite refinement $\{Q_\alpha\}_{\alpha \in A}$ of $\{\tilde{Q}_x\}_{x \in M}$. Since each Q_α is contained in some $\tilde{Q}_x$, each Q_α is compact. Hence there is a partition of unity $\{f_\alpha\}_{\alpha \in A}$ subordinate to $\{Q_\alpha\}_{\alpha \in A}$. Since the support of θ is compact and $\{Q_\alpha\}_{\alpha \in A}$ is locally

finite, the number of Q_α's intersecting the support of θ is finite. Hence $f_\alpha\theta$ is zero except for a finite number of α's. If $f_\alpha\theta \neq 0$, then the support $f_\alpha\theta$ is contained in Q_α. But since $\sum f_\alpha = 1$, we can write $\theta = \sum_{\alpha \in A} f_\alpha\theta$. Thus θ is a sum of a finite number of continuous differential forms of order n, each of whose support is contained in a Q_α, and hence in a cubic coordinate neighborhood Q_x. Suppose θ is expressed in two different ways: $\theta = \sum_{i=1}^{m} \theta_i = \sum_{j=1}^{l} \omega_j$. Here we assume that θ_i, ω_j are continous differential forms of order n whose supports are contained in cubic coordinate neighborhoods. The integrals of θ_i and ω_j on M were defined in (1). If we can show that

$$\sum_{i=1}^{m} \int_M \theta_i = \sum_{j=1}^{l} \int_M \omega_j, \tag{8}$$

then by setting

$$\int_M \theta = \sum_{i=1}^{m} \int_M \theta_i,$$

we can define the integral of θ on M. If K is the union of the supports of θ_i, ω_j ($i = 1, \ldots, m; j = 1, \ldots, l$), then K is compact. Hence there is a C^∞ function f on M which is identically equal to 1 on K (cf. Problem 2 of II, §15). As in the case of θ, f can be expressed as $f = f_1 + \cdots + f_k$, where each f_i is a C^∞ function whose support is contained in a cubic coordinate neighborhood. Since f is identically 1 on K, and since the supports of θ, θ_i, and ω_j are contained in K, we have $f\theta = \theta$, $f\theta_i = \theta_i$, and $f\omega_j = \omega_j$. Hence we have

$$\theta = f_1\theta + \cdots + f_k\theta,$$
$$f_k\theta = f_k\theta_1 + \cdots + f_k\theta_m = f_k\omega_1 + \cdots + f_k\omega_l.$$

Since the supports of $f_k\theta$, $f_k\theta_i$, and $f_k\omega_j$ are all contained in the cubic coordinate neighborhood containing the support of f_k, from the additivity of integrals stated at the end of (1) we get

$$\int_M f_k\theta = \sum_i \int_M f_k\theta_i = \sum_j \int_M f_k\omega_j. \tag{9}$$

On the other hand, since $\theta_i = f\theta_i = \sum_k f_k\theta_i$, and since the support of each $f_k\theta_i$ is contained in the cubic coordinate neighborhood containing the support of θ_i, we have

$$\int_M \theta_i = \sum_k \int_M f_k\theta_i. \tag{10}$$

From (9) and (10) we have

$$\sum_i \int_M \theta_i = \sum_k \int_M f_k \theta.$$

Similarly, we have

$$\sum_j \int_M \omega_j = \sum_k \int_M f_k \theta.$$

Hence (8) is proved, and the integral of θ is well defined.

It is clear from the definition of the integral that the integral has the following properties:

(1) If θ and η are continuous differential forms of order n with compact supports, then

$$\int_M (\lambda \theta + \mu \eta) = \lambda \int_M \theta + \mu \int_M \eta \qquad (\lambda, \mu \in \mathbf{R}).$$

(2) If $-M$ is the oriented manifold with the negative orientation of M,* then

$$\int_{-M} \theta = - \int_M \theta.$$

(3) Let $\{M_\alpha\}_{\alpha \in A}$ be the set of connected components of M, and let θ_α be the restriction of θ to M_α. Then

$$\int_M \theta = \sum_\alpha \int_{M_\alpha} \theta_\alpha.$$

Problem 1.   Prove the properties (1), (2), and (3) above.

Let ω be a volume element of M (i.e., ω is a positive differential form of order n), and let f be a continuous function on M with compact support. The integral of the differential form $f\omega$ of order n on M is called the integral of f with respect to the volume element ω. Often we use the notation dv or dx† instead of ω, and write

$$\int_M f \, dv$$

---

*M and $-M$ have the same differentiable structure, but the positive orientation of M is chosen to be the negative orientation of M.

†dv, dx denote the volume element ω and are not the differentials of v, x here.

for the integral $\int_M f\omega$.

In particular, if M is compact, we can define

$$\int_M dv$$

and the value of this integral is called the *volume* of M (with respect to the volume element $dv = \omega$). It is denoted by $V_\omega(M)$.

Now let g be a Riemannian metric on M. Let g_{ij} be the components of g with respect to the local coordinate system $(x^1, \ldots, x^n)$ and set

$$G = \det(g_{ij}).$$

Since the matrix (g_{ij}) is positive and symmetric, we have $G > 0$. The value depends on the choice of the local coordinate system $(x^1, \ldots, x^n)$. Let $(y^1, \ldots, y^n)$ be another local coordinate system, g'_{ij} the components of g with respect to this local coordinate system, and $G' = \det(g'_{ij})$. Then we have

$$g'_{ij} = \sum_{k,l} \frac{\partial x^k}{\partial y^i} \frac{\partial x^l}{\partial y^j} g_{kl},$$

so that we get $G' = [\det(\partial x^k/\partial y^i)]^2 G$. Since (x^i) and (y^i) are positive local coordinate systems, we have $\det(\partial x^k/\partial y^i) > 0$. Hence

$$(G')^{1/2} = \det\left(\frac{\partial x^k}{\partial y^i}\right) G^{1/2},$$

so that we can define a volume element of M by

$$\omega = G^{1/2} \, dx^1 \wedge \cdots \wedge dx^n.$$

We call this volume element the volume element of the oriented Riemannian manifold M.

Problem 2. Take a positive basis $u_1, \ldots, u_n$ of $T_p(M)$ such that $g_p(u_i, u_j) = \delta_{ij}$, and define a differential form ω of order n on the Riemannian manifold M by $\omega_p(u_1, \ldots, u_n) = 1$. Prove that this ω is the volume element of M defined above.

Now let M and M' be n-dimensional oriented manifolds, Ω' a volume element on M', and φ a diffeomorphism from M onto M'. The differential form $\varphi^*\Omega'$ of order n does not vanish at any point of M. If $\varphi^*\Omega'$ is a positive differential form of order n, then we call φ an *orientation preserving transformation*, while if $\varphi^*\Omega'$ is negative, then we call φ an *orientation reversing transformation*. If M and M' are connected, then $\varphi^*\Omega'$ is either positive or negative, and so a diffeomorphism from an oriented connected manifold M to an oriented connected manifold M' is either orientation preserving or orientation reversing. Let $(x^1, \ldots, x^n)$ be a positive local coordinate system on an open set U of M'. Then $(x^1 \circ \varphi, \ldots, x^n \circ \varphi)$ is a local coordinate system on the open set $\varphi^{-1}(U)$ of M. If φ is orientation preserving, then $(x^1 \circ \varphi, \ldots, x^n \circ \varphi)$ is positive, while if φ is orientation reversing, then $(x^1 \circ \varphi, \ldots, x^n \circ \varphi)$ is negative.

Let us prove the following lemma.

LEMMA 1.   *Let ω be a differential form of order n on M' with compact support, and let φ be a diffeomorphism from M onto M'. If φ is an orientation preserving (resp. reversing) transformation, then*

$$\int_M \varphi^*\omega = \int_{M'} \omega \quad (\text{resp.} \int_M \varphi^*\omega = -\int_{M'} \omega). \tag{11}$$

Proof.   It suffices to prove (11) for the case where the support of ω is contained in a cubic coordinate neighborhood

$$U = \{p|\ |x^i(p)| < 1\}$$

with respect to a positive local coordinate system $(x^1, \ldots, x^n)$. In this case, the support of $\varphi^*\omega$ is contained in $\varphi^{-1}(U)$. $(x^1 \circ \varphi, \ldots, x^n \circ \varphi)$ is a local coordinate system on $\varphi^{-1}(U)$, and if we set $y^i = x^i \circ \varphi$, then we have

$$\varphi^{-1}(U) = \{q|\ |\, y^i(q)| < 1\}.$$

If $\omega_p = h \cdot (dx^1)_p \wedge \cdots \wedge (dx^n)_p$ on U, then, setting $q = \varphi^{-1}(p)$, we have $(\varphi^*\omega)_q = h(\varphi(q))(dy^1)_q \wedge \cdots \wedge (dy^n)_q$. Since $y^i(q) = x^i(p)$, depending on whether $(y^1, \ldots, y^n)$ is positive or negative, we have

$$\int_M \omega = \pm \int_{-1}^{+1} \cdots \int_{-1}^{+1} h(u, \ldots, u)\, du^1 \cdots du^n = \pm \int_M \omega.$$

Hence (11) is proved.

Let f be a continuous function on M' with compact support, and let Ω' be a volume element of M'. Set $\omega = f\Omega'$. Let φ be a diffeomorphism from M onto M'. Then we have $\varphi^*\omega = (f \circ \varphi)\varphi^*\Omega'$. If φ is an orientation preserving (resp. reversing) transformation, then we have

$$\int_{M'} f \cdot \Omega' = \int_M (f \cdot \varphi)\varphi^*\Omega' \quad (\text{resp.} \int_{M'} f \cdot \Omega' = -\int_M (f \cdot \varphi)\varphi^*\Omega').$$

Let Ω and Ω' be volume elements of M and M', respectively, and suppose φ satisfies the condition

$$\varphi^*\Omega' = c \cdot \Omega \quad (c \text{ is a constant}). \tag{12}$$

Here $c \neq 0$, and if φ is orientation preserving (resp. reversing), then $c > 0$ (resp. $c < 0$). Hence, by the formulas above, we obtain

$$\int_{M'} f\Omega' = |c| \int_M (f \circ \varphi)\Omega. \tag{13}$$

In particular, if $|c| = 1$ in (12), i.e., if we have

$$\varphi^*\Omega' = \pm\Omega,$$

then φ is called a *volume preserving transformation* from M onto M'. If $M = M'$ and $\Omega = \Omega'$, then φ is called a volume preserving transformation of M.

If φ is a volume preserving transformation, then, from (13), we have

$$\int_{M'} f\Omega' = \int_M (f \circ \varphi)\Omega. \tag{14}$$

In particular, suppose M is compact and φ is a diffeomorphism of M onto itself, and suppose (12) holds with $\Omega = \Omega'$. Then, in (13), we have $M = M'$, $\Omega = \Omega'$, and $f \equiv 1$ (and hence $f \circ \varphi \equiv 1$), so that we get $|c| = 1$. Hence we have the following lemma.

LEMMA 2. *Let φ be a diffeomorphism of a compact oriented manifold M onto itself. For a volume element Ω of M, if $\varphi^*\Omega = c\Omega$ ($c \in \mathbf{R}$) holds, then $c = \pm 1$, and φ is a volume preserving transformation of M. Hence (14) holds with $M = M'$, $\Omega = \Omega'$.*

Problem 3. Prove that any motion of an oriented connected Riemannian manifold is volume preserving.

PROPOSITION. *Let G be a connected Lie group, M an oriented connected manifold, and suppose that G is a Lie transformation group on M. Then every transformation of G is an orientation preserving transformation.*

Proof. Let t_g denote the transformation of M determined by $g \in G$. If Ω is a volume element of M, then we have

$$t_g^*\Omega = f_g\Omega,$$

where f_g is a function on M. Since M is connected and $t_g^*\Omega$ does not vanish at any point of M, we have $f_g > 0$ or $f_g < 0$. If $f_g > 0$, then t_g preserves orientation, while if $f_g < 0$, then t_g reverses orientation. If we set $G^+ = \{g \in G \mid f_g > 0\}$ and $G^- = \{g \in G \mid f_g < 0\}$, then we see that G^+ and G^- are both open sets of G. Furthermore $G^+ \cap G^- = \varnothing$, $G^+ \cup G^- = G$. Since G is connected, either G^+ or G^- is empty. However G^+ contains the identity element of G, so is nonempty, and we have $G = G^+$. Hence t_g preserves orientation for all $g \in G$.

Problem 4. Prove that all elements of a one-parameter group of transformations $\{\text{Exp } tX | t \in \mathbf{R}\}$ of M are volume preserving transformations if and only if $L_X\Omega = 0$.

§3. Invariant Integration on Lie Groups

Let G be an n-dimensional Lie group and Ω a left invariant differential form of order n on G. The form Ω does not vanish at any point of G, and so we can choose the orientation of G so that Ω is positive. We denote, as usual, by L_g and R_g ($g \in G$), the left translation $x \to gx$ and the right translation $x \to xg$ of G, respectively. Since we have

$$L_g^*\Omega = \Omega,$$

L_g is a volume preserving transformation. Hence, if f is a continuous function on G with compact support, then

$$\int_G (f \circ L_g)\Omega = \int_G f\Omega. \tag{1}$$

It is customary to write dx ($x \in G$) instead of Ω, and write (1) as

$$\int_G f(gx)\, dx = \int_G f(x)\, dx. \tag{1'}$$

Now, as before, we let $A_g = L_g \circ R_{g^{-1}}$, i.e., A_g is the transformation of G defined by $A_g(x) = gxg^{-1}$. Let us show that

$$A_g^*\Omega = \det \text{Ad}(g) \cdot \Omega, \quad R_{g^{-1}}^*\Omega = A_g^*\Omega \tag{2}$$

hold, where $g \to \text{Ad}(g)$ is the adjoint representation of G. The second relation is immediate from $A_g^*\Omega = R_{g^{-1}}^*(L_g^*\Omega) = R_{g^{-1}}^*\Omega$. Since $L_s \circ R_{g^{-1}} = R_{g^{-1}} \circ L_s$ holds for an arbitrary $s \in G$, we have $L_s^*(A_g^*\Omega) = L_s^*(R_{g^{-1}}^*\Omega) = R_{g^{-1}}^*(L_s^*\Omega) = R_{g^{-1}}^*\Omega = A_g^*\Omega$, and hence $A_g^*\Omega$ is also left invariant. The differential forms $A_g^*\Omega$ and $\det \text{Ad}(g) \cdot \Omega$ are both left invariant, and so to prove that they are equal, it suffices to show that

$$(A_g^*\Omega)_e(X_1, \ldots, X_n) = \det \text{Ad}(g) \cdot \Omega_e(X_1, \ldots, X_n)$$

for a basis $\{X_1, \ldots, X_n\}$ of the tangent space $T_e(G)$. However, the left member is equal to $\Omega_e(A_{g*}X_1, \ldots, A_{g*}X_n)$, and since $A_{g*}X_i = \text{Ad}(g)X_i$ by the definition of $\text{Ad}(g)$, we see immediately that the two members of the equality above are actually equal. Hence (2) is proved.

From (2) we have

$$R_g^*\Omega = \det \mathrm{Ad}(g^{-1}) \cdot \Omega, \tag{2'}$$

and hence by (13) of §2, we have

$$\int_M f(xg)\, dx = |\det \mathrm{Ad}(g)| \int_M f(x)\, dx. \tag{3}$$

If $\det \mathrm{Ad}(g) = \pm 1$ holds for any element g of G, then G is called a *unimodular Lie group*. By (2'), this condition is equivalent to the condition that R_g be a volume preserving transformation for all $g \in G$.

If G is unimodular, then, by (3), for any continuous function f on G with compact support,

$$\int_M f(xg)\, dx = \int_M f(x)\, dx \tag{4}$$

holds for an arbitrary g in G.

If G is compact, then, by Lemma 2 of §2 and (2'), G is unimodular.

Problem 1. If G is unimodular and connected, then show that a left invariant volume element is two-sided invariant.

Let ψ denote the map $x \to x^{-1}$ from G onto itself. Suppose that G is unimodular and connected. We shall show that

$$\psi^*\Omega = (-1)^n\Omega \qquad (n = \dim G). \tag{5}$$

Since we have $R_{g^{-1}} \circ \psi = \psi \circ L_g$ for each $g \in G$, we have $L_g^*(\psi^*\Omega) = \psi^*(R_{g^{-1}}^*\Omega)$. Since G is unimodular and connected, Ω is also right invariant (cf. Problem 1), and hence we get $L_g^*(\psi^*\Omega) = \psi^*\Omega$ for any $g \in G$. This shows that $\psi^*\Omega$ is also left invariant. Therefore there exists a real number λ such that $\psi^*\Omega = \lambda\Omega$. Now, for $v_1, \ldots, v_n \in T_e(G)$, we have $(\psi^*\Omega)_e(v_1, \ldots, v_n) = \Omega_e(\psi_{e*}v_1, \ldots, \psi_{e*}v_n) = (-1)^n\Omega_e(v_1, \ldots, v_n)$, because $\psi_{e*}v_i = -v_i$ $(i = 1, \ldots, n)$. Hence we obtain $\lambda = (-1)^n$, and this proves (5). The equality (5) shows that ψ is volume preserving. Using (14) of §2, we conclude that if G is unimodular and connected, then

$$\int_G (f \circ \psi)\Omega = \int_G f\Omega,$$

that is,

$$\int_G f(x^{-1})\, dx = \int_G f(x)\, dx \tag{6}$$

holds for any continuous function f on G with compact support.

Now assume that G is unimodular but not connected. Let $\{G_\alpha\}_{\alpha \in A}$ be the set of connected components of G, and let G_0 denote the connected component of G containing the identity. For a function h on G, we denote its restriction to G_α by h_α. Let the restriction of Ω to G_α be denoted by Ω_α. Then, for any continuous function f on G with compact support, we have

$$\int_G f\Omega = \sum_\alpha \int_{G_\alpha} f_\alpha \Omega_\alpha, \tag{7}$$

$$\int_G (f \circ \psi)\Omega = \sum_\alpha \int_{G_\alpha} (f \circ \psi)_\alpha \Omega_\alpha. \tag{8}$$

The map ψ sends a connected component G_α onto another connected component $G_{\alpha'}$, and $\alpha \to \alpha'$ is a bijection of A onto itself. If we denote the restriction of ψ to G_α by ψ_α, then ψ_α is a diffeomorphism of G_α onto $G_{\alpha'}$, and we have

$$(f \circ \psi)_\alpha = f_{\alpha'} \circ \psi_\alpha.$$

If we can show that

$$\psi_\alpha^* \Omega_{\alpha'} = \pm\Omega_\alpha, \tag{9}$$

i.e., that ψ is volume preserving, then, by (14) of §2, we will have

$$\int_{G_\alpha} (f_{\alpha'} \circ \psi_\alpha)\Omega_\alpha = \int_{G_{\alpha'}} f_{\alpha'}\Omega_{\alpha'}.$$

But the left member is equal to $\int_{G_\alpha} (f \circ \psi)_\alpha \Omega_\alpha$ by the equality before (9). This shows that the right members of (8) and (7) are actually equal, and hence also the left members. Thus we conclude that

$$\int_G f(x^{-1})\, dx = \int_G f(x)\, dx$$

for any unimodular Lie group G, not necessarily connected, and any continuous function f on G with compact support.

To prove (9), let g_α be an element of G_α. Then $G_\alpha = g_\alpha G_0 = G_0 g_\alpha$ and $G_{\alpha'} = \psi(G_\alpha) = G_\alpha^{-1} = g_\alpha^{-1}G_0 = G_0 g_\alpha^{-1}$. The translation L_{g_α} induces a diffeomorphism of $G_{\alpha'}$ onto G_0, and since Ω is left invariant, we have

$$\Omega_{\alpha'} = L_{g_\alpha}^* \Omega_0, \tag{10}$$

and analogously

$$\Omega_\alpha = L_{g_\alpha^{-1}}^* \Omega_0. \tag{11}$$

Now let $y \in G_\alpha$. Then we can write $y = g_\alpha \cdot x$, where $x \in G_0$. Since $\psi_\alpha(y) = y^{-1} = x^{-1} \cdot g_\alpha^{-1} = \psi_0(x)g_\alpha^{-1}$, we get

$$\psi_\alpha = R_{g_\alpha^{-1}} \circ \psi_0 \circ L_{g_\alpha^{-1}},$$

and hence, by (10), we have

$$\psi_\alpha^* \Omega_{\alpha'} = L_{g_\alpha^{-1}}^* (\psi_0^*(A_{g_\alpha}^* \Omega_0)), \tag{12}$$

because $A_{g_\alpha} = L_{g_\alpha} \circ R_{g_\alpha^{-1}}$. By (2), we have $A_{g_\alpha}^* \Omega_0 = \det(\mathrm{Ad}(g_\alpha))\Omega_0$, and hence

$$\psi_0^*(A_{g_\alpha}^* \Omega_0) = \det(\mathrm{Ad}(g_\alpha))\psi_0^* \Omega_0 = (-1)^n \det(\mathrm{Ad}(g_\alpha))\Omega_0$$

by (5), because G_0 is unimodular and connected. Thus from (12) and (11), we get

$$\psi_\alpha^* \Omega_{\alpha'} = (-1)^n \det(\mathrm{Ad}(g_\alpha))\Omega_\alpha,$$

and this proves (9), because $\det \mathrm{Ad}(g_\alpha) = \pm 1$.

§4. Applications of Invariant Integration

Using the invariant integral on a compact Lie group, we can average an arbitrary quantity over the group to produce a quantity that is invariant under the group action. Let us show this by a few examples.

First we prove the following theorem.

THEOREM 1. *Let G be a connected Lie group, H a compact subgroup of G, and set $M = G/H$. Let $\mathfrak{g}$ and $\mathfrak{h}$ be the Lie algebras of G and H, respectively, and let $\mathrm{Ad}_{\mathfrak{g}/\mathfrak{h}}(h)$ $(h \in H)$ be the adjoint representation of H on $\mathfrak{g}/\mathfrak{h}$. Then M is orientable if and only if*

$$\det \mathrm{Ad}_{\mathfrak{g}/\mathfrak{h}}(h) = 1 \tag{1}$$

holds for all $h \in H$. In particular, if H is connected and compact, then G/H is orientable.

Proof. We have shown in Example 5 of §1 that condition (1) is sufficient without the compactness assumption on H. Hence we shall show here that if H is compact and G/H is orientable, then (1) holds. Let Ω be a volume element of M. For $h \in H$, $x \in M$, $u_1, \ldots, u_n \in T_x(M)$, we set

$$F(h; x, u_1, \ldots, u_n) = \Omega_{t_h(x)}((t_h)_* u_1, \ldots, (t_h)_* u_n).$$

If x and $u_1, \ldots, u_n$ are fixed, then F is a continuous function of h, and we have

$$F(hk;x, u_1, \ldots, u_n) = F(h;t_k(x), (t_k)_* u_n, \ldots, (t_k)_* u_n). \qquad (2)$$

Fix x and $u_1, \ldots, u_n \in T_x(M)$ and set

$$\theta_x(u_1, \ldots, u_n) = \int_H F(h;x, u_1, \ldots, u_n) \, dh. \qquad (3)$$

The function θ_x is an alternating n-linear function on $T_x(M)$, and it defines the differential form $\theta: x \to \theta_x$ which is clearly of class C^∞. Since H is compact, H is unimodular, and so we have

$$\int_H F(hk;x, u_1, \ldots, u_n) \, dh = \int_H F(h;x, u_1, \ldots, u_n) \, dh.$$

Hence, from (2) and (3), we have

$$(t_h^* \theta)_x(u_1, \ldots, u_n) = \theta_x(u_1, \ldots, u_n). \qquad (4)$$

and thus, for an arbitrary $h \in H$, we have

$$t_h^* \theta = \theta.$$

Now let us show that θ is a positive differential form of order n. By the proposition in §2, t_g is an orientation preserving transformation for each $g \in G$. Hence, if $(u_1, \ldots, u_n)$ is a positive basis of $T_x(M)$, then $((t_h)_* u_1, \ldots, (t_h)_* u_n)$ is a positive basis of $T_{t_h(x)}(M)$. Then, by the definition of F, for an arbitrary $h \in H$, we have $F(h;x, u_1, \ldots, u_n) > 0$. Hence, by (2), we have $\theta_x(u_1, \ldots, u_n) > 0$, which shows that θ is positive.

Now let o be the point of G/H corresponding to the identity element of G. If we identify $T_o(M)$ and $\mathfrak{g}/\mathfrak{h}$, then we have $(t_h)_{*o} = \mathrm{Ad}_{\mathfrak{g}/\mathfrak{h}}(h)$, so that by (4) we have

$$\theta_o(u_1, \ldots, u_n) = \theta_o(\mathrm{Ad}_{\mathfrak{g}/\mathfrak{h}}(h)u_1, \ldots, \mathrm{Ad}_{\mathfrak{g}/\mathfrak{h}}(h)u_n)$$
$$= \det \mathrm{Ad}_{\mathfrak{g}/\mathfrak{h}}(h) \cdot \theta_o(u_1, \ldots, u_n).$$

Hence $\det \mathrm{Ad}_{\mathfrak{g}/\mathfrak{h}}(h) = 1$ for all $h \in H$, that is, (1) holds.

If H connected and compact, then the image of the map $H \to \mathbf{R}$ given by $h \to \det \mathrm{Ad}_{\mathfrak{g}/\mathfrak{h}}(h)$ is a compact connected subgroup of $\mathbf{R}$, but there is only one such subgroup of $\mathbf{R}$, namely, the subgroup consisting of the single element 1. Hence, if H is connected and compact, then we have $\det \mathrm{Ad}_{\mathfrak{g}/\mathfrak{h}}(h) = 1$ for all $h \in H$, and hence G/H is orientable.

As an application of Theorem 1, we shall prove the following theorem.

THEOREM 2. *The Grassmann manifold $G_{N,n}$, formed by all n-dimensional vector subspaces of an N-dimensional real vector space, is orientable if $N* *is even, and is nonorientable if N is odd. In particular, since the m-dimensional projective space is equal to $G_{m+1,1}$, it is orientable if m is odd, but nonorientable if m is even.*

Proof. Let $N = n + m$, $G = SO(n + m)$ and

$$H = \left\{ \begin{pmatrix} A & 0 \\ 0 & B \end{pmatrix} \middle| A \in O(n), B \in O(m), \det A \cdot \det B = 1 \right\}.$$

H is a compact subgroup of G, and we have

$$G_{N,n} = G/H.$$

If $\mathfrak{g}$ and $\mathfrak{h}$ are the Lie algebras of G and H, respectively, then

$$\mathfrak{g} = \mathfrak{o}(n + m) = \{X \in \mathfrak{gl}(n + m, R); \, {}^tX + X = 0\}$$

$$\mathfrak{h} = \left\{ \begin{pmatrix} C & 0 \\ 0 & D \end{pmatrix} \middle| C \in \mathfrak{o}(n), D \in \mathfrak{o}(m) \right\}.$$

If we set

$$\mathfrak{m} = \left\{ \begin{pmatrix} 0 & F \\ -{}^tF & 0 \end{pmatrix} \middle| F \text{ is an } n \times m \text{ matrix} \right\},$$

then we have

$$\mathfrak{g} = \mathfrak{m} + \mathfrak{h} \quad \text{(direct sum)}.$$

If $h \in H$ and $X \in \mathfrak{m}$, then $\mathrm{Ad}(h)X = hXh^{-1}$, and we have $hXh^{-1} \in \mathfrak{m}$. In fact, if

$$h = \begin{pmatrix} A & 0 \\ 0 & B \end{pmatrix} \quad \text{and} \quad X = \begin{pmatrix} 0 & F \\ -{}^tF & 0 \end{pmatrix},$$

then we have

$$hXh^{-1} = \begin{pmatrix} 0 & AFB^{-1} \\ -B^tFA^{-1} & 0 \end{pmatrix}, \qquad {}^t(AFB^{-1}) = B^tFA^{-1}.$$

Hence, if $\rho(h)$ is the linear transformation of $\mathfrak{m}$ given by $X \to hXh^{-1}$, then, by taking the elements of $\mathfrak{m}$ as representatives of the residue classes of $\mathfrak{g}/\mathfrak{h}$, we can identify the representation of H on $\mathfrak{g}/\mathfrak{h}$ with the representation of H on $\mathfrak{m}$ given by $h \to \rho(h)$.

On the other hand, by the correspondence $X \leftrightarrow F$, $\mathfrak{m}$ is identified with the $(n \cdot m)$-dimensional vector space $M_{n,m}(\mathbf{R})$ formed by the $n \times m$ matrices. By this correspondence, we have $\rho(h)X \leftrightarrow AFB^{-1}$. Hence, if if we set

$$\sigma(h)F = AFB^{-1} \quad (h \in H),$$

then we obtain a representation σ of H on the vector space $M_{n,m}(\mathbf{R})$. This representation is identified with $h \to \mathrm{Ad}_{\mathfrak{g}/\mathfrak{h}}(h)$, and, in particular, $\det \sigma(h) = \det \mathrm{Ad}_{\mathfrak{g}/\mathfrak{h}}(h)$. We shall show in the lemma below that

$$\det \sigma(h) = (\det A)^m (\det B)^{-n}. \tag{5}$$

Since A and B are both orthogonal matrices satisfying $\det A \cdot \det B = 1$, we have either $\det A = \det B = 1$, in which case $\det \sigma(h) = 1$ by (5), or $\det A = \det B = -1$, in which case $\det \sigma(h) = (-1)^N$ by (5). Hence, by Theorem 1, if N is even, then $G_{N,n}$ is orientable, while if N is odd, then $G_{N,n}$ is nonorientable.

To show that (5) holds, it suffices to prove the following general lemma.

LEMMA. Let $M_{n,m}(\mathbf{R})$ be the $(n \cdot m)$-dimensional vector space formed by all $n \times m$ real matrices. For a pair (X, Y), where X is an $n \times n$ matrix and Y is an $m \times m$ matrix, set

$$\sigma(X, Y)F = XF^t Y \quad (F \in M_{n,m}(\mathbf{R})).$$

Then $\sigma(X, Y)$ is a linear transformation of $M_{n,m}(\mathbf{R})$. For this linear transformation, we have

$$\det \sigma(X, Y) = (\det X)^m \cdot (\det Y)^n. \tag{6}$$

Proof. Set $\det \sigma(X, Y) = f(X, Y)$. Let $X = (x_{ij})$, $Y = (y_{st})$, and consider $f(X, Y)$ as a polynomial in variables x_{ij} $(i, j = 1, \ldots, n)$ and y_{st} $(s, t = 1, \ldots, m)$. Since $\sigma(XX', YY') = \sigma(X, Y)\sigma(X', Y')$, we have

$$f(XX', YY') = f(X, Y)f(X', Y'). \tag{7}$$

If X and Y are the diagonal matrices $\alpha 1_n$ and $\beta 1_m$, respectively, then $\sigma(X, Y)$ can be expressed as the $nm \times nm$ diagonal matrix $\alpha\beta 1_{nm}$. Hence we have $f(X, Y) = (\alpha\beta)^{nm}$. But, in this case, $\det X = \alpha^n$, $\det Y = \beta^m$, so that (6) holds. Now for the general case, let Δ_{ij} be the cofactor of x_{ij} in $X = (x_{ij})$, and set $\mathbf{X} = (\Delta_{ij})$. Then $X \cdot {}^t\mathbf{X} = (\det X)1_n$. Define $\mathbf{Y}$ similarly and we have $Y^t\mathbf{Y} = (\det Y)1_m$. Hence by (7) we have

$$f(X, Y)f({}^t\mathbf{X}, {}^t\mathbf{Y}) = (\det X)^{nm}(\det Y)^{nm}.$$

However, $f(X, Y)$, $f({}^t\mathbf{X}, {}^t\mathbf{Y})$, $f(X)$, and $f(Y)$ are polynomials of $n^2 + m^2$ variables (x_{ij}, y_{st}) over $\mathbf{R}$, and $\det X$ and $\det Y$ are irreducible. Hence by the unique factorization property of polynomial rings, we have

$$f(X, Y) = c(\det X)^k(\det Y)^l \quad (c \in \mathbf{R}).$$

If we set $X = x1_n$, $Y = y1_m$, then we get $f(X, Y) = (xy)^{nm}$, while $c(\det X)^k(\det Y)^l = cx^{nk}y^{ml}$. Hence $c = 1, k = m, l = n$, and (6) is proved.

Let us now prove Weyl's theorem and show an application.

THEOREM 3 (Weyl). *Let (ρ, V) be a representation of a compact Lie group G on a real vector space V. Then there is a positive inner product (u, v) $(u, v \in V)$ such that*

$$(\rho(s)u, \rho(s)v) = (u, v)$$

holds for all $s \in G$. That is, there is a G-invariant positive inner product on V.

Proof. Let $\alpha(u, v)$ be an arbitrary positive inner product on V, and set

$$f(t; u, v) = \alpha(\rho(t)u, \rho(t)v).$$

If we fix u and v, then $f(t; u, v)$ is clearly a continuous function of t, and we have

$$f(st; u, v) = f(s; \rho(t)u, \rho(t)v). \tag{8}$$

If we set

$$(u, v) = \int_G f(s; u, v) \, ds,$$

then it is clear that (u, v) is a positive inner product on V. But we have

$$\int_G f(st; u, v) \, ds = \int_G f(s; u, v) \, ds = (u, v),$$

and, by (8), the left member is equal to $(\rho(t)u, \rho(t)v)$. Hence $(\rho(t)u, \rho(t)v) = (u, v)$ holds for all $t \in G$, $u, v \in V$.

Remark 1. If we take an orthonormal base with respect to a G-invariant positive inner product on V, then $\rho(g)$ $(g \in G)$ is expressed by an orthogonal matrix.

Remark 2. If the representation space V is a complex vector space, then we can prove, in a completely analogous way, that there is a G-invariant positive Hermitian inner product. Hence, if we take an orthonormal basis of V, then each $\rho(g)$ is expressed by a unitary matrix.

THEOREM 4. *If G is a Lie group and H a compact subgroup of G, then the homogeneous space G/H admits a G-invariant Riemannian metric. By* *the G-invariance of a Riemannian metric, we mean that each transformation of G is a motion.*

Proof. Let o be the point in M corresponding to the identity element e of G. If we set $\rho(h) = ((t_h)_*)_o$ for $h \in H$, then ρ is a representation of the compact group H on $T_o(M)$. Hence, by Weyl's theorem, there is an H-invariant positive inner product $g_o(u, v)$ on $T_o(M)$.

For an arbitrary point p of M, pick an element k of G such that $t_k(o) = p$, and let the inner product of $T_p(M)$ be given by

$$g_p(u, v) = g_o((t_{k^{-1}})_* u, (t_{k^{-1}})_* v) \qquad (u, v \in T_p(M)).$$

We can prove that the definition of g_p does not depend on the choice of k such that $t_k(o) = p$, and that $g : p \to g_p$ is a G-invariant metric, by the method given in Example 5 of §1.

§5. Stokes' Theorem

Let D be a domain (i.e., an open connected subset) of a manifold M. The domain D is said to have a smooth boundary if for every point p of the set ∂D of the boundary points of D, there are a neighborhood U of p in M and a local coordinate system $(x^1, \ldots, x^n)$ on U such that

$$U \cap \bar{D} = \{q \in U \mid x^n(q) \geqq x^n(p)\}. \tag{1}$$

Let us prove the following lemma.

LEMMA 1. *Let M be an n-dimensional manifold, and let D be a domain in M with a smooth boundary. The boundary ∂D is an $(n - 1)$-dimensional closed submanifold of M. Furthermore, if M is paracompact and orientable, then ∂D is also orientable.*

Proof. The boundary ∂D of D is a closed set of M, and we give it the topology of a subspace. For each point p of ∂D, choose a neighborhood U of M and a local coordinate system on U such that (1) is satisfied. Then we have

$$U \cap \partial D = \{q \in U \mid x^n(q) = x^n(p)\}. \tag{2}$$

In fact, if $q \in U \cap \bar{D}$ and $x^n(q) > x^n(p)$, then we can take a neighborhood V of q sufficiently small so that V is contained in U and $x^n(q') > x^n(p)$ for all $q' \in V$. Then, by (1), $V \subset \bar{D}$, and q is an interior point of D. Hence, if $q \in U \cap \partial D$, then $x^n(q) = x^n(p)$. Conversely, if $x^n(q) = x^n(p)$, then we have $q \in U \cap \partial D$. If we set $f = x^n - x^n(p)$, then, since $df =$

$dx^n \neq 0$, by Theorem 1 in II, §10, ∂D is a submanifold of M of dimension $n - 1$, and $x^1, \ldots, x^{n-1}$, restricted to $U \cap \partial D$ (denoted by the same letters $x^1, \ldots, x^{n-1}$), are local coordinates of ∂D on $U \cap \partial D$.

Now let M be paracompact and orientable, and pick a positive orientation. Since we can choose a positive local coordinate system satisfying (1)*, in what follows we shall only consider positive local coordinate systems. Now let $(\bar{x}^1, \ldots, \bar{x}^n)$ be a (positive) local coordinate system on a neighborhood U' of a point p' of ∂D satisfying

$$U' \cap \bar{D} = \{q \in U' | \bar{x}^n(q) \geq \bar{x}^n(p')\}. \tag{1'}$$

Then, as before, we have

$$U' \cap \partial D = \{q \in U' | \bar{x}^1(q) = \bar{x}^n(p')\}. \tag{2'}$$

Set $x^n(p) = a$ and $\bar{x}^n(p') = a'$. Suppose $U \cap U' \neq \emptyset$, and let

$$\bar{x}^i = \varphi^i(x^1, \ldots, x^n) \qquad (i = 1, \ldots, n)$$

on $U \cap U'$. From (2) and (2'), we have $\varphi^n(x^1, \ldots, x^{n-1}, a) = \bar{a}$. Hence, if $q \in U \cap U' \cap \partial D$, then we have

$$\left(\frac{\partial \bar{x}^n}{\partial x^i}\right)_q = 0 \qquad (i = 1, \ldots, n - 1),$$

and therefore we obtain

$$\frac{D(\bar{x}^1, \ldots, \bar{x}^n)}{D(x^1, \ldots, x^n)_q} = \left(\frac{\partial \bar{x}^n}{\partial x^n}\right)_q \frac{D(\bar{x}^1, \ldots, \bar{x}^{n-1})}{D(x^1, \ldots, x^{n-1})_q}. \tag{3}$$

Let us prove that $(\partial \bar{x}^n/\partial x^n)_q \geq 0$. Let q_k $(k = 1, 2, \ldots)$ be points of $U \cap U'$ such that $x^i(q_k) = x^i(q)$, $i = 1, \ldots, n - 1$, $x^n(q_k) = x^n(q) + \varepsilon_k$ $(\varepsilon_k > 0, \varepsilon_k \to 0)$. From (1) and (2), we have $q_k \in D$, $q_k \to q$. Since $q_k \in D$, we have $\bar{x}^n(q_k) > \bar{x}^n(q)$. Hence

$$\left(\frac{\partial \bar{x}^n}{\partial x^n}\right)_q = \lim_{k \to +\infty} \frac{\bar{x}^n(q_k) - \bar{x}^n(q)}{\varepsilon_k} \geq 0.$$

However, since $D(\bar{x}^1, \ldots, \bar{x}^n)/D(x^1, \ldots, x^n)_q > 0$, from (3) and $(\partial \bar{x}^n/\partial x^n)_q \geq 0$, we conclude that $(\partial \bar{x}^n/\partial x^n)_q > 0$. Hence we get

$$\frac{D(\bar{x}^1, \ldots, \bar{x}^{n-1})}{D(x^1, \ldots, x^{n-1})_q} > 0. \tag{4}$$

Hence, by Theorem 1 of §1, ∂D is orientable.

*If (x^i) is not positive, for example, we can take $(-x^1, x^2, \ldots, x^n)$.

As we have seen in the preceding proof, if $(x^1, \ldots, x^n)$ is a positive local coordinate system satisfying (1), then we can choose a positive orientation on ∂D so that $(x^1, \ldots, x^{n-1})$ is a positive local coordinate system. However, it is more convenient to choose the positive orientation of ∂D so that if *n is even, then* $(x^1, \ldots, x^{n-1})$ *is positive, while if n is odd,* $(x^1, \ldots, x^{n-1})$ *is negative.* Such an orientation of ∂D is called *an orientation of ∂D compatible with the orientation of M.*

Example. Let D be a domain with a smooth boundary in an *n*-dimensional Euclidean space $\mathbf{E}^n$. Then the boundary ∂D is a hypersurface in $\mathbf{E}^n$. We shall denote here by $(y^1, \ldots, y^n)$ the canonical coordinates in $\mathbf{E}^n$, and choose an orientation of $\mathbf{E}^n$ so that $(y^1, \ldots, y^n)$ is a positive coordinate system. On ∂D we consider the orientation compatible with that of $\mathbf{E}^n$. Let $\{v_1, \ldots, v_{n-1}\}$ be a positive basis of $T_p(\partial D)$. A normal vector u of ∂D at p is said to be *outgoing*, if $\{u, v_1, \ldots, v_{n-1}\}$ is a positive basis of $T_p(\mathbf{E}^n) = \mathbf{E}^n$. As we have seen in Theorem 3 of §1, there is a continuous everywhere nonvanishing normal vector field N of ∂D which corresponds to the orientation of ∂D. We shall show that N is an outgoing normal vector field, that is, N_p is outgoing at each point p of ∂D.

To see this, let Ω be the volume element on ∂D, and let $(x^1, \ldots, x^n)$ be a positive local coordinate system on a neighborhood U of $p \in \partial D$ in $\mathbf{E}^n$ satisfying condition (1) of this section. Then, by the definition of the orientation of ∂D, we have

$$\Omega = (-1)^n a \, dx^1 \wedge \cdots \wedge dx^{n-1}$$

on U, where $a > 0$ on $U \cap \partial D$, and by the same letter x^i ($i = 1, \ldots, n-1$) we denote the restriction of x^i to $U \cap \partial D$. Now let F be the alternating *n*-linear function on the underlying vector space of $\mathbf{E}^n$ such that $F(e_1, \ldots, e_n) = 1$, where $e_1, \ldots, e_n$ denote the canonical unit vectors in $\mathbf{E}^n$. Then, by the definition of the normal vector field N (cf. proof of Theorem 3 of §1), we have

$$\Omega(v_1, \ldots, v_{n-1}) = F(N_p, v_1, \ldots, v_{n-1})$$

for any $v_1, \ldots, v_{n-1} \in T_p(\partial D)$. In particular, if $\{v_1, \ldots, v_{n-1}\}$ is a positive basis, then $\Omega(v_1, \ldots, v_{n-1}) > 0$, and hence $F(N_p, v_1, \ldots, v_{n-1}) > 0$, and this shows that $\{N_p, v_1, \ldots, v_{n-1}\}$ is a positive basis of $\mathbf{E}^n$ (Figs. 5.5 and 5.6).

Now let M be an *n*-dimensional oriented manifold. Let D be a domain in M with a smooth boundary, and we suppose that $\bar{D}$ is compact. Let c be a function that is 1 on $\bar{D}$ and 0 outside of $\bar{D}$. If θ is a differential form

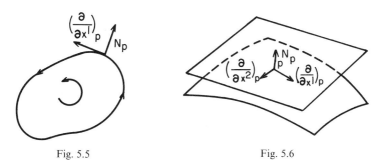

Fig. 5.5   Fig. 5.6

on M of order n, then $c\theta$ has a compact support, and the integral of $c\theta$ on M is defined.* The integral of θ on D is defined by

$$\int_D \theta = \int_M c\theta.$$

THEOREM 1 (Stokes' theorem). *Let M be an oreinted paracompact n-dimensional manifold. Let D be a domain in M with a smooth boundary, and suppose $\bar{D}$ is compact. Let ω be a differential form on M of order $n - 1$, and let i be the injection from ∂D into M. Then*

$$\int_D d\omega = \int_{\partial D} i^*\omega \tag{5}$$

holds. Here, for the integral of the right member, we choose the orientation on ∂D to be compatible with the oreintation of M.

Proof. Choose a locally finite cover $\{U_\alpha\}_{\alpha \in A}$ of M satisfying the following conditions: (1) Each U_α is contained in a cubic coordinate neighborhood and (2) for each α, $U_\alpha \cap \bar{D} = \varnothing$ or $U_\alpha \subset D$ or $U_\alpha \cap \partial D \neq \varnothing$; in case $U_\alpha \cap \partial D \neq \varnothing$, we have

$$U_\alpha \subset \{q|\ |x^i(q)| < \varepsilon_\alpha,\ i = 1, \ldots, n\},$$
$$U_\alpha \cap \bar{D} = \{q \in U_\alpha | x^n(q) \geqq 0\}.$$

Then let $\{f_\alpha\}$ be a partition of unity subordinate to $\{U_\alpha\}$. Since ∂D and $\bar{D}$ are compact, and $\{U_\alpha\}$ is locally finite, the number of α's such that $U_\alpha \cap \partial D \neq \varnothing$ or $U_\alpha \cap \bar{D} \neq \varnothing$ is finite. Hence we have

$$\int_D d\omega = \sum_\alpha \int_D d(f_\alpha\omega) \quad \text{and} \quad \int_{\partial D} i^*\omega = \sum_\alpha \int_{\partial D} i^*(f_\alpha\omega),$$

*The differential form $c\theta$ is not necessarily continuous, but its integral is nevertheless defined as in §2.

so that, to prove (5), it suffices to show that

$$\int_D d(f_\alpha \omega) = \int_{\partial D} i^* f_\alpha \omega.$$

That is, it suffices to prove (5) under the hypothesis that the support of ω is contained in U_α. If $U_\alpha \cap \bar{D} = \varnothing$, then ω and $d\omega$ are 0 on $\bar{D}$, and both members of (5) are equal to zero, so that (5) holds in this case. So let us assume that $U_\alpha \cap \bar{D} \neq \varnothing$, and let $(x^1, \ldots, x^n)$ be a positive local coordinate system on U_α. We assume further that $(x^1, \ldots, x^n)$ satisfies the above conditions in case $U_\alpha \cap \partial D \neq \varnothing$. On U_α we can write ω as

$$\omega = \sum_{j=1}^{n} (-1)^{j-1} a_j \, dx^1 \wedge \cdots \wedge dx^{j-1} \wedge dx^{j+1} \wedge \cdots \wedge dx^n,$$

and $d\omega$ becomes

$$d\omega = \left(\sum_{j=1}^{n} \frac{\partial a_j}{\partial x^j} \right) dx^1 \wedge \cdots \wedge dx^n.$$

If $U_\alpha \subset D$, then, since $U_\alpha \subset \{q \mid |x^i(q)| < \varepsilon_\alpha, \, i = 1, \ldots, n\}$, we have, by the definition of the integral, writing $\varepsilon = \varepsilon_\alpha$ for simplicity,

$$\int_D d\omega = \int_{-\varepsilon}^{\varepsilon} \cdots \int_{-\varepsilon}^{\varepsilon} \left(\sum_{j=1}^{n} \frac{\partial a_j}{\partial x^j} \right) dx^1 \cdots dx^n$$

$$= \sum_{j=1}^{n} \int_{-\varepsilon}^{\varepsilon} \cdots \int_{-\varepsilon}^{\varepsilon} [a_j(x^1, \ldots, x^{j-1}, \varepsilon, x^{j+1}, \ldots, x^n)$$

$$- a_j(x^1, \ldots, x^{j-1}, -\varepsilon, x^{j+1}, \ldots, x^n)] \, dx^1 \ldots dx^{j-1} \, dx^{j+1} \ldots dx^n.$$

However, the support of ω is contained in U_α so that $a_j(x^1, \ldots, \varepsilon, \ldots, x^n) = a_j(x^1, \ldots, -\varepsilon, \ldots, x^n) = 0$. Hence

$$\int_D d\omega = 0.$$

Since we have assumed that $U_\alpha \subset D$, we have $U_\alpha \cap \partial D = \varnothing$, so that $i^* \omega = 0$ and hence

$$\int_{\partial D} i^* \omega = 0.$$

Hence (5) holds for the case $U_\alpha \subset D$ also. Finally, we consider the case where $U_\alpha \cap \partial D \neq \varnothing$. In this case, we have

$$\int_D d\omega = \int_0^\varepsilon \int_{-\varepsilon}^\varepsilon \cdots \int_{-\varepsilon}^\varepsilon \left(\sum_{j=1}^n \frac{\partial a_j}{\partial x^j} \right) dx^1 \cdots dx^n.$$

For the same reason as in the previous case, we have

$$\int_{-\varepsilon}^\varepsilon \frac{\partial a_j}{\partial x^j} dx^j = 0 \qquad (j < n), \quad a_n(x^1, \ldots, x^{n-1}, \varepsilon) = 0,$$

and hence

$$\int_D d\omega = -\int_{-\varepsilon}^\varepsilon \cdots \int_{-\varepsilon}^\varepsilon a_n(x^1, \ldots, x^{n-1}, 0) \, dx^1 \cdots dx^{n-1}.$$

Now $(x^1, \ldots, x^{n-1})$ is a local coordinate system on $U_\alpha \cap \partial D$, and since $x^n = 0$ on $U_\alpha \cap \partial D$, we have

$$i^*\omega = (-1)^{n-1} a_n(x^1, \ldots, x^{n-1}, 0) \, dx^1 \cdots dx^{n-1}.$$

By the definition of the positive orientation of ∂D, if n is even, then $(x^1, \ldots, x^{n-1})$ is positive, while if n is odd, then it is negative. Hence, by the definition of the integral, we have

$$\int_{\partial D} i^*\omega = (-1)^{n-1} \int_{-\varepsilon}^\varepsilon \cdots \int_{-\varepsilon}^\varepsilon a_n(x^1, \ldots, x^{n-1}, 0) \, dx^1 \cdots dx^{n-1}$$

$$\text{(if } n \text{ is even)}$$

$$= -(-1)^{n-1} \int_{-\varepsilon}^\varepsilon \cdots \int_{-\varepsilon}^\varepsilon a_n(x^1, \ldots, x^{n-1}, 0) \, dx^1 \cdots dx^{n-1}$$

$$\text{(if } n \text{ is odd)}$$

and in either case we have

$$\int_{\partial D} i^*\omega = -\int_{-\varepsilon}^\varepsilon \cdots \int_{-\varepsilon}^\varepsilon a_n(x^1, \ldots, x^{n-1}, 0) \, dx^1 \cdots dx^{n-1}.$$

Hence (5) holds in this last case too, and Stokes' theorem is proved.

Problem 1. Let D be a bounded domain in $\mathbf{E}^n$ with a smooth boundary. Let $(y^1, \ldots, y^n)$ be a positive coordinate system of $\mathbf{E}^n$, and let $\omega = \sum (-1)^{i-1} \alpha_i \, dy^1 \wedge \cdots \wedge dy^{i-1} \wedge dy^{i+1} \wedge \cdots \wedge dy^n$. Let Ω be the volume element determined by the Riemannian metric on ∂D, and let $(b^1, \ldots, b^n)$ be the components of the outgoing normal to ∂D of unit length with respect to $(y^1, \ldots, y^n)$. Prove that

$$i^*\omega = \sum_{i=1}^n \alpha_i \cdot b^i \cdot \Omega.$$

(*Hint*: If we take, at each point, a neighborhood U and a positive local coordinate system

$(x^1, \ldots, x^{n-1}, x^n)$ on U with the above property, then on $U \cap \partial D$ we can write $i^*\omega = a_n(x^1, \ldots, x^{n-1}, 0) \, dx^1 \wedge \cdots \wedge dx^{n-1}$. Using this, it suffices to prove that the equality holds on $U \cap \partial D$.)

Example 1. Let (x, y) be a positive coordinate system on the plane, and let D be a domain in the plane bounded by a regular closed curve C (Fig. 5.7). Then D has a smooth boundary C. If we let

$$\omega = P \, dx + Q \, dy,$$

then

$$d\omega = \left(\frac{\partial Q}{\partial x} - \frac{\partial P}{\partial y} \right) dx \wedge dy,$$

so that, by Stokes' theorem, we have

Fig. 5.7

$$\int_C (P \, dx + Q \, dy) = \int_D \left(\frac{\partial Q}{\partial x} - \frac{\partial P}{\partial y} \right) dx \, dy.$$

Example 2. If D is as above, set $z = x + iy$ and consider D as a domain in the complex plane. If $f(z)$ is a complex-valued function of z, then set

$$f(z) = u(x, y) + iv(x, y).$$

Consider the complex differential form $\omega = f(z) \, dz = (u + iv) \, (dx + i \, dy) = u \, dx - v \, dy + i(v \, dx + u \, dy)$, and set $\omega_1 = u \, dx - v \, dy$, $\omega_2 = v \, dx + u \, dy$. Then we have $\omega = \omega_1 + i\omega_2$,

$$d\omega = d\omega_1 + i \, d\omega_2,$$

$$d\omega_1 = -\left(\frac{\partial u}{\partial y} + \frac{\partial v}{\partial x} \right) dx \wedge dy \quad \text{and} \quad d\omega_2 = \left(\frac{\partial u}{\partial x} - \frac{\partial v}{\partial y} \right) dx \wedge dy.$$

By Stokes' theorem, we have

$$\int_D f(z) \, dz = \int_C \omega_1 + i \int_C \omega_2 = \int_D d\omega_1 + i \int_D d\omega_2.$$

Observe that we have the following equivalent statements:

$$d\omega = 0 \leftrightarrow d\omega_1 = 0 \quad \text{and} \quad d\omega_2 = 0$$

$$\leftrightarrow \frac{\partial u}{\partial x} = \frac{\partial v}{\partial y} \quad \text{and} \quad \frac{\partial u}{\partial y} = - \frac{\partial v}{\partial x}$$

$$\leftrightarrow f(z) \text{ is holomorphic.}$$

Hence, if $f(z)$ is holomorphic on D and continuous on $\bar{D}$, then

$$\int_C f(z)\, dz = 0.$$

This is Cauchy's theorem.

Example 3. Let D be a bounded domain in the 3-dimensional Euclidean space $\mathbf{E}^3$ with a smooth boundary, and let (x, y, z) be a positive coordinate system in $\mathbf{E}^3$. Set

$$\omega = P\, dy \wedge dz + Q\, dz \wedge dx + R\, dx \wedge dy.$$

Then we have

$$d\omega = \left(\frac{\partial P}{\partial x} + \frac{\partial Q}{\partial y} + \frac{\partial R}{\partial z} \right) dx \wedge dy \wedge dz.$$

Hence we can write Stokes' theorem as

$$\iiint_D \left(\frac{\partial P}{\partial x} + \frac{\partial Q}{\partial y} + \frac{\partial R}{\partial z} \right) dx\, dy\, dz = \int_{\partial D} i^*\omega.$$

If (L, M, N) are the components of an outgoing normal vector to ∂D of unit length, then, by Problem 1, the integral in the right member is

$$\int_{\partial D} (PL + QM + RN)\, dv$$

where dv is the volume element of ∂D determined by the Riemannian metric on ∂D.

THEOREM 2. *Let M be an oriented compact n-dimensional manifold, and let ω be a differential form on M of order $n - 1$. Then we have*

$$\int_M d\omega = 0.$$

Proof. If we consider M as a domain in itself, then the conditions of Stokes' theorem are satisfied, and since ∂M is empty, the integral of $d\omega$ on M is equal to 0 by Stokes' theorem. (If such an argument does not appeal to the reader's taste, then a direct proof can be obtained by repeating part of the proof of Stokes' theorem.)

Now let M be an oriented connected compact n-dimensional manifold, and consider the n-dimensional de Rham cohomology group $H^n(M)$

(cf. III, §6 for notations). If $\eta \in B^n(M)$, then, by Theorem 2, we have

$$\int_M \eta = 0.$$

Since n is the dimension of M, for an arbitrary differential form ω of order n, we have $d\omega = 0$. Hence $Z^n(M) = D^n(M)$. On the other hand, if ω is a volume element of M, then, by the definition of the integral, we have

$$\int_M \omega > 0.$$

Hence $\omega \notin B^n(M)$, $\omega \in Z^n(M)$, so that the dimension of $H^n(M)$ is positive. However, the following theorem holds.

THEOREM 3. *If M is an oriented compact connected n-dimensional manifold, and ω is a differential form of order n on M such that*

$$\int_M \omega = 0,$$

then there is a differential form θ of order $n - 1$ such that $d\theta = \omega$, that is, $\omega = B^n(M)$.

We shall omit the proof of this theorem,* but, using the theorem, we see that for an orientable compact connected n-dimensional manifold M,

$$\dim H^n(M) = 1$$

holds. In fact, if ω is a volume element of M, and η a differential form of order n on M, then there is a real number λ such that

$$0 = \int_M \eta - \lambda \int_M \omega = \int_M (\eta - \lambda\omega).$$

Then, by Theorem 3, we have $\eta = \lambda\omega + d\theta$. So $Z^n(M)/B^n(M)$ is spanned by the residue class containing ω. Hence $H^n(M)$ is of dimension 1.

Problem 2. Let M be a nonorientable compact connected n-dimensional manifold. Prove that $H^n(M) = (0)$, using the manifold $\tilde{M}$ given in Problem 5 of §1.

---

*For an elementary proof of Theorem 3, see, for example, S. Sternberg, *Lectures on Differential Geometry*, Mathematics Series, Prentice-Hall, Englewood Cliffs, New Jersey, 1964, p. 120. This book is a good reference book for Chapter V.

§6. Degree of Mappings

Let M be an oriented compact connected n-dimensional manifold. A volume element ω satisfying

$$\int_M \omega = 1$$

is called a *fundamental differential form of order n* of M. Let θ be an arbitrary differential form on M of order n, and let

$$\alpha = \int_M \theta.$$

Then the value of the integral of the differential form $\theta - \alpha\omega$ on M is 0, and hence, by Theorem 3 of §5, there is some η such that $\theta - \alpha\omega = d\eta$. That is, in the n-dimensional cohomology group $H^n(M)$, we have $[\theta] = \alpha[\omega]$. Here $[\theta]$ and $[\omega]$ denote the elements in $H^n(M)$ represented by θ and ω, respectively.

Now let M and M' be oriented compact connected n-dimensional manifolds, and let ω and ω' be fundamental differential forms of order n of M and M', respectively. If f is a C^∞ mapping from M to M', then there is a real number α such that $[f^*\omega'] = \alpha[\omega]$. We denote this α by $\deg(f)$, and call it the *degree* of f. That is,

$$\deg(f) = \int_M f^*\omega', \tag{1}$$

where ω' is a fundamental differential form of order n on M'. If η is an arbitrary differential form of order n on M', then

$$\int_M f^*\eta = \deg(f)\int_{M'} \eta \tag{2}$$

holds. In fact, if we let

$$\beta = \int_{M'} \eta,$$

then we can write $\eta = \beta\omega' + d\xi$, so that $f^*\eta = \beta f^*\omega' + df^*\xi$. Hence, by Theorem 2 of §5, we obtain

$$\int_M f^*\eta = \beta \int_M f^*\omega' = \deg(f)\,\beta.$$

Note that (2) shows that the definition of the degree of f does not depend on the choice of the fundamental differential form of order n on M'. In fact, take η to be another fundamental differential form of order n on M', i.e., let $\int_{M'}\eta = 1$. Then we have $\int_M f^*\eta = \deg(f)$ from (2).

Let us prove the following theorem.

THEOREM. *Let M and M' be oriented compact connected n-dimensional manifolds, and let f be a C^∞ map from M to M'. If q is a regular value* of f, then $f^{-1}(q)$ is a finite set, and*

$$\deg(f) = \sum_{p\in f^{-1}(q)} [\text{signature of } \det(f_*)_p]. \tag{3}$$

Here $\det(f_*)_p$ is the determinant of the matrix of the differential $(f_*)_p$ of f with respect to a positive local coordinate system at each p and a positive local coordinate system at q. If $f^{-1}(q)$ is empty, i.e., if $q \notin f(M)$, then we define the right member of (3) to be 0. It follows from (3) that, in particular, the degree of f is an integer.

Proof. By Sard's theorem (II, §8), there is always a regular value q of f. First let us assume that $f^{-1}(q) \neq \varnothing$. Since M is compact and $f^{-1}(q)$ is a closed set of M, $f^{-1}(q)$ is also compact. If $p \in f^{-1}(q)$, then, since p is a regular point of f, by the inverse function theorem, we can take a sufficiently small neighborhood U_p of p such that $f|U_p$ is a homeomorphism from U_p onto a neighborhood of q. In particular, f is 1:1 on U_p, so that $f^{-1}(q) \cap U_p - \{p\}$. Since $f^{-1}(q)$ is compact, this shows that $f^{-1}(q)$ is a finite set. Since $f^{-1}(q)$ is finite, we can choose a neighborhood U_p of each $p \in f^{-1}(q)$ so that U_p is homeomorphic by $f|U_p$ to a fixed cubic coordinate neighborhood V of q, and so that $U_p \cap U_{p'} = \varnothing$ for $p, p' \in f^{-1}(q)$, $p \neq p'$. By taking V sufficiently small, we have $f^{-1}(V) = \cup_{p\in f^{-1}(q)} U_p$. In fact, let $\{V^{(i)}\}$ be a decreasing sequence of neighborhoods of q contained in V such that $\cap_i V^{(i)} = \{q\}$, and let $U_p^{(i)} = f^{-1}(V^{(i)}) \cap U_p$. If $f^{-1}(V^{(i)}) \neq \cup_{p\in f^{-1}(q)} U_p^{(i)}$ for all i, then there is an infinite sequence $\{q_i\}$ of points q_i in $V^{(i)}$ converging to q, such that $q_i = f(p_i)$ and $p_i \notin \cup_{p\in f^{-1}(q)} U_p^{(i)}$. Since $f|U_p$ is 1:1, actually we have $p_i \notin \cup_{p\in f^{-1}(q)} U_p$. Since M is compact, we can pick a subsequence $\{p_{i_s}\}$ of $\{p_i\}$ such that $\{p_{i_s}\}$ converges to p_0. We have $p_0 \notin \cup_{p\in f^{-1}(q)} U_p$ and $f(p_0) = \lim f(q_{i_s}) = q$, a contradiction.

Let $(y^1, \ldots, y^n)$ be a positive local coordinate system on V, and let h be a C^∞ function on M' such that h is positive on a neighborhood of

*Cf. II, §8.

q contained in V and 0 outside of V, and set $\eta = h \, dy^1 \wedge \cdots \wedge dy^n$ on V and $\eta = 0$ outside of V. Then η is a differential form on M' of order n, and we have

$$\int_{M'} \eta = \int_{-\varepsilon}^{\varepsilon} \cdots \int_{-\varepsilon}^{\varepsilon} h \, dy^1 \cdots dy^n,$$

where $V = \{q' \mid |y^i(q)| < \varepsilon\}$. The support of $f^*\eta$ is contained in $\cup_{p \in f^{-1}(q)} U_p$, and since $U_p \cap U_{p'} = \varnothing$ (for $p \neq p'$), writing the restriction of $f^*\eta$ to U_p as η_p, we get

$$\int_M f^*\eta = \sum_{p \in f^{-1}(q)} \int_M \eta_p, \tag{4}$$

by the definition of the integral. On the other hand, if we write f_p for the restriction of f to U_p, and let $y^i \circ f_p = x^i$, then $(x^1, \ldots, x^n)$ is a local coordinate system on U_p, and

$$\int_M \eta_p = \pm \int_{-\varepsilon}^{\varepsilon} \cdots \int_{-\varepsilon}^{\varepsilon} h \, dx^1 \cdots dx^n,$$

where, in the right member, we take $+$ if $(x^1, \ldots, x^n)$ is positive, and $-$ if $(x^1, \ldots, x^n)$ is negative. Hence we have

$$\int_M \eta_p = (\text{signature of } \det(f_*)_p) \times \int_{M'} \eta,$$

and, from (2) and (4), we conclude that $\deg(f) = \sum_{p \in f^{-1}(q)}$ (signature of $\det(f_*)_p$).

For the case where $f^{-1}(q) = \varnothing$, i.e., when $q \notin f(M)$, we pick a neighborhood V of q such that $V \cap f(M) = \varnothing$, and consider a differential form η of order n similar to that in the previous case. Then we have $f^*\eta = 0$, so by (2), we obtain $\deg(f) = 0$.

Remark. If $f(M) \neq M'$, then, from the theorem, we get $\deg(f) = 0$.

COROLLARY. *Let M, M' and f be as in the theorem, and suppose that $\det(f_*)_p \geqq 0$ at each point p of M. Then we have either $\det(f_*)_p = 0$ at all points p of M, or $f(M) = M'$.*

Proof. It suffices to show that $\det(f_*)_p = 0$ at all $p \in M$, if $f(M) \neq M'$. From the theorem we have $\deg(f) = 0$. If there is a point p such that $\det(f_*)_p > 0$, then a sufficiently small neighborhood U of p is mapped homeomorphically by f onto a neighborhood $f(U)$ of $f(p)$.

By Sard's theorem the open set $f(U)$ of M' contains a regular value of f. Then by the theorem we have $\deg(f) > 0$, a contradiction.

Problem 1. Let f be a diffeomorphism of M. Then show that $\deg(f) = \pm 1$.

Problem 2. Let f and g be two C^∞ mappings from M to M'. The maps f and g are said to be *homotopic* if there is a continuous map f from $[0, 1] \times M$ into M' satisfying (1) for each fixed t, $f(t, p)$ is a C^∞ map from M to M', and (2) $f(0, p) = f(p), f(1, p) = g(p)(\forall p \in M)$. Show that, if f and g are homotopic, then $\deg(f) = \deg(g)$.

Problem 3. If G is a connected compact Lie group, find the degree of the map from G to G given by $g \to g^2$.

§7. Divergences of Vector Fields. Laplacians

Let M be an oriented paracompact n-dimensional manifold, Ω a volume element of M, and X a vector field on M. The Lie derivative L_X of Ω with respect to X is a differential form of order n on M, so that $L_X\Omega$ is equal to Ω times a C^∞ function. Denote this function by div X, so that we have

$$L_X\Omega = (\text{div } X)\,\Omega. \tag{1}$$

The function div X is called the *divergence* of the vector field X with respect to the volume element Ω. In particular, if Ω is the volume element determined by a Riemannian metric,* then div X is called the divergence of the vector field X (with respect to the Riemannian metric).

From the formula $L_X = d \cdot i(X) + i(X) \cdot d$ (III, §4) and $d\Omega = 0$, we obtain

$$L_X\Omega = d(i(X)\Omega). \tag{2}$$

Hence, from Stokes' theorem, we get the following Green's theorem.

THEOREM 1.   *Let M be an oriented paracompact manifold, Ω a volume element of M, D a domain in M with a smooth boundary, and suppose that $\bar{D}$ is compact. Then we have*

$$\int_D \text{div } X \cdot \Omega = \int_{\partial D} i^*(i(X)\Omega). \tag{3}$$

---

*Every paracompact manifold admits a Riemannian metric (cf. Example 1 of II, §14). Conversely, every Riemannian manifold is paracompact (cf. Kobayashi-Nomizu, *Foundations of Differential Geometry*, Vol. 1, p. 60).

In particular, if M is compact, then

$$\int_M \text{div } X \cdot \Omega = 0. \tag{4}$$

Now let M be a Riemannian manifold, let f be a C^∞ function on M, and set $X = \text{grad } f$. That is, X is a vector field determined by

$$df(Y) = g(X, Y),$$

where Y is an arbitrary vector field on M. We denote by Δf the divergence of the vector field grad f with respect to g, that is, we set

$$\Delta f = \text{div grad } f. \tag{5}$$

The operator $\Delta : f \to \Delta f$ on M is called the *Lapalcian* of the Riemannian manifold M. From Green's theorem and the definition of the Laplacian, we get the following theorem.

THEOREM 2. *Let M be an oriented Riemannian manifold, f a C^∞ function on M, and D a domain in M with a smooth boundary. Suppose $\bar{D}$ is compact. Then*

$$\int_D \Delta f \, dv = \int_{\partial D} i^*(i(X)\Omega), \qquad X = \text{grad } f, \tag{6}$$

*where $dv = \Omega = G^{1/2} \, dx^1 \cdots dx^n$ is the volume element determined by the Riemannian metric, and we have**

$$i(X)\Omega = \sum_{i,j=1}^n (-1)^i \, G^{1/2} \, g^{ij} \frac{\partial f}{\partial x^j} \, dx^1 \wedge \cdots \wedge dx^{i-1} \wedge dx^{i+1} \wedge \cdots$$
$$\wedge \, dx^n.$$

In particular, if M is compact, then we have

$$\int_M \Delta f \, dv = 0. \tag{7}$$

For an n-dimensional oriented Riemannian manifold M, let us find the expressions for div X and Δf with respect to a local coordinate system. Let us write

$$X = \sum_i \xi^i \frac{\partial}{\partial x^i}$$

*(g^{ij}) denotes the inverse matrix of (g_{ij}), i.e., $\sum_{j=1} g^{ij} \cdot g_{jk} = \delta_k^i$ for all i and k.

with respect to a positive local coordinate system $(x^1, \ldots, x^n)$. Then we have

$$(L_X \Omega) \left(\frac{\partial}{\partial x^1}, \ldots, \frac{\partial}{\partial x^n} \right) = X \left(\Omega \left(\frac{\partial}{\partial x^1}, \ldots, \frac{\partial}{\partial x^n} \right) \right)$$
$$- \sum_{i=1}^{n} \Omega \left(\frac{\partial}{\partial x^1}, \ldots, \left[X, \frac{\partial}{\partial x^i} \right], \ldots, \frac{\partial}{\partial x^n} \right).$$

On the other hand, since

$$\left[X, \frac{\partial}{\partial x^i} \right] = - \sum \frac{\partial \xi^j}{\partial x^i} \frac{\partial}{\partial x^j},$$

$$\Omega \left(\frac{\partial}{\partial x^1}, \ldots, \frac{\partial}{\partial x^n} \right) = G^{1/2}, \qquad L_X \Omega = (\text{div } X) \Omega,$$

we have

$$(\text{div } X) G^{1/2} = \sum_{i=1}^{n} \xi^i \frac{\partial G^{1/2}}{\partial x^i} + \sum_{i=1}^{n} \frac{\partial \xi^i}{\partial x^i} G^{1/2}.$$

Hence we obtain

$$\text{div } X = \sum_{i=1}^{n} \frac{\partial \xi^i}{\partial x^i} + \frac{1}{2} \sum_{i=1}^{n} \xi^i \frac{\partial \log G}{\partial x^i}. \tag{8}$$

In particular, if $X = \text{grad } f$, then $\xi^i = \sum_{j=1}^{n} g^{ij} \partial f / \partial x^j$, and hence we have

$$\Delta f = \sum_{i,j=1}^{n} g^{ij} \frac{\partial^2 f}{\partial x^i \partial x^j}$$
$$+ \sum_{i,j=1}^{n} \left(\frac{\partial g^{ij}}{\partial x^i} + \frac{1}{2} g^{ij} \frac{\partial \log G}{\partial x^i} \right) \frac{\partial f}{\partial x^j}. \tag{9}$$

We have defined div X and Δf by (1) and (5), respectively, when the Riemannian manifold M is orientable, using a volume element. Equations (8) and (9) were derived from (1) and (5). If M is not orientable, we use (8) and (9) to define div X and Δf, respectively, with respect to a local coordinate system. It can be checked by a straightforward computation that these definitions do not depend on the choice of the local coordinate system.

Example.   If M is n-dimensional Euclidean space and $(x^1, \ldots, x^n)$ is the standard coordinate system, then we have $g_{ij} = g^{ij} = \delta_{ij}$, $G = 1$, so that we get

$$\text{div } X = \sum_{i=1}^{n} \frac{\partial \xi^i}{\partial x^i},$$

$$\Delta f = \sum_{i=1}^{n} \frac{\partial^2 f}{(\partial x^i)^2}.$$

A function f, on a Riemannian manifold M, satisfying $\Delta f = 0$, is called a *harmonic function* on M.

To prove a theorem on harmonic functions on a connected Riemannian manifold, we compute Δf^2, using formula (9), and obtain

$$\Delta f^2 = 2f \Delta f + 2 \sum_{i,j} g^{ij} \frac{\partial f}{\partial x^i} \frac{\partial f}{\partial x^j}$$

$$= 2f \Delta f + 2 \sum_{s,t} g_{st} \left(\sum_i g^{si} \frac{\partial f}{\partial x^i} \right) \left(\sum_j g^{tj} \frac{\partial f}{\partial x^j} \right).$$

Hence

$$\Delta f^2 = 2f \Delta f + 2 \| \text{grad } f \|^2,$$

where $\| \ \ \|$ is the length of a tangent vector with respect to the Riemannian metric. If $\Delta f = 0$, then

$$\Delta f^2 = 2 \| \text{grad } f \|^2.$$

If M is compact and orientable, then, from Theorem 2, we have

$$\int_M \| \text{grad } f \|^2 \, dv = \frac{1}{2} \int_M \Delta f^2 \, dv = 0.$$

Since $\| \text{grad } f \| \geq 0$, the above equality gives $\| \text{grad } f \| = 0$, i.e., grad $f = 0$, and f is a constant. Thus we obtain the following theorem.

THEOREM 3.   *If M is a compact connected orientable Riemannian manifold, then any harmonic function on M is a constant.*

Remark.   This theorem is valid without the orientability assumption (see Corallary 3 to Theorem 4).

Now if M is orientable, a vector field X, satisfying div $X = 0$, is called a *volume preserving* infinitesimal transformation. If X is a complete vector field on M, then by the theorem in III, §5, X is volume preserving if and only if the one-parameter group of transformations Exp tX generated by X leaves the volume element Ω invariant, i.e., if and only if (Exp tX)*$\Omega = \Omega$ for all real values t (cf. Problem of §2).

Problem 1.   Find a necessary and sufficient condition that an infinitesimal affine transformation in Euclidean space $\mathbf{E}^n$ be volume preserving.

Problem 2.   Show that an infinitesimal motion in a Riemannian space M is volume preserving.

PROPOSITION.   *Let M be an oriented compact manifold, Ω a volume element of M, and X a volume preserving infinitesimal transformation on M. For an arbitrary C^∞ function f on M, we have*

$$\int_M Xf \, dv = 0 \qquad (dv = \Omega).$$

Proof.   Set $\omega = f\Omega$. Then we have

$$L_X \omega = Xf \cdot \Omega + f L_X \Omega, \qquad L_X \Omega = \text{div } X \cdot \Omega.$$

By hypothesis, we have div $X = 0$, and hence $L_X \omega = Xf \cdot \Omega$. But, from $L_X = i(X) \cdot d + d \cdot i(X)$ and $d\omega = 0$, we have $L_X \omega = d(i(X)\omega)$. Hence we get $Xf \cdot \Omega = d(i(X)\omega)$. By Theorem 2 in §5, we conclude that

$$\int_M Xf \, \Omega = 0.$$

A linear mapping L from $C^\infty(M)$ to $C^\infty(M)$ is called an *elliptic differential operator* on M if, for a local coordinate system $(x^1, \ldots, x^n)$ and any $f \in C^\infty(M)$, we can write

$$Lf = \sum_{i,j=1} a^{ij} \frac{\partial^2 f}{\partial x^i \partial x^j} + \sum_{i=1} b^i \frac{\partial f}{\partial x^i} \qquad (a^{ij} = a^{ji}),$$

where a^{ij} are the components of a symmetric contravariant tensor field A of order 2, and the symmetric matrix (a^{ij}) is positive definite at each point. This definition does not depend on the choice of the local coordinate system.

From (8), we see that the Laplacian Δ is an elliptic differential operator. Let us prove one theorem on elliptic differential operators.

THEOREM 4.   *Let U be a domain in Euclidean space $\mathbf{E}^n$, and let*

$$L = \sum_{i,j} a^{ij} \frac{\partial^2}{\partial x^i \partial x^j} + \sum_i b^i \frac{\partial}{\partial x^i}$$

be an elliptic differential operator. Let f be a C^2 function on U, and suppose

$$L(f) \geq 0$$

holds on U. If for a point $p_0 \in U$, $f(p_0) \geq f(q)$ holds for all $q \in U$, then f is equal to a constant, namely $f(p_0)$, on U.

Before proving this theorem, we shall give some corollaries that follow immediately from this theorem.

COROLLARY 1 (The Maximum Principle). *Let D be a domain in a manifold M such that the closure $\bar{D}$ is compact. Let f be a continuous function on $\bar{D}$ which is of class C^2 on D, and suppose it satisfies $Lf = 0$. Since f is continuous on $\bar{D}$, it achieves a maximum value M_0 and a minimum value m_0 on $\bar{D}$. If f is not a constant, then $m_0 < f(q) < M_0$ for all $q \in D$. That is, if $Lf = 0$ and f is not a constant, then f does not take maximum value or minimum value inside D.*

Proof of Corollary 1. Suppose p_0 is a point in D such that $f(p_0) = M_0$. Then taking a sufficiently small neighborhood U of p_0 and local coordinates on U, we see immediately from Theorem 4 that $f(q) = M_0$ at all points q of U. Hence the set A of points $p \in D$, satisfying $f(p) = M_0$ is an open set in D. On the other hand, A, by its definition, is clearly a closed set in D, and, since A is not empty, we get $A = D$ by the connectivity of D. That is, $f \equiv M_0$. If there were a point $p_0 \in D$ such that $f(p_0) = m_0$, then $-f$ takes a maximum value $-m_0$ at p_0, and we get $-f \equiv -m_0$, i.e., $f \equiv m_0$.

In particular, if M is compact, we get Corollaries 2 and 3.

COROLLARY 2. *Let L be an elliptic differential operator on a compact manifold M, and let f be a C^2 function satisfying $Lf = 0$. Then f is a constant.*

COROLLARY 3. *A harmonic function on a compact Riemannian manifold is a constant.*

Proof of Theorem 4. Put $f(p_0) = M_0$, and let $A = \{p \in U | f(p) = M_0\}$. Then A is a nonempty closed subset of U. If A is open, then, since U is connected, we get $A = U$, and this implies $f = M_0$, which completes the proof. Therefore we assume that A is not open, and we shall show that this leads to a contradiction.

Let $U - A = V$. Then V is a nonempty open set of U. For each $p \in \mathbf{R}^n$ and $r > 0$, we shall denote by $B_r(p)$ the open ball in $\mathbf{R}^n$ of radius r and centered at p, i.e., $B_r(p) = \{x \in \mathbf{R}^n | d(p, x) < r\}$, where $d(p, x)$ denotes the Euclidean distance between p and x.

We first show that there exist a point $q_0 \in V$ and a positive real number R such that $B_R(q_0) \subset V$ and $\overline{B_R(q_0)} \cap A$ consists of a single point. In fact, since A is not open by our assumption, there is a point a in A such that $B_\varepsilon(a) \not\subset A$ for any $\varepsilon > 0$. Taking ε sufficiently small, we may assume $\overline{B_\varepsilon(a)} \subset U$, and let q be a point of $B_{\varepsilon/2}(a)$ such that $q \notin A$. Let d be the distance between q and A, i.e., $d = \inf_{x \in A} d(q, x)$. Then, since A is closed and $a \in A$, d is positive and $d < \varepsilon/2$. Then there is a point p_1 of A such that $d(p_1, q) = d$. Then $B_d(q)$ is contained in $V = U - A$. Let q_0 be a point on the line segment $\overline{qp_1}$ different from q and p_1. Let $R = \overline{q_0 p_1}$. Then q_0 and $B_R(q_0)$ satisfy our condition, that is, $B_R(q_0) \subset V$ and $\overline{B_R(q_0)} \cap A = \{p_1\}$.

To simplify our computation, we assume, by translating the coordinate system in $\mathbf{R}^n$, that q_0 is the origin 0 of $\mathbf{R}^n$. Now choosing $R_1 > 0$ $(R > R_1)$ sufficiently small, we may assume that $\overline{B_{R_1}(p_1)} \subset U$ (Fig. 5.8). Let $S_1 = \partial B_{R_1}(p_1)$, and put

$$F_0 = \{x \in S_1 | d(0, x) \geq R\},$$
$$F_1 = \{x \in S_1 | d(0, x) \leq R\}.$$

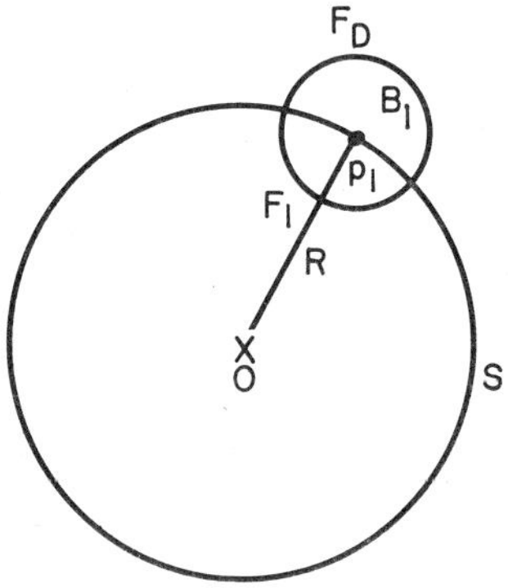

Fig. 5.8

Then $S_1 = F_0 \cup F_1$ and $F_1 \cap A = \varnothing$. Therefore, by the definition of A and M_0, there exists $\varepsilon_0 > 0$ such that

$$f \leq M_0 - \varepsilon_0 \quad \text{on } F_1; \quad f \leq M_0 \quad \text{on } F_0. \tag{10}$$

Let

$$r(x) = \{(x^1)^2 + \cdots + (x^n)^2\}^{1/2},$$

and

$$g(x) = e^{-\alpha r^2} - e^{-\alpha R^2} \quad (\alpha > 0).$$

Then, by a simple computation, we obtain

$$L(g) = e^{-\alpha r^2}\{4\alpha^2 \sum_{i, j} a^{ij} x^i x^j - 2\alpha \sum_i (b^i x^i + a^{ii})\}.$$

Since the matrix (a^{ij}) is positive definite, taking $\alpha > 0$ sufficiently large, we may assume that

$$L(g) > 0 \quad \text{on } \overline{B_{R_1}(p_1)} \tag{11}$$

holds. On the other hand, since $r^2 \geqq R^2$ on F_0, we have $g \leqq 0$, and $g = 0$ only on $F_0 \cap \partial B_R(0)$. Similarly, we have $g \geqq 0$ on F_1. Put

$$h = f + \delta g,$$

where δ is a positive constant. Then, by choosing δ sufficiently small, because of (10) we may assume $h < M_0$ on F_1. On F_0 we have $f \leqq M_0$, $g \leqq 0$ and $g = 0$ only on $F_0 \cap \delta B_R(0)$. However, we have $\partial B_R(0) \cap A = \{p_1\}$, and hence $F_0 \cap \partial B_R(0) \cap A = \varnothing$. Thus we have $f < M_0$ on $F_0 \cap \partial B_R(0)$, and hence $h < M_0$ on F_0. Thus we have shown that, for δ small enough, we have

$$h < M_0 \quad \text{on } S_1. \tag{12}$$

On the other hand, since $p_1 \in A$ and $r(p_1) = R$, $h(p_1) = f(p_1) + \delta g(p_1) = M_0$. Therefore the function h must attain its maximum on $\overline{B_{R_1}(p_1)}$ at an interior point q of $B_{R_1}(p_1)$. Then at q we have

$$\frac{\partial h}{\partial x^i}(q) = 0, \qquad i = 1, \ldots, n,$$

and the Hessian matrix

$$H(q) = \left(\frac{\partial^2 h}{\partial x^i \partial x^j}(q) \right)$$

must be nonpositive.* Then we have

$$L(h)(q) = \sum_{i,j} a^{ij}(q) \frac{\partial^2 h}{\partial x^i \partial x^j}(q) = \text{Tr } A(q) \cdot H(q),$$

where $A(q) = (a^{ij}(q))$. Since the matrix $A(q)$ is positive definite and $H(q)$ is nonpositive, we have $\text{Tr } A(q) \cdot H(q) \leqq 0$, and hence $L(h)(q) \leqq 0$. On the other hand, we have $L(h) = L(f) + \delta L(g)$. But $L(f) \geqq 0$ by the hypothesis, and $L(g) > 0$ on $\overline{B_{R_1}(p_1)}$ by (2), so that, in particular, we have $L(h)(q) > 0$. Thus we have arrived at a contradiction, and this shows that A must be open.

---

*A symmetric $n \times n$ matrix (h_{ij}) is said to be nonpositive if $\sum_{i,j} h_{ij} x_i x_j \leqq 0$ for any $(x_1, \ldots, x_n) \in R^n$.

Bibliography

[1] C. Chevalley, *Theory of Lie Groups I*, Princeton Univ. Press, Princeton, New Jersey, 1946.

[2] S. S. Chern, *Complex Manifolds Without Potential Theory*, Van Nostrand Mathematical Studies 15, Van Nostrand, Princeton, New Jersey, 1967.

[3] H. Flanders, *Differential Forms*, Academic Press, New York, 1963.

[4] K. Nomizu, *Lie Groups and Differential Geometry*, Publications of the Mathematical Society of Japan, 1956.

[5] S. Kobayashi and K. Nomizu, *Foundations of Differential Geometry*, I and II, Wiley-Interscience, New York, 1963 and 1969.

[6] S. Helgason, *Differential Geometry and Symmetric Spaces*, Academic Press, New York, 1962.

[7] J. Milnor, *Morse Theory*, Annals of Mathematics Studies, Princeton Univ. Press, Princeton, New Jersey, 1963.

[8] L. Pontryagin, *Topological Groups*, 2nd ed., translated from the Russian by A. Brown, Gordon and Breach, New York, 1966.

[9] S. Sternberg, *Lectures on Differential Geometry*, Prentice-Hall, Englewood Cliffs, New Jersey, 1964.

[10] I. M. Singer and J. A. Thorpe, *Lecture Notes on Elementary Topology and Geometry*, Scott, Foresman and Co., 1967.

[11] A. Weil, *Introduction à l'étude des variétés kählériennes*, Hermann, Paris, 1958.

INDEX

301